Evolving Neural Crest Cells

Evolutionary Cell Biology

Series Editors
Brian K. Hall – Dalhousie University, Halifax, Nova Scotia, Canada
Sally A. Moody – George Washington University, Washington DC, USA

Editorial Board
Michael Hadfield – University of Hawaii, Honolulu, USA
Kim Cooper – University of California, San Diego, USA
Mark Martindale – University of Florida, Gainesville, USA
David M. Gardiner – University of California, Irvine, USA
Shigeru Kuratani – Kobe University, Japan
Nori Satoh – Okinawa Institute of Science and Technology, Japan
Sally Leys – University of Alberta, Canada

Science publisher
Charles R. Crumly – CRC Press/Taylor & Francis Group

Cells in Evolutionary Biology: Translating Genotypes into Phenotypes – Past, Present, Future
Edited by Brian K. Hall and Sally Moody

Deferred Development: Setting Aside Cells for Future Use in Development in Evolution
Edited by Cory Douglas Bishop and Brian K. Hall

Cellular Processes in Segmentation
Edited by Ariel Chipman

Cellular Dialogues in the Holobiont
Edited by Thomas Bosch and Michael G. Hadfield

Evolving Neural Crest Cells
Edited by Brian F. Eames, Daniel Meulemans Medeiros, Igor Adameyko

Evolving Neural Crest Cells

Edited by
Brian F. Eames,
Daniel Meulemans Medeiros,
and Igor Adameyko

CRC Press
Taylor & Francis Group
Boca Raton London New York

CRC Press is an imprint of the
Taylor & Francis Group, an **informa** business

First edition published 2020
by CRC Press
6000 Broken Sound Parkway NW, Suite 300, Boca Raton, FL 33487-2742

and by CRC Press
2 Park Square, Milton Park, Abingdon, Oxon, OX14 4RN

First issued in paperback 2022

ISBN: 978-0-367-52274-2 (pbk)
ISBN: 978-1-138-63081-9 (hbk)
ISBN: 978-1-315-20919-7 (ebk)

DOI: 10.1201/b22096

Typeset in Times
by Deanta Global Publishing Services, Chennai, India

Visit the Taylor & Francis Web site at
http://www.taylorandfrancis.com

and the CRC Press Web site at
http//www.crcpress.com

Contents

Editors

Dr. Igor Adameyko is currently a professor and a department chair at the Center for Brain Research of the Medical University of Vienna, Austria. For the last 14 years, his research has focused on developmental, biomedical, and evolutionary aspects of the neural crest cells. Dr. Adameyko is known for the discovery of multipotency of nerve-associated Schwann cell precursors, a population of neural crest derived cells with exceptional plasticity and distribution in the body. These studies introduced a radically new concept for developmental biology in that defined precursor pools existing in a highly specialized niche use nerves as conduits to migrate and differentiate through temporally and spatially delineated nerve-Schwann cell communication.

Dr. Adameyko graduated from Lobachevsky University in Russia, where he also earned his PhD studying the development of the mammalian heart. During that time, Dr. Adameyko spent three years as an intern at Dartmouth Medical School in the lab of Prof. Tevosian. After defending his PhD thesis, he moved to the Karolinska Institutet in Sweden, where he started a postdoc under the supervision of Prof. Ernfors—the leading expert in neural crest and sensory biology. In 2012, Dr. Adameyko started his independent lab at Karolinska, which he still keeps simultaneously with other appointments in Vienna.

Dr. Brian F. Eames is associate professor in Anatomy, Physiology, and Pharmacology at the University of Saskatchewan's College of Medicine in the Canadian Prairies. For the past 25 years, he has studied gene programs of cartilage and bone cell differentiation in the contexts of embryonic development, evolution, and most recently, tissue engineering. The focus of many of these studies is the head skeleton, derived from neural crest cells, but his most significant contributions to understanding of this fascinating cell type were done with Dr. Richard Schneider at University of California, San Francisco. Inter-specific neural crest transplants between quail and duck demonstrated the wonderful dominance of cranial neural crest in tissue interactions with brain epithelia, mesoderm, and ectodermal epithelia, imparting species-specific information on patterning, often associated with changes to the timing of skeletal tissue formation.

Dr. Eames studied HIV during his undergrad at University of North Carolina at Chapel Hill, becoming fascinated with understanding molecular mechanisms of evolution. He switched fields from molecular virology to skeletal development during his PhD studies in Biomedical Sciences at UCSF (under Dr. Jill Helms). After postdocs at UCSF and the University of Oregon (under Dr. Chuck Kimmel), Dr. Eames set up his lab on skeletal cell differentiation, evolution, and regeneration at the University of Saskatchewan.

Dr. Daniel Meulemans Medeiros is an associate professor at the University of Colorado, Boulder. Dr. Medeiros became interested in evolution as a child on the island of Oahu, where he was forced to come to terms with the fact that he would never find a dinosaur fossil there, no matter how deep he dug. He felt much better when he learned birds were dinosaurs and subsequently began raising small parrots, learning basic genetics in the process. As an undergraduate at the University of Hawaii working with Prof. Gert De Couet, he was introduced to the emerging field of evolutionary developmental biology, which uses modern genetic and molecular tools to better understand how genetic and genomic changes lead to macroevolutionary novelties. Dr. Medeiros earned his Ph.D. with Prof. Marianne Bronner at the California Institute of Technology. Prof. Bronner further encouraged Dr. Medeiros' interest in evolution, and introduced him to neural crest cells, one of the best examples of a bona fide evolutionary novelty in animals. Dr. Medeiros started his own lab at in 2008, and has published over 30 original research articles and several literature reviews on vertebrate head skeleton development and evolution, focusing on neural crest-derived skeletal tissues. Current projects in the Medeiros lab exploit new methods for monitoring and perturbing developmental gene expression in organisms occupying key phylogenetic nodes for understanding vertebrate evolution, including the jawless vertebrate lamprey, the vertebrate-like invertebrate chordate, amphioxus, zebrafish, and the African clawed frog.

Contributors

Philip B. Abitua
Department of Molecular and Cellular
  Biology
Harvard University
Cambridge MA

Per Erik Ahlberg
Department of Organismal Biology
Uppsala University
Uppsala, Sweden

Mansour Alkobtawi
Université Paris Saclay
Orsay, France

and

Institut Curie Research Division
PSL Research University
Orsay, France

Marianne Bronner
Division of Biology and Biological
  Engineering
California Institute of Technology
Pasadena, CA

Anne H. Monsoro-Burq
Université Paris Saclay
Orsay, France

and

Institut Curie Research Division
PSL Research University
Orsay, France

and

Institut Universitaire de France
Paris, France

Tatjana Haitina
Department of Organismal Biology
Uppsala University
Uppsala, Sweden

David Jandzik
Ecology and Evolutionary Biology
University of Colorado
Boulder, CO

and

Zoology
Comenius University
Bratislava, Slovakia

Shigeru Kuratani
Laboratory for Evolutionary
  Morphology
RIKEN Center for Biosystems
  Dynamics Research (BDR)
Kobe, Japan

and

Evolutionary Morphology Laboratory
RIKEN Cluster for Pioneering
  Research (CPR)
Kobe, Japan

David W. McCauley
Department of Biology
University of Oklahoma
Norman, OK

Patsy Gómez-Picos
Anatomy, Physiology, and
  Pharmacology
University of Saskatchewan
Saskatoon, Canada

Yi-Hsien Su
Institute of Cellular and Organismic
    Biology (ICOB)
Academia Sinica
Taipei, Taiwan

Jr-Kai Yu
Institute of Cellular and Organismic
    Biology (ICOB)
Academia Sinica
Taipei, Taiwan

and

Marine Research Station
ICOB
Academia Sinica
Yilan, Taiwan

Joshua R. York
Department of Biology
University of Oklahoma
Norman, OK

Kevin Zehnder
Department of Biology
University of Oklahoma
Norman, OK

Introduction
Tribute to the Neural Crest

This book represents a tribute to our favorite cell type: the neural crest. Neural crest cells are thought to be a vertebrate innovation, and one that helped to promote the success of vertebrates as predators, leading to expansion of their "New Head" as proposed by Gans and Northcutt in 1983. Neural crest cells arise at the neural plate border as the central nervous system (CNS) is forming, then emigrate via undergoing an epithelial-to-mesenchymal transition (EMT), as detailed in several chapters (e.g., introduced by Alkobtawi and Monsoro-Burq). These multipotent progenitors undergo some of the most extensive migrations of any embryonic cell type and differentiate into an enormous variety of derivatives, including cartilage and bone of the face, neurons and glia of the peripheral nervous system, and melanocytes of the skin. In terms of the New Head of vertebrates, interactions between neural crest cells and the adjacent CNS as well as mesoderm and associated epithelia enabled enlargement of the brain, reorganization and remodeling of the pharynx, and expansion of novel sensory systems. For this reason, the neural crest is sometimes referred to as the "fourth germ layer", promoting elaboration of the vertebrate body plan.

In addition to their origin at the neural plate border, migratory ability, and multipotency to form many derivatives, neural crest cells also are characterized by a shared molecular signature. For example, genes characteristic of newly specified neural crest cells include transcription factors like the *FoxD3*, *Snai2*, and *SoxE* genes. These are part of a gene regulatory program that imbues the neural crest with its unique properties. Interactions between these transcription factors together with signaling pathways (primarily FGFs, BMPs, and WNTs) comprise a gene regulatory network (GRN) that controls various steps in neural crest formation. This starts with induction at the neural plate border, whereby distinct sets of transcription factors in the neural plate border module activate the neural crest specification program. In turn, neural crest specifier genes initiate EMT and the migration program. Finally, differentiation genes activate the various neural crest differentiation programs, resulting in formation of specific derivatives.

Several chapters in this book (e.g. Alkobtawi and Monsoro-Burq; Eames, Jandzik and Gomez-Picos) discuss gene regulatory networks that drive neural crest development and how these GRNs may have evolved via cooption of existing gene batteries responsible for differentiation of cell types like chondrocytes, melanocytes, neurons and glia. Cooption of these programs may have enabled cells at the CNS border to acquire multipotency and migratory ability. A good example of this is provided by chondrocytes, as discussed in Eames and colleagues. Whereas chondrocytes exist in nonvertebrate chordates, they are not of neural crest origin. With their advent in vertebrates, neural crest cells appear to have coopted parts of GRN subcircuits, which in invertebrates were involved in terminal chondrocyte differentiation, to produce craniofacial cartilage and bone in the vertebrate lineage.

Modifications to the vertebrate body plan are thought to be a direct consequence of small changes in this developmental GRN, as evidenced by deep conservation of the core neural crest GRN as discussed by York, Zehnder and McCauley. While there is some variation in individual transcription factors between species, this is likely to reflect restructuring of the GRN in individual clades. Consistent with the overall conservation of the neural crest GRN, there are remarkable similarities in morphology, migratory patterns, and neural crest derivatives across vertebrates, as discussed by Kuratani. From an evolutionary perspective, examining different neural crest GRN modules is useful to help identify subcircuits for reconstructing the evolutionary history of neural crest cells.

Interestingly, most neural crest–derived cell types can be found in invertebrates. Thus, although only vertebrates have a *bona fide* neural crest characterized by multipotency and migratory ability, the rudiments and gene regulatory recipes for making neural crest cells can be inferred from comparative studies of multiple organisms spanning the deuterstome tree. It has become increasingly clear, as discussed in several chapters (Yu and Su; York, Zehnder and McCauley; Abitua), that there are many indications of cells at the neural plate border in invertebrate chordates as well as other invertebrates that share properties with vertebrate neural crest cells. Studies of these cells may provide important clues as to how gene regulatory network (GRN) modules may have been coopted to promote elaboration of the neural crest. For example, analysis of neural crest–like cell types in nonvertebrate chordates has revealed intriguing embryonic cell types in ascidians that possess some neural crest characteristics and gene expression profiles. As discussed in the chapters by Abitua and Yu and Su, the a9.49 cell lineage of *Ciona intestinalis* arises from the neural plate border and forms pigmented sensory cells of the otolith and the ocellus. Moreover, misexpression of the transcription factor *twist* results in making these cell migratory. Similarly, the caudal neural plate border of *Ciona* contains a precursor to bipolar tail neuron (BTNs) that shares molecular markers and migratory ability reminiscent of vertebrate neural crest cells, including coexpression of neural plate border and neural crest markers like *Snail* and *Msx*. Thus, neural crest–like cells are likely to have evolved from these neural plate border lineages.

While neural crest cells are multipotent and often referred to as *stem cells*, in fact they are a transient population that rapidly undergoes differentiation into numerous cell types including glia and Schwann cells of the peripheral nervous system. For the Schwann cell lineage, the progenitor cells first form cells that have been termed *Schwann cell precursors* (SCPs) and are retained along peripheral nerves. As discovered by Adameyko and discussed in his chapter, these cells retain broad developmental potential and contribute to diverse cell types including melanocytes, parasympathetic and enteric neurons, Schwann cells, and likely the same cell types that early embryonic neural crest cells can form.

Just as vertebrates have continued to evolve, so has the neural crest, acquiring novel additions to its differentiation repertoire during the course of vertebrate evolution. Examples include acquisition of jaws, sympathetic neurons, and much of the enteric nervous system. How this elaboration may have occurred is nicely discussed in this chapter by Adameyko. For example, lamprey apparently lack a vagal neural crest which in gnathostomes forms most of the enteric nervous system. Instead, trunk neural crest–derived Schwann cell precursors appear to give rise to enteric nervous system (ENS) neurons in lamprey, suggesting that Schwann cell precursors may represent an ancient evolutionarily conserved source of cells that contribute to the peripheral nervous system, which were then coopted to the embryonic neural crest in stem gnathostomes. Importantly, SCPs are likely to have played a major role in promoting the elaboration of the neural crest to take on new cell types.

Studies of many organisms, ranging from basal metazoans to basal vertebrates, have been key in promoting understanding of changes that have driven evolution of the neural plate border and ultimately formation of the neural crest, as nicely summarized in the chapter by York, Zehnder and McCauley. Cyclostomes have been particularly useful for understanding the origin of craniofacial structures. As jawless vertebrates, lamprey and hagfish are the closest living approximation from which to glean clues as to the nature of the vertebrate ancestor. Kuratani nicely illustrates their usefulness in understanding panvertebrate traits and how they contribute to understanding the origin of craniofacial structures and patterning events. Finally, in order to understand vertebrate origins, it is important to combine and compare what we know from comparative embryological studies to information found in the fossil record. This can be done by analyzing changes in neural crest–derived structures through comparative analysis of fossils, as beautifully reviewed in the chapter by Ahlberg and Haitina, by viewing paleontology while being mindful of the underlying embryology.

In summary, this book covers neural crest cell formation and evolution, providing insights into how regulatory changes enabled evolutionary and morphological change. Neural crest cells are likely to be the most rapidly evolving population in the vertebrate embryo, such that more derivative cell types come under the umbrella of the neural crest as more differentiation gene batteries come under the control of neural crest transcription factors. Studying their continuing evolution gives important clues not only into the history of vertebrates but also into what may lie in store for the changing vertebrate body form in the distant future.

Marianne Bronner
Director, Beckman Institute
Albert Billings Ruddock Professor
Division of Biology and Biological Engineering 139-74
California Institute of Technology
Pasadena, California

1 The Neural Crest, A Vertebrate Invention

Mansour Alkobtawi and Anne H. Monsoro-Burq

CONTENTS

ABBREVIATIONS

BMP:	bone morphogenetic protein
CIL:	contact inhibition of locomotion
ECM:	extracellular matrix
EMT:	epithelial-to-mesenchymal transition
ENS:	enteric nervous system
ES:	embryonic stem cells
FGF:	fibroblast growth factor
HNK-1 (NC1):	monoclonal antibody recognizing a carbohydrate epitope at the cell surface of migrating NCC in chick embryos; HNK-1 is also present on many other cell types

MHB:	midbrain–hindbrain boundary
MMP:	matrix metalloprotease
NB:	neural border
NC:	neural crest
NCC:	neural crest cells
NC-GRN:	neural crest gene regulatory network
NNE:	nonneural ectoderm
NP:	neural plate
PNS:	peripheral nervous system
QȼPN:	quail nonchick perinuclear antigen
SCP:	Schwann cell progenitor
WNT:	protein of the Wingless-Int family

NOTE ON GENE/PROTEIN NAMES NOMENCLATURE

This chapter mentions studies using a variety of model organisms, such as fish, frog, chick, human, and mouse. The gene and protein nomenclature is not (yet) homogeneous for each of these species. We have here adopted the following guidelines:

For a result obtained in multiple species, we adopt the mouse nomenclature (gene: *Snai2*; protein: SNAI2).

For results linked to a single species, the nomenclature of this model is used:

Fish and frog: gene: *snai2*; protein: Snai2
Chick and human: gene: *SNAI2*; protein: SNAI2

KEY DEFINITIONS

NB: neural border (also called *neural plate border* elsewhere): the *neural border* denomination is preferred because the domain involved in neural crest formation includes a region of the dorsal ectoderm negative for *sox2* gene expression, a canonical panneural marker (see Figure 1.3; thus the neural border is not completely included in the neural plate). We advocate naming this region the *neural border*; however, *neural/nonneural border* is more accurate, though less practical.

Specifier, e.g. *NB specifier* or *NC specifier*: A factor X plays a role of "specifier" for a given tissue or cell type if X is expressed in this tissue, and if X function is necessary for the induction and development of this tissue, but X is not necessary for the development of the progenitor populations. The need for factor X's function is usually assessed by loss-of-function experiments. Loss of X activity will impact the tissue of interest and its derivatives. For example, loss of a NB specifier will prevent NB and NC formation, while loss of a NC specifier will affect NC development but not formation of the NB.

Immediate-early targets: Target genes that can be transcribed and activated without the need for *de novo* protein synthesis, usually minutes after stimulation (e.g. in Hong and Saint-Jeannet, 2007; Plouhinec et al., 2014). Usually these target genes prove to be direct targets (e.g. by binding of the protein of interest to regulatory sequences of this target, using chromatin immunoprecipitation experiments). This technique was used prior to availability of the chromatin immunoprecipitation technique.

1.1 THE NEURAL CREST: DISCOVERY AND FATES

The neural crest was initially discovered as a layer of cells lying between the ectoderm and the neural tube of chick embryos, later forming sensory ganglia. Intense research spanning over a century has been dedicated to tracing the neural crest cells from their formation to their differentiation. This has demonstrated the highly multipotent nature of the neural crest and its importance in generating key vertebrate-specific features. More recently, the cellular and molecular steps of development of many neural crest derivatives have been explored. These studies have led to the construction of one of the most advanced vertebrate gene regulatory networks. Using knowledge from model organisms, this introduction chapter reviews the evolution of ideas, strategies, and experimental discoveries leading to our current view of neural crest biology. The following parts summarize the formation of the main neural crest derivatives and highlight the major molecular actors of their development.

1.1.1 THE NEURAL CREST: AN ANATOMICAL DEFINITION FOR A HIGHLY DYNAMIC AND MULTIPOTENT CELL POPULATION

The neural crest (NC) was first described in the chick embryo 150 years ago as a band of cells located dorsal to the neural tube under the superficial ectoderm by the renowned German embryologist and anatomist Wilhelm His (Dixon, 1904), who named it the "*Zwischenstrang*", literally, the "intermediate cord" (His, 1868). Using morphological criteria, he also proposed that the NC formed the sensory ganglia (also named the *dorsal root ganglia*), the first NC derivative described. NC was further described in amphibians, then in the main vertebrate models (Horstadius, 1950; Raven, 1931). For example, during the second day of egg incubation in chick embryos, neural crest cells (NCC) can be visualized morphologically on scanning electron microscopy images as they emerge from the recently closed neural tube (Figure 1.1; trunk NCC are highlighted in purple). Thus, a key and defining feature of the vertebrate NC is its delamination from an area located at the frontier between neural and non-neural domains of the ectoderm. In the cephalic area, NC delamination takes place either prior to neural tube closure, during neural fold elevation (e.g. in mouse and frog embryos [Nichols, 1981; Sadaghiani and Thiebaud, 1987]) or after fusion of the neural folds (e.g. in chick embryos). In the trunk, the NC emigrates after neural tube closure (Collazo et al., 1993; Martinez-Morales et al., 2011). The temporal dynamics of NCC exit from the neural epithelium also varies according to different species. In chick embryos, the trunk NC emigrates about 4.5 hours after neural tube closure, while in frogs the trunk NC emigrates more than 12h after the end of neurulation (Collazo et al., 1993; Martinez-Morales et al., 2011).

A typical epithelial-to-mesenchymal transition (EMT) allows NCC to exit the neural epithelium and to acquire cell motility. NCC then migrate actively along defined embryonic routes and settle in various tissues and organs, to form over 30 distinct differentiated cell types, including cell types that are otherwise formed by

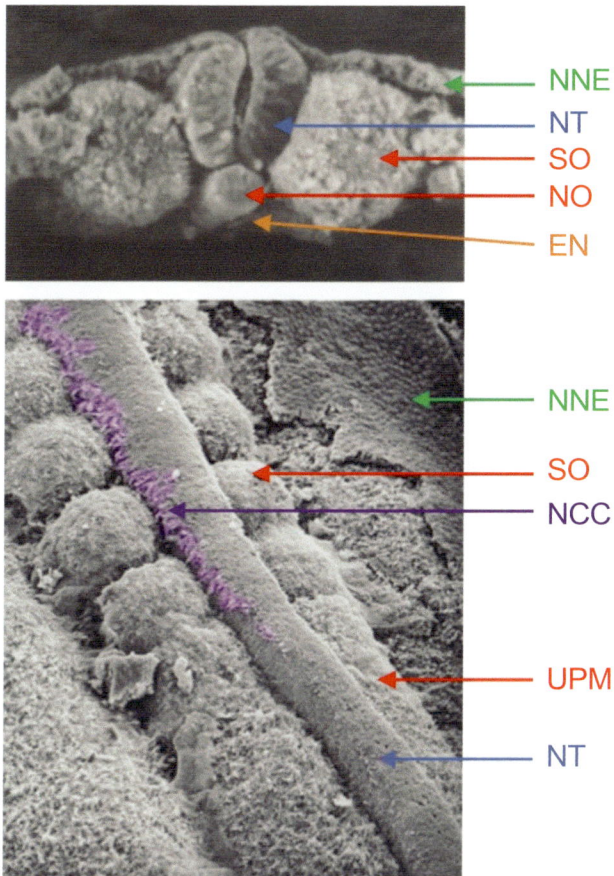

FIGURE 1.1   Neural crest formation in chick embryos. A: Scanning electron micrograph showing a cross-section of the closing neural tube, at the trunk level of a 2-days old chick embryo. B: Scanning electron micrograph done after peeling off the superficial ectoderm. The emigrating trunk neural crest cells, highlighted in purple, are seen dorsal to the neural tube, starting approximately at the level of the second and third last formed somites. ("The Nogent Institute collection" Courtesy of N. Le Douarin and Pierre Coltey.) NNE: non-neural ectoderm. NT: trunk neural tube, NO: notochord, SO: somite, EN: endoderm, UPM: unsegmented paraxial mesoderm, NCC: neural crest cells.

the mesoderm germ layer, such as chondrocytes, osteocytes, and connective cells. For this reason, the NC has also been named the *fourth germ layer* and the mesenchymal cell types formed by this ectoderm-derived cell population are called *ectomesenchyme* (previously also called *mesectoderm*). The identification of NC derivatives scattered throughout the body has proven one of the most daunting and challenging tasks in developmental biology over the last 50 years, yet rare or scattered NC derivatives are still being discovered (Etchevers et al., 2001; Forni et al., 2011). Furthermore, the formation of these various derivatives from an apparently homogeneous cell population characterizes the NC as a highly multipotent cell population. The molecular basis of NC multipotency, as well as whether individual cells are multipotent or partially lineage-restricted prior to EMT, remains debated. The differences in fate and migration behavior between cranial and trunk NCC are described in the paragraphs below. Briefly, a major difference between NCC forming in cranial areas and trunk NCC is that only the anterior NCC possess the capacity to form ectomesenchyme *in vivo* (Simoes-Costa and Bronner, 2016), with one exception known so far, the denticles in the skate (Gillis et al., 2017). In addition, migration paths and individual cell migratory behavior differ along the anterior-posterior embryonic axis.

1.1.2   THE NEURAL CREST GENE REGULATORY NETWORK: ONE OF THE MOST ADVANCED VERTEBRATE DEVELOPMENTAL GENE NETWORKS

Studies of neural crest formation rely upon the availability of reliable means for tracking the NC-derived cells; cell labeling techniques and molecular markers have been used to identify pre-migratory as well as migrating and differentiated NCC. While many very useful molecular tools have been devised, it is important to point out that, to this day, there is no marker exclusively expressed by NCC and their derivatives throughout NC development. All the frequently called "NC markers" are indeed strongly enriched in NCC but are also expressed by other cell types in the developing embryo. This creates the need to systematically use a combination of several markers to formally identify NCC, in particular when no formal lineage tracing results support a cell's NC origin. This is of particular importance for the studies of NC development and evolution in novel animal species. In the next paragraph, we detail the various strategies used to label NCC and their progeny in the chronological order of their discovery.

The first widely used molecular marker for neural crest identification is the antigen recognized by the HNK-1/NC1 monoclonal antibodies, raised against the quail ciliary ganglion and a human leukemia cell line respectively (Tucker et al., 1984; Vincent et al., 1983). These antibodies recognize the same carbohydrate epitope, present on multiple glycoproteins, such as neural cell adhesion molecule (N-CAM) and Myelin-associated glycoprotein (MAG) (Holley and Yu, 1987; Tibboel et al., 1987). HNK-1 immunostaining is strongly detected on chick migrating NCC, from day 1 to day 3, but also reacts on other cell types, such as the neural tube and the perichondrium in older chick embryos. Later on, HNK-1 immunoreactivity is retained in the peripheral nervous system but lost in other NC derivatives such as melanocytes and mesenchymal derivatives. HNK-1 immunoreactivity is also found on NCC in dog, pig, human, and rat embryos (transiently), but not in frog and mouse embryos. In mice, HNK-1 labels parts of the developing neural tube (Holley and Yu, 1987; Tibboel et al., 1987; Tucker et al., 1988). This illustrates the caution needed when comparing species using a given NC marker, or when characterizing NC differentiated from stem cells in culture. Many studies using HNK-1 have significantly advanced our understanding of chick NCC migration both *in vitro* and *in vivo* as described below. However, because the epitope recognized by HNK-1 is not maintained on all NCC later on and is not exclusively found on NC-derived cells, HNK-1 cannot be used unequivocally to follow the fates of NCC on the long term.

Between 1992 and 1995, with the emergence of molecular embryology and *in situ* hybridization techniques, the first NC-enriched marker genes were discovered. In particular, *Snai1/2* genes (initially named *snail* and *slug* after their *Drosophila melanogaster* ortholog *Snai*) turned out to be strongly expressed in the pre-migratory neural crest in frog, mouse, fish, and chick, as well as in the mesoderm during gastrulation (Essex et al., 1993; Hammerschmidt and Nusslein-Volhard, 1993; Mayor et al., 1995; Nieto et al., 1994, 1992; Smith et al., 1992; Thisse et al., 1995). As the search for cell-type-specific developmental markers was intensified in vertebrate embryos, a myriad of genes was described to be coexpressed with *Snai* genes and to play roles at various stages of NC formation (see below). However, to date, none of these genes stands as a specific and exclusive NC marker. With these new molecular tools in hand, an extensive body of research using experimental embryology and genetic approaches has explored the tissue interactions controlling NC induction at the edge of the neural plate, the parameters of NC emigration, and the control of NC differentiation into diverse lineages.

Finally, over the last 15 years, there has been a huge effort to integrate these results into a comprehensive panspecies network, called the *neural crest gene regulatory network* (NC-GRN, [Meulemans and Bronner-Fraser, 2004]). In particular, using the simultaneous manipulation of several pathways and gene expression to study their epistatic relationships, or using epigenomic approaches to understand NC-specific gene expression, the cascade of interactions forming the current core of the NC-GRN is one of the most elaborate networks described in vertebrate development. Its conservation between species, from earlier-derived vertebrates to mammals, suggests that this network is a robust starting point to explore the emergence of the NC lineage in chordate evolution and beyond (Sauka-Spengler et al., 2007).

1.2 THE FATES OF THE NEURAL CREST

1.2.1 A Challenge in Developmental Biology: How to Trace a Dynamic and Highly Invasive Embryonic Cell Population

Pioneer works by W. His proposed a NC origin for the dorsal root sensory ganglia based on the examination of histological sections from chick embryos (His, 1868). However, defining all NC derivatives with accuracy has been and remains a major challenge, first because the NCC migrate over long distances and disperse throughout the entire body, and because there are no obvious morphological criteria to distinguish them from the surrounding cells during and after migration. Since L.S. Stone identified the ability of NC to form skeleton, using heterotopic transplantations in *Rana palustris* (Stone, 1929), numerous strategies have been used to define the major NC derivatives, mainly in amphibian and chick embryos, chosen for their amenability to experimental embryology approaches.

Early studies have mainly used two complementary approaches *in vivo:* extirpation or labeling of the NC primordium. Removal or destruction of the neural folds or neural tube, at precise levels of the neuraxis, followed by examination of missing structures at subsequent time points, has allowed to deduce most trunk NC derivatives in frog and chick embryos to be deduced (for example, see [Hammond and

Yntema, 1947; Horstadius, 1950; Weston, 1970; Yntema and Hammond, 1954]).
However, these studies were not optimal because, on one hand, they potentially disrupted essential morphogenetic interactions between embryonic tissues that could
indirectly affect NC formation and, on the other hand, they did not consider possible
regulatory mechanisms restoring some of the NC derivatives after the surgery. To
minimize these issues, the second series of approaches used the labeling of pieces of
neural fold or neural tube, combined with grafting experiments. Two main strategies
were used: either detection of intrinsic markers such as differences in nucleus and
nucleolus size between two closely related species, or experimental labeling with
vital dyes or radiolabeling. The labeled tissue from the donor is then grafted into a
stage-matched host embryo, either from the same species or from a compatible one,
after removal of the equivalent tissue fragment from the host embryo. For radiolabeling the prospective NCC, a neural tube fragment taken from a donor chick embryo is
labeled with tritiated thymidine *in vitro* and grafted into a stage-matched chick host.
This technique, introduced by J. Weston, has allowed for the following of the migration of early trunk NC, while intermingled with somitic mesenchyme, addressing
debated issues such as the NC origin of Schwann myelinating cells and sympathetic
ganglia (Weston, 1963). D. Noden has used similar approaches to define the migration patterns of cephalic NC and their contribution to cranial ganglia (Noden, 1975).
In amphibians, P. Chibon used radiolabeled neural fold fragments from the amphibian *Pleurodeles Waltlii* to establish the fate map of cranial and trunk NC: he confirmed the NC origin of cranial pigment cells and cranial skeleton, including several
elements which had remained debated or unknown from studies in the chick, such
as the tooth odontoblasts. He also confirmed trunk NC contribution to spinal sensory
ganglia and to the Rohon-Béard cells, and established NC origin for Schwann cells,
sympathetic ganglia, dorsal fin mesenchyme, and meninges in amphibians. At that
time, the NC origin of adrenal medulla chromaffin cells was suggested but could
not be assessed with certainty, as the radioactive labelling was lost in differentiated
chromaffin cells at metamorphosis stage (Chibon, 1967, 1964). In these studies, the
urodele cranial ganglia were found to derive from the ectodermal placodes rather
than from the neural fold (Chibon, 1967).

The alternative labelling techniques, using classical vital dyes such as neutral red,
or fluorescent lipophilic dyes such as DiI® and DiO® (Molecular Probes), have the
advantage of being easier and less invasive, since they avoid the need for neural tube/
fold grafting, thus minimizing potential technical effects of the surgery. However,
neither radiolabeling nor vital dye staining allow unambiguous and long-term tracing of a highly dividing cell population such as the NC because the staining is lost
after a few days, and because the staining may diffuse to adjacent cells (diffusion or
cytolysis followed by uptake of the radioactive molecules by the adjacent cells was
described, see references in Weston, 1963).

In 1969, N. Le Douarin discovered a permanent species-specific cell marker,
using the difference in interphase nucleolar chromatin organization between cells
from two bird species with compatible embryonic development: the Japanese quail
(*Coturnix japonica*) and the chick (*Gallus gallus*). The chick nucleolar heterochro-
matin is dispersed, while the quail nucleus contains a large and dense nucleolus,
allowing a clear distinction of cells from either species using conventional DNA

dyes such as Feulgen-Rossenbeck staining. This permanent and species-specific cell marker has been fully exploited to establish the current NC fate maps in birds. This technique relies on grafting a small piece of either cephalic neural fold or trunk neural tube taken from a quail embryo into a chick recipient of the same developmental stage, after surgical extirpation of an equivalent piece of the host neural fold or neural tube (the various procedures are described in Le Douarin and Kalcheim, 1999). By grafting neural tube fragments as small as a single rhombomere along the entire anterior–posterior axis, these experiments virtually mapped all neural crest derivatives in the chick embryo. Since then, either Feulgen-Rossenbeck staining or antibodies raised against quail or chick cells, such as the monoclonal antibodies Q¢PN and 8F3, respectively (Carlson, B.M. and Carlson, J.A.; Halfter, W.M.; DSHB Hybridoma bank), were widely used to follow NCC as they form either large cell populations such as in the peripheral nervous system and the craniofacial dermis and skeleton, or scattered and discrete cell populations such as individual pigment cells, adrenal medulla neuroendocrine cells, brain pericytes, and heart outflow tract cells.

More recently, transgenesis in mouse, fish, and frog and electroporation in chick, combined with confocal and bi-photon microscopy, has allowed the following of NCC formation *in vivo*, either after fixation at various developmental stages, or by live videomicroscopy. Although, as stated above, there is no perfect pan-NC and NC-restricted marker so far, several powerful molecular tools have been devised, using the regulatory sequences of the main NC specifiers to drive reporter genes expression in fish, frog, chick, and mouse. The differences between populations of NCC labelled in these different transgenic reporter lines have been directly compared (Table 1.1; Chen et al., 2017). In some lines, the anteriormost NCC is labeled while other lines better trace the posterior NCC; other lines allow labeling pre-migratory NC or NC from delamination to differentiation into selected NC derivatives. Table 1.1 lists examples in fish: for example, *sox10* promoter elements driving *gfp, membrane-rfp, or kaede* expression, and *foxd3-gfp* lines (Gfrerer et al., 2013; Gilmour et al., 2002; Kwak et al., 2013; Williams and Bohnsack, 2017); examples in mouse such as *Tfap2-IRES-Cre* (Macatee et al., 2003); *Wnt1-Cre/Wnt1-Cre2* (two lines with different properties, the most recent one being more accurate; Danielian et al., 1998; Lewis et al., 2013); *Sox10-Cre*, (Matsuoka et al., 2005), *Cdx2-Cre* (Coutaud and Pilon, 2013), *Mef2C-10N-LacZ* (Aoto et al., 2015); and examples in frog, *pax3-gfp, snai2-gfp* and *sox10-gfp* (Alkobtawi et al., 2018). Strictly speaking, these transgenic lines trace NC specifically only if the driver is selectively expressed by NCC. These tools have highlighted several novel derivatives and have helped to detail the contribution of NC in organs already known to be NC targets (Barraud et al., 2010; Forni et al., 2011; Muller et al., 2008). In addition, they have allowed the refined analysis of various parameters of NC development such as migration characteristics (Richardson et al., 2016).

Moreover, recent developments in imaging techniques allow tracking of small groups or single NC cells *in vivo*, for example, using iontophoresis of fluorescent dextrans or by combining electroporation of plasmids encoding photoconvertible dyes followed by laser-directed photoconversion and *in ovo* imaging (Bhattacharyya et al., 2008; Kulesa et al., 2013). These high-resolution approaches are usually dedicated to studies of migratory behavior of NCC out of the neural tube (McKinney et al., 2013; Richardson et al., 2016; Wynn et al., 2013).

TABLE 1.1

Selected transgenic lines to label the neural crest

Line	Species	Pre-migratory NC	Migrating NC	Derivative (N, G, M, O, C, Ms, other)	Non-NC expression	References
Wnt1-Cre	Mouse	✓	✓	N, G, M, O, C, P, Ms	Neuroepithelium	(Chai et al., 2000; Danielian et al., 1998)
Wnt1-Cre2	Mouse	✓	✓	N, G, M, O, C, P, Ms	Neuroepithelium	(Chen et al., 2017)
Sox10-Cre	Mouse	ND	✓	N, G, M		(Matsuoka et al., 2005)
Cdx2-Cre	Mouse	✓	✓	N, G, M	Neural tube, mesoderm	(Coutaud and Pilon, 2013)
Tfap2a-ires-Cre	Mouse	✗	✓	ND	Caudal ectoderm over pharyngeal arches 3-6	(Macatee et al., 2003)
Pax3-Cre	Mouse	✓	✗	N, G, M,	Neural tube	(Degenhardt et al., 2010; Li et al., 2000)
Pax3-Gfp	Mouse	ND	ND	N, G, M	Somites	(Relaix et al., 2005)
Pax7-Cre	Mouse	✓	✓	N,G, C	Neural tube, muscle, olfactory epithelium, frontonasal mesenchyme	(Murdoch et al., 2012)
Ht-PA-Cre	Mouse	✗	✓	N, G, M, O, C, P, Ms		(Pietri et al., 2003)
P0-Cre	Mouse	✗	✓	N, G, C, Ms	Notochord	(Yamauchi et al., 1999)
Tyrosinase-Cre	Mouse	✗	✓	N, G, M, C	Brain	(Tonks et al., 2003)
sox10-gfp	Frog	✗	✓	ND	Otic vesicle	(Alkobtawi et al., 2018)
pax3-gfp	Frog	✓	✗	ND	Neural tube	(Alkobtawi et al., 2018)
snai2-gfp	Frog	✓	✓	N, G, C		(Li et al., 2019)
sox10-gfp /mRFP / kaede	Fish	✓	✓	G	Oligodendrocytes, otic epithelium	(Dougherty et al., 2012; Kirby et al., 2006)

Name	Organism			Description	Tissue	Reference
Tg (sox10-cre)	Fish	✓	✓	N,G, M, C,	Oligodendrocytes, otic epithelium	(Rodrigues et al., 2012)
foxd3-gfp; Tg (foxd3-gfp)^zf15 (foxd3^zdf1)	Fish	✓	✓	foxd3^zdf1 N, G, M: xanthophore precursors and iridophores, but not in terminally differentiated melanophores		(Gilmour et al., 2002; Stewart et al., 2006)
Tg(foxd3-citrine)^(ct110a)	Fish	✓	✓	N, G, M, C	Anterior neural tube	(Hochgreb-Hägele and Bronner, 2013)
Tg(−4.9 sox10:egfp)	Fish	✓	✓	N, G, M, C	Otic epithelium	(Carney et al., 2006)
fugu-tyrp1 (pt102)	Fish	✗	✓	M	ND	(Zou et al., 2006)
mitfa-gfp (Tg(mitfa:gfp)^w47)	Fish	ND	✓	M	ND	(Curran et al., 2009)

Legend:

N: peripheral neuron; G: peripheral glia; M: melanocyte; O: osteocyte; C: chondrocyte; P: pericyte; Ms: Mesenchymal cells; ND: not described; ✗: NO; ✓: YES

Last, recent advances in transcriptomics have been applied to NC studies in fish, frog, chick, and mouse; NC and related samples have been either microdissected from embryos or sorted out by fluorescence activated cell sorting (FACS), allowing RNA sequencing of small sample (small-RNAseq) or single cell RNA sequencing (scRNA-seq). For example, in frog embryos, complementary fragments of the developing ectoderm have been dissected out to map the dorsal ectoderm transcriptionally, by small-RNAseq, during stages of neural border and neural crest induction (Plouhinec et al., 2017). This work provides a tool to explore gene expression and putative relationships (synexpression) between genes along the anterior–posterior and dorsal–ventral axes of the ectoderm, thus predicting if a particular gene/set of genes is enriched in the frog NC region. In zebrafish and mouse, selected *foxd3* and *wnt1* enhancer elements driving fluorescent reporter expression have been used to isolate NCC populations and individual cells at different stages of early development: either pre-migratory or during migration and initial differentiation, followed by small-RNAseq and scRNA-seq (Lukoseviciute et al., 2018; Soldatov et al., 2019; Williams et al., 2019). In mouse neural crest, single cell analysis proposes a model of cell fate acquisition, in which multipotent neural crest cells progress through a sequence of decisions that progressively restricts cell fates. Prior to migration, NC cells exhibit a common transcriptional state; then a series of lineage-restriction events starts during migration. For example, as the trunk NC cells migrate, a first decision separates sensory neuro-glial fates from others (Soldatov et al., 2019). Another study used postembryonic stages in zebrafish to dissociate the various skin cell types and study pigment cell lineages after the end of migration (Saunders et al., 2019). Finally, spatial transcriptomics, detecting dozens of genes on the same sample with a single cell resolution, allows the validation of gene coexpression and large gene signatures, for example in premigratory NCC located in the dorsal midbrain in chick (Lignell et al., 2017). Altogether, these analyses provide a deep molecular characterization of genes expressed in NCC at various stages of their development, heralding a novel era for the building of the NC-GRN.

In summary, classical and molecular embryology studies conducted during the past century and a half have defined the NC fate map in multiple model vertebrate embryos, but also in nonmodel organism species such as the earlier-derived vertebrate, the lamprey *Petromyzon marinus* (Green et al., 2017; Modrell et al., 2014). Advances in transcriptome and epigenome analyses and refinement at the scale of individual cells will certainly provide a new depth of understanding for each NC derivative in the near future.

1.2.2 THE MAIN NEURAL CREST DERIVATIVES

In all the vertebrates described so far, NCC generate a common array of different cell types. This includes pigment cells, cells of the peripheral nervous system (PNS) such as the Schwann myelinating cells, neurons and glial cells forming the sensory, sympathetic and enteric ganglia, the cranial skeletal cells including chondrocytes, osteocytes, and other ectomesenchymal cell types such as connective cells, smooth muscle cells, adipocytes, cardiac cells populating the outflow tract, pericytes in the brain, and secretory cells such as the adrenal medulla chromaffin cells (Figure 1.2; Table 1.2).

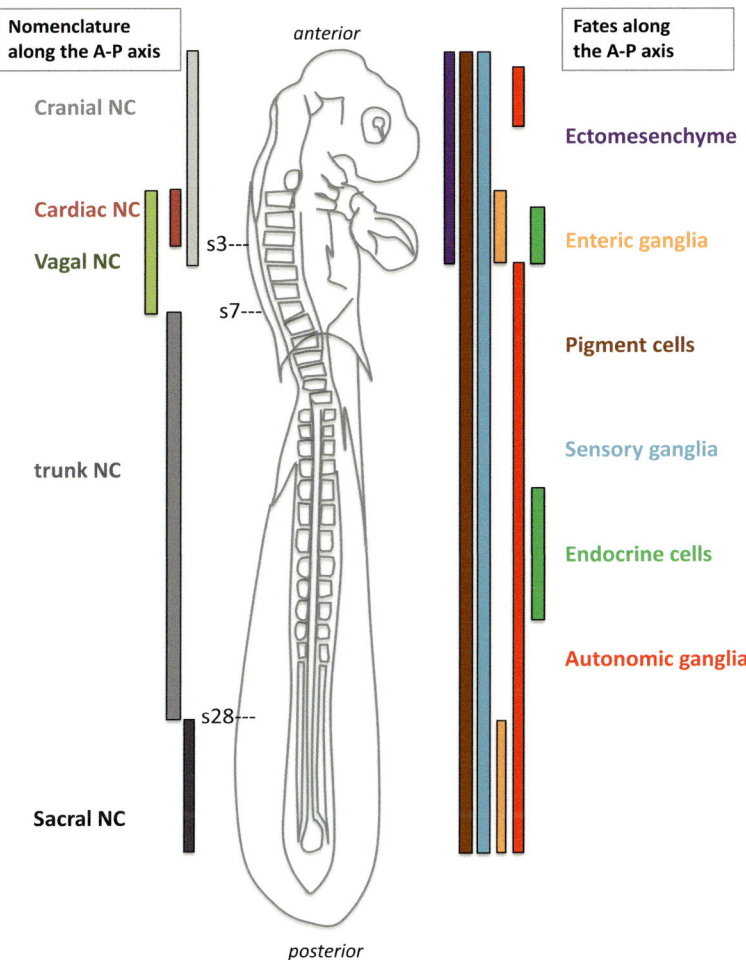

FIGURE 1.2   Position and fates of the neural crest cells along the anterior–posterior axis. Schematic drawing of a chick embryo after 2 days of incubation (24-somite stage). On the left, the nomenclature used for labeling the NC along the anterior-posterior (A-P) axis is indicated for the cranial (or cephalic) NC (i.e. from the posterior diencephalon rostrally to somite 4 caudally) thus including the cardiac NC (i.e along somites 1-3), for the vagal NC (i.e. along somites 1-7), for the trunk NC (from somites 7 to 28) and for the sacral NC (posterior to somite 28, at the level of secondary neurulation). On the right side, the fates adopted by the NC at each level of origin along the A-P axis are indicated. (Modified from Le Douarin et al., 2004.)

In addition to the derivatives directly generated by the early NC, the contribution of Schwann cell progenitor cells as a secondary source of multipotent progenitors in development, tissue homeostasis, and repair in adults was highlighted for generating pigment cells, neurons, and ectomesenchyme derivatives, among others (Adameyko et al., 2009; Dyachuk et al., 2014; Espinosa-Medina et al., 2017, 2014; Furlan et al., 2017; Laranjeira et al., 2011; Petersen and Adameyko, 2017; Uesaka et al., 2015).

TABLE 1.2
List of the main NC Derivatives (see text for references)

General fate	Details on NC-derived cell types	Notes
Peripheral nervous system (PNS)	Glial cells	Generate myelinating, non-myelinating (Remak) and terminal Schwann cells
	Olfactory ensheathing cells (OECs)	
	GnRH-1 neurons, olfactory and vomeronasal cells	Mixed origin NC and ectoderm (placode)
	Neurons and glia of the enteric ganglia of the myenteric and submucosal plexuses	From vagal NC in all vertebrates and sacral NC in amniotes
	Neurons and glia of the dorsal root (spinal) ganglia	Sensory ganglia
	Neurons and glia of the sympathetic ganglia	Autonomous noradrenergic neurons and glia
	Schwann cell progenitors: a neural crest-derived multipotent population generating later subsets of neuronal and other derivatives.	Mouse SCP-derived parasympathetic neurons (autonomic neurons) starting from E12.5
		Mouse: SCP-derived Enteric neurons in ENS
		Lamprey ENS
		Neuroendocrine cells of the adrenal medulla (chromaffin cells) from SCPs starting from E11.5 in the mouse embryo
Head ectomesenchyme	Chondrocytes, Osteocytes	Calvaria and visceral skeleton,
	Mesenchymal cells, adipocytes	
	Cardiac septum cells	
	Cardiac outflow tract cells	
	Smooth muscle cells around vessels	
	Pericytes, choroidal cells, meningeal cells	
	Schwann cell progenitors: a neural crest-derived multipotent population generating later subsets of mesenchymal derivatives.	Dental mesenchyme dental mesenchyme stem cells (MSCs) during mouse incisor tooth self-renewal process
Neck ectomesenchyme	Mesenchymal cells of the neck and shoulder skeleton chondroblasts/osteoblasts	Mixed NC and mesoderm origin, Pectoral girdle in fish
Trunk ectomesenchyme	Osteoblasts of the trunk skeleton	Caudal fin lepidotrichia osteoblasts in fish
	Fin mesenchymal cells	Dorsal and ventral fin in frog

(Continued)

TABLE 1.2 (CONTINUED)

List of the main NC Derivatives (see text for references)

General fate	Details on NC-derived cell types	Notes
Pigment cell	Melanocytes	In all vertebrates, Skin and other locations: Inner ear melanocytes; heart melanocytes; inner organs in some species.
	Xanthophores	In fish, reptiles, amphibians.
	Iridophores	In fish, reptiles, amphibians.
	Leucophores, erythro-iridophores	In fish
	Cyanophores	In fish: rare true blue pigment in vertebrates
	Melanocytes from Schwann Cell Precursors	Mouse embryo
Secretory cells	Carotid body glomus cells	Dopaminergic, hypoxia sensitive
	Chromaffin cells of the adrenal medulla	Dopaminergic, hypoxia sensitive
	Catecholaminergic cells associated with zebrafish pharyngeal arch blood vessels	

Schwann cell progenitor cells can thus be defined as multipotent neural-crest-derived cells associated with nerves. During late embryogenesis or in adults, Schwann cell progenitor cells provide cell types which, in the early embryo, are formed directly by migratory NC. This allows large areas of the growing embryo to be populated, as well as providing a distribution of these cells in coordination with the growing innervation (reviewed in Furlan and Adameyko, 2018).

In amniotes, the most striking difference between the possible fates of neural crest cells lies between cranial and trunk neural crest, as only cranial NC possesses ectomesenchymal potential. In contrast, there is evidence of some mesenchymal NC derivatives in fish and amphibians. The paragraphs below summarize the main aspects of pigment cells, PNS derivatives, ectomesenchyme, cardiac, and vascular NC development.

1.2.2.1 Pigment Cell

A variety of pigmented cells arise from the NC in vertebrates: apart from the pigmented cells of the retinal pigmented epithelium and of the pineal gland, formed by the central nervous system, all other chromatophore cells are of neural crest origin. The melanocytes form brown-black melanin or yellow-red pheomelanin in all vertebrates (reviewed in Simon et al., 2009). Reptiles, fishes, and amphibians also possess other types of pigmented cells such as the shiny blue, gold, or silver iridophores (forming reflective platelets of crystallized guanine [Morrison, 1995], the xanthophores and erythrophores, containing yellow, red, and orange pigment from carotenoids and pteridins [Matsumoto, 1965; Ziegler, 2003]). Fishes additionally display white leucophores (crystallized guanine randomly oriented), violet erythro-iridophores (crystallized purines and carotenoids, [Goda et al., 2011]), and, in rare species such as sturgeonfish, a true blue pigment in cyanophores (Goda et al., 1994).

Pigmented cells are usually found in the dermis or the epidermis, and sometimes in internal organs (Han et al., 2015; Lecoin et al., 1995). The pigment is formed and accumulates in specialized organelles moving around in the cell along microtubules, allowing rapid changes in color for camouflage or other behaviors (Goda, 2017). In the skin, melanosomes are transferred towards the surrounding keratinocytes to extend the pigmented surface coverage and optimize protection against ultraviolet irradiation (Bouchard and Garcia, 1987; Laurent-Gengoux et al., 2018).

During early organogenesis stages, the NCC fated to become melanocytes migrate out from the dorsal part of the neural tube and follow a superficial migration route located between the surface ectoderm and the dermomyotome, named the *dorsolateral pathway*. This is the exclusive route for melanoblasts in mammals and chick. Melanoblast migration depends on chemoattractants and their receptors, such as ENDOTHELIN3 and its receptor EDNRB2, EPHRINS and EPH receptors, and SDF1-CXCL12 (Belmadani et al., 2009; Kawasaki-Nishihara et al., 2011; Pla et al., 2005; Santiago and Erickson, 2002; Serbedzija et al., 1990). In fish, xanthophore precursors also follow this path. In contrast, in fish, iridophores and some melanoblasts/melanophores also migrate along an inner route between the neural tube and the somite, the *medial pathway*, also followed by neuron and glial precursors. In addition, as mentioned above, in mammals, during later stages of organogenesis, the SCP generate a secondary wave of melanoblasts (Adameyko et al., 2009).

Melanogenesis is the best-studied pigment synthesis pathway. It involves an incompletely understood gene regulatory network, activated by cooperation between the transcription factors PAX3 and SOX10 and the signaling pathways WNT and CREB. This GRN controls the expression of *MITF*, the key regulator of melanogenesis (Dorsky et al., 2000; Minchin and Hughes, 2008; Wan et al., 2011). In turn, MITF and its partners control the activation of the main enzymes of melanin biosynthesis, TYR (*tyrosinase*), TYRP2 (*Dct*), and TYRP1 (reviewed in Galibert et al., 1999; Levy et al., 2006; Seberg et al., 2017).

1.2.2.2   Neurons and Glia of the Peripheral Nervous System

Both cranial and trunk NCC generate neurons and glia of the peripheral nervous system. At trunk levels, both sensory and autonomous ganglia are formed exclusively by NCC, while in the head, ganglia have a dual neural crest and ectodermal placode origin (D'Amico-Martel, 1982; D'Amico-Martel and Noden, 1983; Schlosser and Ahrens, 2004). A complex interplay between the placode-derived neuroblasts and the neural crest coordinates the formation of cranial ganglia and nerves (Coppola et al., 2010; Freter et al., 2013). In addition, the segmentation of the posterior brain into distinct rhombomeres along the anterior–posterior axis shapes the streams of migrating NC, resulting in precise and segmented positioning of the cranial ganglia and nerves (Köntges and Lumsden, 1996). In brief, cranial NCC form sensory neurons primarily in the proximal sensory ganglia, while placodes contribute to the majority of sensory neurons in the distal ganglia (D'Amico-Martel and Noden, 1983; Thompson et al., 2010). NCC also generate Schwann and satellite glia in cranial sensory ganglia along the nerves that innervate the sensory organs, and glia that ensheathe the placode-derived sensory neurons (e.g. olfactory ensheathing glial cells [Barraud et al., 2010; Forni et al., 2011]). Finally, cranial NCC also form the neurons and glia in parasympathetic ganglia, either directly or by a later Schwann Cell Progenitor contribution (Dyachuk et al., 2014).

At trunk levels, NCC migrate through the *ventral-lateral (medial) pathway* between the neural tube and the somites to form a vast array of neurons and glia: in the autonomous ganglia close to the dorsal aortas, in the sensory (or dorsal root) ganglia lateral to the neural tube, and in the enteric nervous system (ENS). The segmentation of the paraxial mesoderm imposes the segmented pattern of the sensory and autonomous PNS: chick and mouse NCC migrate within the anterior half of the somitic mesenchyme, while fish NCC migrate along the medial path at mid-somite level. In contrast, the ENS is not segmented: the vagal NCC enter the gut wall and undergo an extensive migration to populate the whole digestive tract, complemented by a sacral NCC contribution in amniotes (Burns and Le Douarin, 1998; Le Douarin and Teillet, 1973; Nishiyama et al., 2012; Olden et al., 2008; Yntema and Hammond, 1954; Young et al., 2004). In addition, nerve roots also provide a contribution to ENS via Schwann Cell Progenitors (Uesaka et al., 2015).

On the molecular aspects, the SoxE family transcription factor SOX10 is essential for the development of all PNS cells (Southard-Smith et al., 1998). *Sox10* starts to be expressed in pre-migratory NC and is maintained during migration. SOX10 directly controls expression of the myelin genes in Schwann cells (Britsch et al., 2001; Lee et al., 2008; Peirano et al., 2000), activates neurogenic cascades, and is downregulated during final neuron differentiation. In the sensory lineage, SOX10 activates *Neurogenin1/2* and *NeuroD*. These interact with ISL1 and BRN3a, which in turn control *Runx1/ret* and *Cgrp* in nociceptors, *Runx3* in proprioceptors, and *TrkB* in mecanosensory neurons (Carney et al., 2006; Chen et al., 2006; Dykes et al., 2011; Eng et al., 2007; Kramer et al., 2006; Nakamura et al., 2008; Perez et al., 1999; Sun et al., 2008). In the autonomic lineage, SOX10 and BMP signaling from the dorsal aortas activate *Ascl1* and *Phox2b* early expression (Groves et al., 1995; Guillemot et al., 1993; Johnson et al., 1990; Pattyn et al., 1999; Schneider et al., 1999; Sommer et al., 1995). In turn, ASCL1 and PHOX2b indirectly control *Tyrosine hydroxylase* and *Dopamine beta hydroxylase* expression in noradrenergic neurons (Howard et al., 2000; Kim et al., 1998; Lo et al., 1998; Stanke et al., 1999; Tsarovina et al., 2004). In the enteric nervous system, RET/GDNF and Endothelin 3/Endothelin Receptor B signaling are essential to control *Sox10* and *Zeb2 (Sip1)* expression early on, followed by *Ascl1, Phox2b,* and *Hand2* expression, ensuring the survival, proliferation, and differentiation of the myenteric (Auerbach's) and submucosal (Meissner's) plexuses (Blaugrund et al., 1996; Bondurand et al., 2006; Chalazonitis et al., 2001; Herbarth et al., 1998; Nagy and Goldstein, 2006; Uesaka and Enomoto, 2010; Yan et al., 2004).

1.2.2.3   Secretory Cells

The neural crest also contributes neurosecretory cells such as the chromaffin cells of the adrenal medulla and glomus cells in the carotid body (Hockman et al., 2018; Le Douarin, 1975; reviewed in Lumb and Schwarz, 2015; Pardal et al., 2007). Both adrenal chromaffin cells and carotid glomus cells are hypoxia- (low O_2) and hypercapnia- (high CO_2) sensitive dopaminergic cells. Their formation requires Ascl1 and Phox2b transcription factors, among others (Dauger et al., 2003; Huber et al., 2005, 2002; Kameda, 2005). Both cell types may descend indirectly from the neural crest by contribution of the multipotent *Sox10*-positive or Plp1-positive glial stem cells at later stages (Furlan et al., 2017; Hockman et al., 2018). The neural crest origin of calcitonin-producing cells in the thyroid, deduced from quail-chick chimeras in birds, is debated, as in mice these cells derive from endoderm cells (Johansson et al., 2015; Polak et al., 1974).

1.2.2.4   Neural Crest Forming Bone, Cartilage, Connective Tissue, and Smooth Muscle Cells

The cephalic neural cells differentiate into osteocytes, chondrocytes, smooth muscle myoblasts, and adipocytes, which are cell types formed by the mesoderm at trunk levels. Hence, cranial NCC shape most of the facial and skull skeleton as well as the dermis and head and neck muscles (Dupin et al., 2010; Le Lièvre and Le Douarin, 1975; Matsuoka et al., 2005; Noden, 1983; Stone, 1929). This contribution is tightly regulated to define the species-specific craniofacial features, and is also a source of subtle interindividual variation within a given species. Such variation allows facial recognition, a key element in the evolution of social groups in primates (Parr, 2011). On the other hand, altered regulation during the formation of facial structures can lead to craniofacial defects that impair physiological functions (e.g. swallowing, hearing) and affect social integration (De Oliveira Bastos et al., 2008).

More precisely, NCC form the entire ventral pharyngeal skeleton (except the columella in the ear) and dorsally the rostralmost part of the skull (Couly et al., 1998; Le Lievre, 1978; Le Lièvre and Le Douarin, 1975; Noden, 1983; Stone, 1929). The posterior part of the skull is formed by the mesoderm. The boundary between NC and mesoderm-derived skeletal elements varies according to species. In chick, most of the skull, including frontal and parietal bones, are NC-derived (Couly et al., 1993). In fish, the boundary is located within the frontal bones (Kague et al., 2012). In mouse, the boundary is located at the level of the coronal suture, between the frontal and parietal bones (Chai and Maxson, 2006; Couly et al., 1993; Jiang et al., 2002). NCC also form muscle and connective tissues in the face and pharyngeal arches derivatives (e.g. thymus, periocular mesenchyme, part of the external ear). More recently, novel NC skeletal derivatives have been identified, including cells of the gill pillars in fish, otic vesicle elements, and, in the tooth, cells forming dentin, pulp, and cementum (Chai et al., 2000; Freyer et al., 2011; Li et al., 2011; Mongera et al., 2013; Thompson and Tucker, 2013; Wang et al., 2011). In amniotes, trunk NCC are devoid of ectomesenchyme potential (Le Lièvre and Le Douarin, 1975), with the possible exception of turtles, in which bones of the plastron form by intramembranous ossification (i.e. without a cartilaginous template) similarly to skull bones (Vincent et al., 2003). In the turtle Trachemys scripta, although without formal lineage tracing data, it has been proposed that a late-forming NCC population generates melanoblasts dorsally and cells in the somatopleura mesenchyme ventrally, the tissue from which plastron bones arise (Cebra-Thomas et al., 2013; Rice et al., 2017). In nonamniote vertebrates, a limited ectomesenchymal potential seems to be retained in the trunk, although this is debated; fish lepidotrichia and part of the fin mesenchyme in amphibian larvae may come from NCC (Collazo et al., 1993; Kague et al., 2012; Lee et al., 2013; Sobkow et al., 2006). Other cell types with a potential NC origin, such as fish scales, have been excluded (Mongera and Nusslein-Volhard, 2013).

The molecular basis for this difference in NCC potential along the anterior–posterior axis is incompletely understood. In chick embryos, the expression of cranially enriched genes such as *SOX8*, *TFAP2b*, and *ETS1* is necessary (although not sufficient) to endow NCC with cartilage-forming potential (Simoes-Costa and Bronner, 2016). During induction and migration, the early skeletogenic cranial NC shares developmental programs with the nonectomesenchymal NC. After reaching the pharyngeal arches, specific molecular regulations are activated, with *twist* expression in the prospective ectomesenchyme and *foxd3* expression in the nonectomesenchymal crest (Das and Crump, 2012; Mundell and Labosky, 2011; Soldatov et al., 2019).

Cranial NCC migrate around the eye and towards the midface and along three major streams towards the pharyngeal arches. Each pharyngeal stream expresses a defined Hox code along the anterior–posterior axis. The first arch (mandibular) is filled with Hox-negative NCC. The second (hyoid) arch is populated by NCC migrating from the rhombomeres 3 to 5, expressing Hox paralog group 2 genes. Last, NCC emerging from rhombomeres 5 to 7 invade the third (branchial) arch and express Hox paralog group 3/4 genes. Hox gene expression is essential to establish the identity of each pharyngeal arch (Couly et al., 2002; Crump et al., 2006; Santagati et al., 2005).

During pharyngeal arch patterning, key crosstalk between the epithelia, whether ectoderm or endoderm-derived, and the NC-derived mesenchyme controls the development of the skeletal elements (Couly et al., 2002; David et al., 2002; Jheon and Schneider, 2009; Takahashi et al., 1991). Along the dorsal–ventral axis, the various cartilage and bone elements are patterned by a GRN involving secreted endothelins and DLX1-6 transcription factors (Clouthier et al., 1998; Depew et al., 2002; Miller et al., 2003; Talbot et al., 2010). Midfacial structures are formed by the fusion of frontonasal, palate, and maxillary processes, under the control of Sonic Hedgehog (SHH) and FGF8 signaling, mediated by the transcription factors DLX, HAND2, and MSX. Numerous human congenital malformations, including facial clefting, result from mild to severe failure during this process. Finally, the skull vault differentiates according to a carefully controlled process allowing the coordinated growth of the brain and the bones by maintenance of opened sutures. The complex balance between proliferation and differentiation at the suture is ensured by multiple secreted or paracrine signals (FGFs, WNTs, BMP, retinoic acid, EPH-EPHRINS), triggering the action of transcriptional regulators such as MSX, DLX, and TWIST1 (Ciurea and Toader, 2009; Dennis et al., 2012; Holleville et al., 2007; Wada et al., 2005).

1.2.2.5   Cardiac NCC and Vascular NCC

The NC also forms smooth muscle cells and pericytes of the face and forebrain as well as heart vasculature. Anteriorly to the otic vesicle level, the NC forms the pericytes around capillaries, in particular in the forebrain meninges, the retinal choroid, and the facial capillaries (Etchevers et al., 2001). Between the otic vesicle and the third somite level, the rhombencephalic NC is also called the cardiac NC because the first wave of migrating cells colonizes the heart outflow tract (Kirby et al., 1983; Le Lièvre and Le Douarin, 1975; Lo et al., 1997). These cells form the aorticopulmonary septum and the aorticopulmonary cushions and partly make up the wall of posterior pharyngeal arches (PA) arteries. At the end of heart morphogenesis, NCC form the majority of smooth muscle cells in the aortic arch (i.e. in the left and right common carotid arteries, the subclavian artery, the ductus arteriosus, the aorta, and the pulmonary trunk). In the PA arteries, NCC form a subset of the vasculature as differentiated from smooth muscle cells and pericytes surrounding the endothelial cells. The FGF8 ligand, as well as ROBO/SLIT, EPH/EPHRIN signals, and the transcription factor PAX3, are important for the guidance and survival of the cardiac NC (Conway et al., 1997a, 1997b; Kirby and Hutson, 2010; Smith et al., 1997). Sympathetic innervation and vascular tree morphogenesis occur in a coordinated manner (neurovascular congruence), although the molecular mechanisms underlying this coordination remain incompletely understood. By providing capillary innervation and muscle cells in the vessel walls, NC derivatives are key elements in the control of the cranial blood flow.

In conclusion, NCC form a myriad of different cell types throughout the body. In the last 10 years, the elucidation of the gene regulatory networks involved in the differentiation of each kind of derivative has progressed tremendously and highlighted the high evolutionary conservation of these mechanisms. Recently, the discovery of the multipotent Schwann cell precursors, providing a secondary, long-term source of NC derivatives, adds further impact to the role of NC in embryo and adult.

1.3   NEURAL CREST MULTIPOTENCY

One of the most fascinating yet unsolved questions is how the unique degree of NC multipotency is achieved at the cellular and molecular levels. As a cell population, the NC is clearly multipotent. The long-standing debated question is whether the NC comprises truly multipotent progenitors or is formed by an array of partially fate-restricted ones, and at which stage of NC development such cells would be present *in vivo*. The second pending question is to define which molecular mechanisms control multipotency in the NC.

In vitro analyses conducted over three decades, mainly in N. Le Douarin's group, have described a detailed tree of sequential fate restrictions in the NC lineage. By the study of a single NC cell plated in isolation and the identification of its progeny after differentiation, dual, triple, or more multipotent progenitors of melanoblasts, glia, neurons, osteocytes, smooth muscle cells, and adipocytes have been identified (Baroffio et al., 1991, 1988; Sieber-Blum and Cohen, 1980). The rare single cells able to form all these cell types *in vitro* have been described recently (Calloni et al., 2009; reviewed in Dupin et al., 2018). Clonal analysis of NCC taken during their migration or after aggregation in the ganglia (sensory and enteric PNS) has defined the restricted potency of these older cells (Deville et al., 1994, 1992). As all *in vitro* studies, these analyses isolate the cells from their normal microenvironment and rely upon the use of a permissive/instructive cell culture medium. Although the experimental *in vitro* environment could potentially be insufficient for the differentiation of a given lineage, it can also unveil the potential of the isolated cells, such as an ectomesenchymal potential in trunk NCC from quail embryos (Calloni et al., 2007; Coelho-Aguiar et al., 2013). Finally, compared to the original cell culture approach, a three-dimensional culture of isolated NC progenitors (*crestospheres*) has been devised to identify their stem-cell potential, including self-renewal (Kerosuo et al., 2015). Long-term propagation of premigratory cells able to differentiate into multiple cell types *in vitro* has thus been obtained, confirming the existence of multipotent NC stem cells, at least *in vitro*.

Following these analyses *in vitro*, the next question is to identify and localize such multipotent and committed NC progenitors *in vivo*. Whether multipotent cells are found in the premigratory NC population or whether fate-restricted cells are already spatially arranged in the dorsal neural tube has been debated (Bronner-Fraser and Fraser, 1988; Collazo et al., 1993; Krispin et al., 2010; Nitzan et al., 2013; Serbedzija et al., 1994). Single-cell labeling in the trunk dorsal neural tube of chick embryos demonstrates that about half of the labeled cells give rise to more than one derivative (Bronner-Fraser and Fraser, 1988). In the same line of observations, recent *in vivo* clonal labeling strategies and high-resolution live imaging techniques have not identified obvious fate restrictions or spatial arrangement in the premigratory trunk NC progenitors as they are located in the dorsal quadrant of the spinal cord in chick and mouse embryos and move dorsally to exit the neural tube (Baggiolini et al., 2015; McKinney et al., 2013). Rather, the results suggest that NC fates are not fixed at early premigratory and migratory stages, and that a significant proportion of NCC are multipotent prior to EMT and as they initiate migration. However, a few cells seem already lineage-restricted early on, and the proportion of such cells seems higher in fish than in amniotes (Krispin et al., 2010; Raible and Eisen, 1994). As the current techniques cannot ensure the labeling and tracing of all cells in a given embryo, the exact proportion of multipotent versus restricted progenitors remains difficult to assess *in vivo*. The apparent discrepancies between the different studies will probably be understood in the future, perhaps with single-cell analyses (Lignell et al., 2017; Soldatov et al., 2019). Remaining questions include the developmental time point for each lineage restriction as well as the potential maintenance of lineage-restricted yet undifferentiated progenitors beyond this stage. These cells could continue to proliferate and provide NC derivatives in the longer term during development. Alternatively, as mentioned above, multipotency would be retained by Schwann cell precursors, which would in turn be the long-term source of NC derivatives in late embryos and at postembryonic stages (Motohashi et al., 2011; Singh et al., 2016).

The molecular regulations underlying NC multipotency remain incompletely understood. Frog and chick NC express many markers of multipotency (e.g. *Sox2, c-Myc, Oct4*) and other genes such as *Snail* and *Sox5*, which are also involved in multipotency in blastula cells (Buitrago-Delgado et al., 2015). Based on these common gene sets, it has been suggested that the NCC retain a multipotent state similar to blastula-stage multipotent cells, rather than reacquiring a multipotency program after neurulation. However, these genes are not exclusively found in multipotent cells and thus cannot be taken as a "multipotency signature" accounting for the unique ability of NC to form multiple derivatives (Briggs et al., 2018). To understand the mechanisms controlling NC multipotency, VENTX2/NANOG gene family was recently identified as a vertebrate-specific innovation, acting downstream of NB specification, essential for the expression of multipotency markers in the early NC, such as *pou5f3.2* and *tert,* and needed for the formation of the ectomesenchyme lineage but not for the sensory neuronal lineage (Scerbo and Monsoro-Burq, 2020). This study suggests that during their specification, NC progenitors acquire extended potency compared to adjacent cells, rather than retaining multipotency character from a blastula stage. In chick embryos, earlier in development at the open neural plate stage, a fraction of cells located at the neural/nonneural border coexpress markers of neural plate, NC, and placode progenitors, in variable proportion (Roellig et al., 2017). In terms of differentiation, the functional meaning of such gene coexpression is yet to be defined. Finally, SOX10 seems a key element in the NC multipotency GRN, as it is essential for the progenitor induction in multiple NC lineages and because its prolonged expression limits the commitment of several lineages while maintaining immature progenitors (Kim et al., 2003).

1.4   UPSTREAM OF NEURAL CREST EMERGENCE: THE PATTERNING OF THE NEURAL BORDER

In the last 15 years, many studies have focused on the mechanisms driving NC induction in the ectoderm. In this paragraph and the next one, we will detail the actors and the main steps for neural border and neural crest induction. The main players in NC fate acquisition have been alluded to above. As mentioned in introduction, the neural crest cells are specified within the posterior part of the neural (plate) border (NB) at the end of gastrulation. The NB itself is specified and patterned between the non-neural ectoderm (prospective epidermis) and the neural plate (prospective central nervous system) during early gastrulation, in a similar timing as the neural plate (Basch et al., 2006; De Crozé et al., 2011; Monsoro-Burq et al., 2005). The NB lies above the paraxial and intermediate mesoderm. The anterior edge of the neural plate, forming later the anterior neural fold, generates the preplacodal domain rather than NC (Schlosser and Ahrens, 2004; Streit, 2002). In the posterior part of the NB, in addition to the NC, three other ectodermal lineages are also generated: the posterior ectodermal placodes, the prospective dorsal quadrant of the central nervous system, and the nonneural ectoderm. At the time of NB specification, the progenitors of these four lineages are intermingled (Steventon et al., 2009). In the following part, we detail the current understanding of NB formation followed by NC induction within the neural border.

1.4.1   NEURAL BORDER PATTERNING AND THE CORRESPONDING GENE REGULATORY NETWORK

Multiple *in vitro* and *in vivo* studies have defined the essential role of tissue interactions during NB and NC induction. NB induction in the ectoderm is controlled by the combination of different factors secreted by the surrounding nonneural ectoderm, neural plate and underlying mesoderm. In frog and chick embryos, when early neural plate explants are isolated *in vitro,* they do not form NC, whereas NC markers expression is induced when neural explants are cultured adjacent to nonneural ectoderm fragments. In those pieces of tissue, as observed in the embryo, the NC markers are detected at the neural/nonneural tissue boundary (Mancilla and Mayor, 1996; Selleck and Bronner-Fraser, 1995). Similarly, when early frog paraxial/intermediate mesoderm is juxtaposed to pluripotent ectoderm taken from the blastocoel roof ectoderm, expression of NB and NC markers is robustly activated in the ectoderm (Bang et al., 1999; Bonstein et al., 1998; Monsoro-Burq et al., 2003). Moreover, *in vivo* excision of the paraxial mesoderm in frog embryos results in the loss of neural crest (Bonstein et al., 1998; Marchant et al., 1998; Steventon et al., 2009). Thus, all the tissues immediately surrounding the prospective NB are interacting with the ectoderm at gastrulation stages to specify this novel ectoderm territory. Recent experiments using chick blastula-stage ectoderm even suggest that the bias towards NB and NC fate could occur prior to gastrulation (Prasad et al., 2019).

The interactions between the neural plate, the nonneural ectoderm and the mesoderm are mediated by three main groups of signals conserved between species. At gastrula stage, WNT and BMP signals secreted by the nonneural ectoderm and the underlying mesoderm maintain the nonneural ectodermal fate and repress the neural fate (Faure et al., 2002; Garcia-Castro et al., 2002). The axial mesoderm and the medial part of the neural plate secrete diffusible BMP antagonists, such as Chordin, Noggin, Follistatin, and Cerberus, allowing the expression of neural genes (Plouhinec et al., 2013). As a result, a BMP gradient is generated in the ectoderm: high BMP activity in the nonneural ectoderm promotes the epidermal fate and represses the neural fate; moderate BMP levels induce NB genes such as *pax3*, *zic3*, *msx1*, and *hes4*; and a low BMP activity induces a neural fate and represses both NB and ectoderm fates (Brugger et al., 2004; Garnett et al., 2012; Marchant et al., 1998; Mayor et al., 1995; Nichane et al., 2008b; Schumacher et al., 2011; Suzuki et al., 1997; Tribulo et al., 2003). In chick and frog embryos, BMP4 is the main effector of BMP activity acting during NB induction (Patthey et al., 2009; Steventon et al., 2009). In zebrafish embryos, BMP signaling is established by the cooperation of BMP2b and BMP7 (Nguyen et al., 1998; Schmid et al., 2000).

However, the modulation of BMP signaling alone is not sufficient to induce NB and NC markers robustly, suggesting that BMP signaling cooperates with other signaling pathways in the embryo (LaBonne and Bronner-Fraser, 1998). Accordingly, interfering with canonical WNT signaling prevents NB induction (Chang and Brivanlou, 1998; LaBonne and Bronner-Fraser, 1998; Saint-Jeannet et al., 1997). In all model species studied, canonical WNT ligands (WNT6, WNT7b and WNT8) are involved in NB and NC induction, although they come from different sources according to the species considered; for example, in frog and zebrafish, Wnt8 originates from the paraxial mesoderm and Wnt7b from the surface ectoderm, while in chick, WNT6 originates from the nonneural ectoderm (Bang et al., 1999; Chang and Hemmati Brivanlou, 1998; Elkouby et al., 2010; Garcia-Castro et al., 2002; Hong et al., 2008; Lekven et al., 2001; Lewis et al., 2004; Schmidt et al., 2007; Steventon et al., 2009; Wilson et al., 2001). As a result, a broad ectoderm area including the NB responds to WNT–βcatenin signaling *in vivo* (Borday et al., 2018). The activation of WNT signaling alone is not sufficient to induce NB markers in the ectoderm but when combined with a low dose of BMP antagonists to establish moderate levels of BMP signaling, NB/NC marker expression is potently activated in the nonneural ectoderm *in vitro* (Chang and Hemmati Brivanlou, 1998; De Crozé et al., 2011; LaBonne and Bronner-Fraser, 1998; Monsoro-Burq et al., 2005; Saint-Jeannet et al., 1997).

In addition to WNT and BMP signaling, FGFs also participate in neural border induction, likely by the action of FGF ligands secreted by the mesoderm (Christen and Slack, 1997; Monsoro-Burq et al., 2003; Villanueva et al., 2002). Therefore, a finely tuned combination of BMP, WNT, and FGF signaling is established at the neural border, between the nonneural ectoderm area (with high BMP and WNT signaling) and the neural plate area (with absence of BMP signaling), resulting in the establishment of the NB territory. The integration of these multiple inputs on the regulatory sequences of target genes is only starting to be explored; for example, separate enhancers of fish genes *pax3a* and *zic3a* are shown to respond differentially to WNT, FGF, and BMP signals at the neural border (Garnett et al., 2012).

Finally, on top of the main signaling pathways mentioned above, WNT/PCP, Hedgehog, and retinoic acid signaling also participate in NB induction. For example, Ror2, a receptor of the WNT/PCP pathway, upregulates the NB expression of *Gdf6*, another BMP ligand at the neural border (Schille et al., 2016). Ror2 therefore modulates the levels of BMP signaling in NB ectoderm. Similarly, retinoic acid acts as a posteriorizing signal involved in the anterior–posterior axial patterning of the neural plate and the NB. Retinoic acid signaling thus participates in the regulation of neural crest markers expression in frog embryos (Villanueva et al., 2002). Other signaling pathways, such as Indian Hedgehog or its effector Gli2, also take part in the patterning of this paraxial part of the dorsal ectoderm (Agüero et al., 2012; Cerrizuela et al., 2018).

All the signaling pathways mentioned above trigger the expression of the NB specifiers defined as factors necessary for NB patterning and, consequently, for the subsequent steps of the NC-GRN (see Definitions, Meulemans and Bronner-Fraser, 2004). The activity of each individual NB specifier is essential for NC cell induction. To date, most identified NB specifiers are transcription factors, controlling the expression of target genes acting in the NC-GRN. Molecularly, the NB is characterized by the overlapping expression of several NB specifiers; a few are restricted to the NB domain, such as *pax3* and *pax7*, while others are expressed in broader territories encompassing the NB and non-neural ectoderm on one hand (e.g. *tfap2*, *hairy2*, *msx1*), or the NB and neural plate on the other hand (e.g. *zic1* and *gbx2*) (Figure 1.3, Milet and Monsoro-Burq 2012; Simoes-Costa and Bronner 2015). Once activated, these NB specifiers engage collaboratively in the regulation of one another, which maintains/reinforces their continuous expression (De Crozé et al., 2011; Li et al., 2009; Nichane et al., 2008a). We summarize below the roles of the best-studied NB specifiers in neural border and neural crest patterning in multiple vertebrate animal models: fish, frog, chick, and mouse.

1.4.1.1   Tfap2a (Transcription Factor Activating Enhancer-Binding Protein 2 Alpha)

Tfap2a is a transcription factor belonging to a family of four members: Tfap2a, b, c, e. The earliest roles of Tfap2a during NB and NC induction have been mainly studied in frog embryos. *Tfap2a* is expressed in frog non-neural ectoderm at the onset of gastrulation (Figure 1.3; Luo et al., 2003). *Tfap2a* is then the earliest NB specifier known to be upregulated in the NB region at mid-gastrula stage, as an immediate-early target of WNT signaling (De Crozé et al., 2011; Luo et al., 2003). The activity of Tfap2a is important at all subsequent stages of the NB/NC-GRN, from NB specification to later stages of NC development. Firstly,

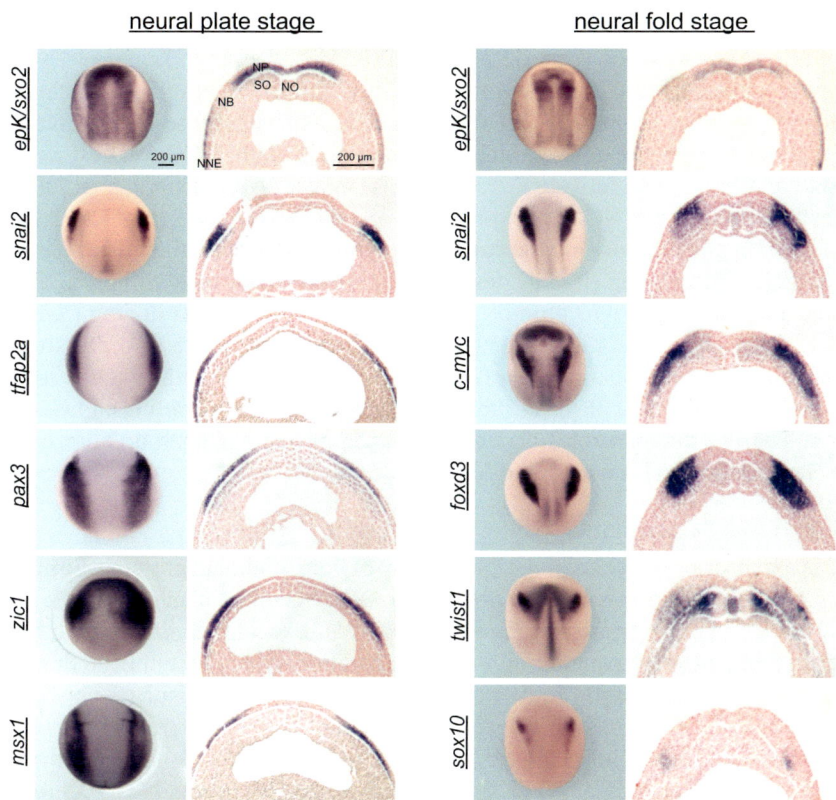

FIGURE 1.3   Expression pattern of selected genes during neural border and neural crest early development. Whole-mount *in situ* hybridization and their corresponding cross-sections are shown for the markers commonly used for neural plate (NP), neural border (NB), non-neural ectoderm (NNE) and neural crest in *Xenopus laevis* embryos, dorsal views, anterior up. On sections, somites (SO) and notochord (NO) are also indicated. During neural plate and neural fold stages, the double staining for epidermal keratin *epk* (labels the NNE) and *sox2* (labels the neural plate) leaves unlabeled the intermediate region corresponding to the NB region where the NC marker *snai2* is appears. During neural plate stage, *tfap2a, pax3, zic1* and *msx1* are expressed at the NB region. During neural fold stage, *c-myc, foxd3, twist* and *sox10* are strongly expressed in the premigratory NC.

Tfap2a cooperates with WNT signaling to activate the other NB specifiers, for example *pax3*, by direct regulation, as well as *msx1* and *hes4* by indirect regulation. Secondly, Tfap2a activity is required during NC induction for *snai2* and *foxd3* activation. Finally, Tfap2a is needed for NC migration (De Crozé et al., 2011). In fish and chick embryos, the role of Tfap2a during NB induction is still incompletely described. Due to zebrafish genome duplication, the effects of early Tfap2a depletion are compensated by paralog genes activity. The double mutation of both Tfap2a and Tfap2c is needed to reveal a NB phenotype in fish (Li and Cornell,

2007). In mouse, as in frog, *Tfap2a* is expressed in the nonneural ectoderm and in the NB at early stages (Schorle et al., 1996). Mice lacking *Tfap2a* die prenatally, with severe craniofacial abnormalities and failure of cranial tube closure. In contrast to what is observed in frog and fish embryos, those mouse mutants still express *Pax3* in the dorsal brain and in the emigrating NCC, suggesting that alternative mechanisms activate *Pax3* in cranial mouse NCC (Schorle et al., 1996). Interestingly, in human NCC derived from hES cells grown *in vitro*, TFAP2a acts as a pioneering factor, opening the chromatin around NC genes allowing the initiation of their transcription (Rada-Iglesias et al., 2012). Consistent with these observations, Tfap2a binds *sox10* promoter in fish and *pax3* promoter in frog (Arduini et al., 2009; De Crozé et al., 2011). As in frog, zebrafish Tfap2a also plays a late role during NC cell diversification (Arduini et al., 2009). Together, these results demonstrate that Tfap2a plays a crucial role at multiple stages of NB and NC formation. These sequential roles depend on Tfap2a partners Tfap2b or Tfap2c, acting as heterodimers (Rothstein and Simoes-Costa, 2020).

1.4.1.2   Gbx2: Gastrulation Brain Homeobox 2

Gbx2 is a transcription factor of the Gbx family of homeodomain proteins. Similarly to *tfap2a*, *gbx2* is expressed in the prospective NB/NC domain and the in the nonneural ectoderm of frog gastrulas (Li et al., 2009). Unlike *tfap2a*, *gbx2* expression is posterior to the prospective midbrain-hindbrain boundary and absent from the most anterior part of the embryo (Tour et al., 2002). A direct comparison of *tfap2a* and *gbx2* patterns would be useful for exploring whether they may coregulate a subset of genes at the NB. As shown for *tfap2a*, *pax3* and *msx1*, *gbx2* is also an immediate-early target of canonical WNT signaling, regulated by βcatenin/TCF3 binding sites present on *gbx2* promoter (Li et al., 2009). Gain and loss of function experiments show that Gbx2 acts upstream of *pax3* and *msx1*, regulates their expression at the neural border, and is required for early NC cell induction and specification (Li et al., 2009). During chick development, *GBX2* is first expressed at early gastrula stage (HH4) in the central part of the epiblast, followed by expression in the neurectoderm (Niss and Leutz, 1998). In zebrafish, cross-sections at gastrula stage show that *gbx2* is expressed in the endomesoderm but not in the ectoderm; later on at 100% epiboly stage, fish *gbx2* is detected at the border of the neural plate and in the ectoderm (Rhinn et al., 2003). In mice, *gbx2* is expressed in the neurectoderm and underlying mesoderm at E7.5 stage (Wassarman et al., 1997). Later on, mouse *Gbx2* is observed in the pharyngeal arches (Byrd and Meyers, 2005). *Gbx2* knockout mice present NC pharyngeal arches defects, especially in fourth arch derivatives (Byrd and Meyers, 2005). In sum, frog studies show a crucial role of Gbx2 during early neural border establishment and a role in the NC-GRN is supported by the mouse mutant phenotype.

1.4.1.3   Msx1 and Msx2: Muscle Segment Homeobox 1 and 2

Msx1 and Msx2, previously named Hox7 and Hox8 respectively, are homeodomain-containing transcription factors. *Msx1* and *msx2* genes were identified originally as vertebrate orthologs of the *Drosophila* gene *msh* expressed in fly segmental muscles (Davidson, 1995). *Msx* genes are BMP signaling immediate-early responsive genes, maternally expressed in the whole ectoderm (Suzuki et al., 1997; Tribulo et al., 2003). *Msx1* gene expression is downregulated in areas with low BMP signaling, thus *msx1* expression is cleared from the neural plate ectoderm during gastrulation (Suzuki et al., 1997). In mid-gastrula stage frog embryos, *msx1* is expressed at the NB and in the non-neural ectoderm (Figure 1.3). At this stage, *msx1* expression is controlled both by FGF8 and by canonical WNT signaling (Monsoro-Burq et al., 2005). Epistasis experiments show that *msx1* regulation by FGF8 does not seem to involve intermediary WNT signals. Moreover, Msx1 can rescue NC induction after FGF8 depletion but not after the loss of WNT signals, suggesting that Msx1 is not sufficient for NC induction by WNT signals. Finally, Pax3 is able to compensate for the depletion of Msx1, indicating that Pax3 acts downstream of Msx1 (Monsoro-Burq et al., 2005). In chick embryos, *msx1* is expressed in the NC cells (Suzuki et al., 1991). In zebrafish, MsxB, MsxC and MsxE act redundantly; their individual inhibition only causes mild defects, while the inhibition of all three blocks early NC cells differentiation (Phillips et al., 2006). In mouse embryos, *Msx1* and *Msx2* expression starts in late gastrulas at E7.5. However, potential early roles in NB/NC induction are not described. *Msx1/2* double mutant mice show increased apoptosis in cranial and cardiac NC cell derivatives, resulting in craniofacial and heart abnormalities (Ishii et al., 2005). Also consistent with a function in craniofacial formation, *Msx1* and *Msx2* are upregulated in *Hdac3* knockout mice which display severe craniofacial abnormalities (Singh et al., 2013). This positions Msx1 and Msx2 as important players acting during NCC induction and later steps of differentiation and craniofacial development.

1.4.1.4   Zic1: Zinc Finger Protein 1

Zic genes form a family of five zinc finger transcription factors, which are vertebrate homologs of the *Drosophila* pair-rule gene odd-paired (*opa*) (reviewed by Merzdorf, 2007). In frog, *zic1* (previously named *opl*) is first expressed in the neurectoderm during gastrulation; then, during neurulation, its expression becomes restricted to the NC-forming NB and the preplacodal anterior neural border (Figure 1.3; Sato et al., 2005). Zic1 overexpression expands the NC domain, while Zic1 knockdown abolishes expression of the NC markers *snai2* and *foxd3* (Sato et al., 2005). Gain-of-function of Zic1 alone promotes the placodal fate, but when Zic1 is combined with Pax3, they trigger NC formation (Hong and Saint-Jeannet, 2007; also discussed in the Pax3 chapter below). Zic1 depletion does not prevent the expression of the NB marker Pax3 (*pax3* expression is rather expanded; Sato et al., 2005). In chick, *Zic1* is expressed in the developing nervous system, ear, and somites, but its potential expression at neural-plate stage has not yet been described (Rhodes and Merzdorf, 2006). In mouse embryos, *Zic* genes start to be expressed early on during nervous-system development around E7.5 and are thus compatible with a potential role during NB formation (Inoue et al., 2004). *Zic2* and *Zic5* mutant mice have fewer NCC, resulting in severe defects of NC derivatives (Elms et al., 2003; Inoue et al., 2004). In contrast, no NC development defect was mentioned in *Zic1* mutant mice, which were studied for cerebellar development and behavioral alterations (Aruga et al., 1994; Ogura et al., 2001). In order to reveal potential functional redundancy between Zic1-5, double-mutant mice have been generated. Compound *Zic1* and *Zic3* mouse mutants present a severe phenotype in the central nervous system, but no NC defect was described (Inoue et al., 2007).

1.4.1.5 Pax3/7: Paired Box Gene 3/7

Pax3 and *Pax7* belong to a family of nine *PAX* genes, homologs of the Drosophila melanogaster *paired* segmentation gene. In addition to the paired DNA-binding motif, Pax3 and Pax7 possess a homeodomain DNA-binding domain and a conserved octapeptide region mediating protein–protein interactions (reviewed in Monsoro-Burq, 2015). In *Xenopus laevis*, *pax3* expression appears in the prospective NB ectoderm as gastrulation starts and remains mostly restricted to the NB and dorsal neural tube later on (Figure 1.3; De Crozé et al., 2011). As mentioned above, frog *pax3* is directly regulated by Tfap2 in combination with WNT signaling. The relative expression levels of Pax3 and Zic1 in ectoderm cells regulate the induction of three distinct cell populations arising at the NB: the NC cells, the ectodermal placodes, and the hatching gland. The hatching gland is an amphibian-specific ectoderm structure, secreting an enzyme needed to degrade the vitelline envelope prior to embryo hatching. High levels of Pax3 alone promote the hatching gland fate, whereas Zic1 high expression promotes the ectodermal placodes fate (Hong and Saint-Jeannet, 2007). In contrast, the balanced coexpression of these two transcription factors in frog blastula-stage pluripotent ectoderm is necessary and sufficient to trigger all the developmental steps of NC formation, including NC induction, NCC migration along NC characteristic routes, and progenitor cell differentiation into multiple derivatives (Hong and Saint-Jeannet, 2005; Milet et al., 2013; Monsoro-Burq et al., 2005). Similarly, in early chick gastrulas, *Pax7*, which is a *Pax3* paralog, is expressed in the NB region and is crucial for NCC induction and specification (Basch et al., 2006). Interestingly, in *Xenopus laevis*, *pax7* is not expressed in the early NB/NC domain but plays an essential role in the paraxial mesoderm during NC induction (Maczkowiak et al., 2010). This is an example of subtle species-to-species variation in the NB/NC-GRN, where paralog gene usage has been swapped during evolution. In zebrafish, the activation of *pax3a* and *zic3a* in the NB is controlled by multiple enhancers, which respond differentially to WNT, FGF, and BMP signals to drive gene expression in specific areas of the NB (Garnett et al., 2012). In mice, a *Pax3* mutation has appeared spontaneously in the Splotch mouse line. Heterozygous Splotch mice present perturbed cranial NCC migration and abnormal pharyngeal arches. Moreover, homozygous Splotch mice die *in utero* from cardiac malformations linked to cardiac NCC developmental defects (Epstein et al., 2000; Li et al., 1999). At cranial levels, however, NCC development is not obviously dependent upon *Pax3/7* function: in mouse double mutants lacking both *Pax3* and *Pax7*, NCC form, migrate, and undergo normal early development (Zalc et al., 2015). The difference between the strict need for PAX3 or PAX7 during early NC induction in frog, fish, and chick and the normal NC development in mouse lacking both genes is a striking example of model-specific variation in the vertebrate NB/NC-GRN.

1.4.1.6 Hes4 (Previously Named Hairy2b) Class B Basic Helix-Loop-Helix Protein 4

Hes4 is a transcription factor of the Enhancer of Split bHLH family. In frog, *hes4* is expressed in the non-neural ectoderm and enriched at the neural border (De Crozé et al., 2011).

Hes4 acts downstream of FGF and BMP signaling at the NB and maintains undifferentiated NCC survival and proliferation by a cell-autonomous mechanism (Nagatomo and Hashimoto, 2007; Nichane et al., 2008b). During NC cell differentiation, Hes4 binds to the FGF4/Stat3 complex and activates the Id3 non-cell autonomously using Delta-Notch signaling. In turn, Id3 promotes NC cell proliferation and differentiation (Nichane et al., 2008a). Therefore, Hes4 plays an important role in regulating the sequence of events that maintains NCC progenitors in an undifferentiated state, then for the switch promoting NCC proliferation and differentiation. In mouse and other embryonic models, the role of HES4 during early NB border patterning is not yet described.

In summary, this paragraph has detailed the function of the main neural border specifiers in multiple model species. TFAP2a, GBX2, MSX1/2, ZIC1, PAX3/7, and HES4 are critical regulators of the NB specification in several species, indicating a global conservation of the vertebrate neural border GRN. However, this comparison also highlights some species-specific variations, suggesting that additional or alternative pathways may be used, especially in mammals. Epistasis analyses have been useful to understand how these factors control the spatiotemporal sequence of NB/NC induction. In contrast, little is known about their potential protein–protein interactions or their cooperation on target gene regulatory elements. In this brief description, some other NB specifiers described in individual species have not been included, such as AXUD1 in chick or Znf703 in frog (Janesick et al., 2019; Simões-Costa et al., 2015). The list of NB specifiers examined in multiple model species will probably be extended in the upcoming years. In addition to transcription factors, it may include other types of protein function such as signaling factors and epigenetic and metabolic modifiers.

The current NB-GRN has also been described in earlier-derived vertebrates such as the sea lamprey *Petromyzon marinus*, showing a global evolutionary conservation over 500 million years (Sauka-Spengler et al., 2007). For example, homologs of *MsxA*, *ZicA*, and *Pax3/7* are expressed in early lamprey embryos and play the role of NB specifiers as described in frog embryos (Medeiros, 2013; Nikitina et al., 2011; Nikitina and Bronner-Fraser, 2009). In contrast to the conservation of a similar genetic NB-GRN in vertebrate models, the timing of neural border specification differs greatly from one model to another. This is, in part, linked to the different duration of embryonic development in the various species. In non-amniote vertebrates, such as frog and zebrafish, NB induction initiates at early gastrula stage and takes place within a few hours of gastrulation (Arduini et al., 2009; Hong and Saint-Jeannet, 2007; Sato et al., 2005). In amniotes (e.g. chick and mouse), but also in earlier-derived anamniote lamprey embryos, NB induction was described to last longer during gastrulation and beyond (Nikitina 2009; Khudyakov and Bronner-Fraser, 2009).

While NC cells are unique to vertebrates, the molecular actors described so far in the NB network are also conserved in invertebrate chordate models. Cephalochordates, such as amphioxus, and urochordates, such as ascidians, express homologs of *Pax3/7* and *Msx* in cells located between the neural plate and epidermis (Imai et al., 2004; Medeiros, 2013). These progenitors generate cell types similar to some vertebrate NCC derivatives such as pigment cells and migratory neuronal progenitors (Stolfi et al., 2015). These data suggest that at least some of the early steps in the NB/NC-GRN were set up in the common chordate ancestor.

1.4.2   NEURAL CREST INDUCTION AT THE NEURAL
BORDER AND ITS MAIN REGULATORS

Once activated, the NB specifiers trigger the expression of another set of transcription factors: the NC specifiers, each essential for NC induction and further development, but not for NB formation (Meulemans and Bronner-Fraser, 2004). This group includes *Snai1* and *Snai2* (also called *Snail1, Snai2*); *Foxd3*; *Twist1*; *Sox8, 9,* and *10* (soxE genes); and *c-Myc*. In addition to their role in pre-migratory and early migrating NC cells (Figure 1.4), they can also be expressed and have a significant role during later stages for NCC survival or during NC cell differentiation (e.g. *Sox10*). Similarly, some neural border specifiers also act as *bona fide* neural crest specifiers independently of their earlier role (*e.g.* Tfap2a [De Crozé et al., 2011]).

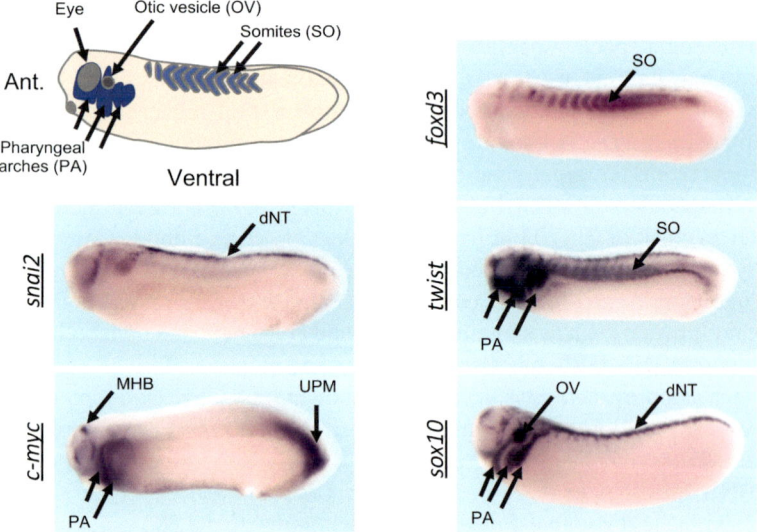

FIGURE 1.4   During NC migration, NC specifiers *twist1* and *sox10* mark migrating NCC while other early markers are no longer expressed in NC. During organogenesis in frog tailbud stages, the cranial NCC migrate into the mandibular, hyoid, third and fourth pharyngeal arches (PA). The frog *mandibular stream* nomenclature includes the cells migrating around the eye and towards midface, corresponding to frontonasal streams in chick and mammals. At trunk levels, the neural crest lies on top of the neural tube undergoing EMT (at anterior trunk levels) or is still located in the dorsal part of the neural tube (dNT) prior to EMT (at posterior trunk levels). At this stage, *snai2* expression has faded out in cranial NC after EMT, weakly labelling the dorsalmost part of the migrating streams. Instead, *snai2* robustly marks the trunk NC. At cranial levels, *c-myc* is expressed in pharyngeal arches, stronger in the second arch. *C-myc* is also found at the mid-hindbrain boundary (MHB) and in the unsegmented paraxial mesoderm (UPM). In tailbud stage embryos, *foxd3* is no longer expressed in the neural crest, but labels the somites and anterior UPM. *Twist1* and *sox10* are strongly expressed in all NC streams. *Twist1* is also expressed in the somites and trunk NC. *Sox10* also labels the otic vesicle and the dorsal premigratory trunk neural crest. dNT: dorsal neural tube, BA: branchial arches, SO: somite, MHB: Midbrain-Hindbrain boundary, UPM: unsegmented paraxial mesoderm, OV: otic vesicle.

NC specifiers thus mainly control induction and expansion of the NCC population, cell proliferation and maintenance of multipotency, and the EMT and differentiation into distinct cell types. In addition to the simplified hierarchical NC-GRN proposed initially by Meulemans and Bronner-Fraser (2004), feedback regulatory loops, epigenetic modifications, the reiterated use of a given NC specifier at several different stages, and metabolic regulations render the NC-GRN more complex and still incompletely understood (De Crozé et al., 2011; Figueiredo et al., 2017; Plouhinec et al., 2014; Strobl-Mazzulla et al., 2010). Below, we highlight the main roles of the best-described NC specifiers.

1.4.2.1   Snai1 and Snai2: Snail Family Zinc Finger 1, 2

Snail1 and Snail2 are the best-studied NC specifiers. These zinc-finger transcriptional repressors are involved in neural crest specification and EMT. In *Xenopus*, at the neurula stage, both genes are expressed in the pre-migratory NC (Figure 1.3). In contrast, in chick embryos only *Snai2*, and in mouse embryos only *Snai1*, is expressed in NC progenitors prior to EMT (Nieto, 2018). Snai1/2 overexpression enlarges the NC population, while their inhibition blocks NC specification and migration. *Snai2* gene regulatory regions contain LEF/TCF and Smad1 binding sites, which are directly regulated by WNT and BMP signaling (Vallin et al., 2001). In addition, *snai1* and *snai2* are direct targets of Zic1 and Pax3 respectively, during frog neurulation (Bae et al., 2014; Plouhinec et al., 2014). In turn, Snai1 and Snai2 directly repress *cdh1* (*e-cadherin*) expression, which is essential to trigger the EMT. For this repression, frog Snai2 recruits the Polycomb repressive complex 2 (PRC2) in order to directly downregulate *cdh1* expression (Tien et al., 2015). Similarly, in chick, SNAI2 interacts with another cofactor, LMO4, to repress *Cdh1* during EMT (Ferronha et al., 2013). Also in chick, SNAI2 interacts with SIN3A, HDAC, and PHD12, forming a complex that is able to deacetylate histone H3 on *Cad6b* promoter, resulting in CAD6B downregulation during NCC migration (Strobl-Mazzulla and Bronner, 2012). SNAI1 and SNAI2 do not seem to be essential for NC formation in mouse embryos since *Snai1* or *Snai2* mutant NCC do form, migrate, and differentiate normally, potentially by a redundant activity of mouse-specific *Snai3* gene (Bradley et al., 2013; Jiang et al., 1998; Murray and Gridley, 2006).

1.4.2.2   FoxD3: Forkhead Box D3 (Winged Helix Transcription Factor)

FoxD3 belongs to the Forkhead protein family. It encodes a transcriptional repressor characterized by a DNA-binding domain and transcriptional activation or repression domain (Wijchers et al., 2006). *Foxd3* is expressed in the pre-migratory (Figure 1.3) and migratory NC cells of many vertebrates. It is required for NC development, in particular for the maintenance of NC multipotency. During early development in chick, FOXD3 prevents precocious NC emigration and inhibits melanogenesis in NC cells, allowing the differentiation of other derivatives (Kos et al., 2001; Mundell and Labosky, 2011; Nitzan et al., 2013). In frog, blockade of Foxd3 activity, using a dominant-negative form of the protein, disturbs early NC formation. Interestingly, this phenotype can be rescued by Snai2 gain-of-function, suggesting that Snai2 mediates or mimics part of Foxd3 activity in early NCC (Sasai et al., 2001). In fish embryos, Foxd3 does not seem to be required for NC induction, but it is important for later differentiation of several NC derivatives, as *foxd3* morpholino-mediated knockdown affects the differentiation of jaw cartilage, peripheral neurons, glia, and only irido-phores. In contrast, other pigment cell lineages, melanophores and xanthophores, are not affected by *foxd3* depletion, suggesting a specific role for Foxd3 in restricting fate choices of NC progenitors. Finally, *Foxd3* mutant mice show increased apopto-sis in NCC, resulting in severe loss of multiple NC derivatives such as the branchial arches and the enteric nervous system (Teng et al., 2008). At 9.5 dpc, *Foxd3* expres-sion is restricted to the presumptive cranial and dorsal root ganglia and is absent in cranial and cardiac NC mesenchyme. Downregulation of mouse *Foxd3* guides NC cells towards a mesenchymal fate (Mundell and Labosky, 2011) suggesting that, in addition to the role in NC multipotency and self-renewal, FOXD3 is also required for lineage decision by inhibiting mesenchymal differentiation.

1.4.2.3   Twist1: Twist Basic Helix-Loop-Helix Transcription Factor 1

The basic helix-loop-helix (bHLH) transcription factor TWIST1 is important for NC induction and survival and is also a major regulator of EMT. In frog embryos, Twist1 is an early NC specifier (Figure 1.3), required for fate determination and cranial NC formation. Interestingly, at later stages, Twist1 loss of function results in reduced craniofacial cartilage preceded by a decreased expression of *sox9* which marks the developing chondrocytes, and an increased expression of the cranial glial cell maker *foxd3*, suggesting that Twist1 favors the cartilage fate (Lander et al., 2013). Twist1 interacts directly with Snai1 and Snai2 proteins via its WR domain. When Twist1 is phosphorylated by GSK3b, Snai1 and Snai2 activity is inhibited (Lander et al., 2013). In mouse embryos, *Twist1* is first detected in the NC domain and the head mesenchyme that is populated by the cranial NC cells (Gitelman, 1997; Soo et al., 2002). TWIST1 is important for the directionality of NC migration and for the proper differentiation of the first branchial arch into bone, muscle, and teeth (Soo et al., 2002). In addition, recent scRNAseq data in mouse and experimental manipulation of *Twist1* expression in chick demonstrates that TWIST1 regulates and promotes the mesenchymal fate choice (Soldatov et al., 2019). Finally, *Twist1-/-* mouse embryos present malformed pharyngeal arches, illustrating its important role at all steps of the NC-GRN for NC EMT, migration, and differentiation.

1.4.2.4   Sox8, 9, 10: SoxE family, Sex-Determining Region Y: SRY-Box 8,9 10

SoxE family members SOX8, SOX9, and SOX10 are transcriptional activators essential for NC specification, maintenance, and survival. However, each SoxE fac-tor displays distinct features in the NC-GRN. In frog, *sox8* and *sox9* are among the NC specifiers expressed earliest in the NB region, prior to *snai2* and *foxd3*, whereas *sox10* is expressed later in the pre-migratory NC cells (Figure 1.3, Alkobtawi et al., 2018; Hong and Saint-Jeannet, 2005; Spokony et al., 2002). Both *sox9* and *sox10* are activated by the canonical WNT pathway, and *sox10* expression is controlled by Sox9 and Snail2 (Aoki et al., 2003; Hong and Saint-Jeannet, 2005; Spokony et al., 2002). *Sox8* loss of function affects NC induction, resulting in a loss of NC deriva-tives. Similarly, time-controlled inhibition of Sox9 has evidenced its early func-tion. At neurula stage, Sox9 blockade expands the neural plate and prevents NC progenitor induction. In contrast, after the induction stage, Sox9 inhibition does not affect cranial NC EMT and migration (Lee et al., 2004). This result limits the role of Sox9 to the early NC specification stage in frog neurulas. Later on, at differentiation stages, Sox9 is critical for chondrocyte differentiation (Lefebvre and de Crombrugghe, 1998). Frog *sox10* is expressed starting at the mid-neurula stage and is maintained throughout subsequent steps of NC formation until differentiation. Although Sox10 is critical for the formation of all peripheral nervous system NC derivatives, Sox10 overexpression at neural plate stages leads to an increased number of pigment cells. This suggests that Sox10 also biases early NC towards a pigment cell lineage fate, in agreement with its important role in the control of melanogenesis and *mitf* expression as mentioned above (Aoki et al., 2003; O Donnell et al., 2006; Spokony et al., 2002). In zebrafish, *sox8* is not detected in the NC cells or their derivatives, whereas *sox9b* is expressed early in the NC progenitors and *sox10* is expressed later in the pre-migratory NC cells. Fish *sox10* mutations do not perturb NC induction but rather affect NCC migration (Dutton et al., 2001). In chick and mouse embryos, *Sox9* and *Sox10* are expressed prior to *Sox8* expression (Cheng et al., 2000; Cheung and Briscoe, 2003; Southard-Smith et al., 1998). In chick, SOX9, ETS1 and c-MYB activate *Sox10* expression in cranial NC (Betancur et al., 2010). In mouse, SOX9 seems to play later roles than in the frog. The conditional inactivation of *Sox9* using the *Wnt1-Cre* driver line (Table 1.2) prevents SOX9 activity after NC induction and prior to EMT. This inactivation results in an abnormal craniofacial formation, including cleft palate, leading to respiratory distress and therefore death at birth (Mori-Akiyama et al., 2003). Moreover, unlike in frog and fish, *FoxD3* expression appears earlier than *Sox9* in mice and is unaffected by Sox9 mutation (Cheung et al., 2005). Finally, *Sox10* knockout mice show disrupted NC development and increased NC cell death, implicating SOX10 as essential player in the survival of NCC (Mollaaghababa and Pavan, 2003). In sum, the regulation of SoxE family gene spatiotemporal expression and function varies significantly between species. This is another example of the species-specific usage of paralog genes at different steps of the NC-GRN.

1.4.2.5   C-Myc: Protooncogene

The protooncogene c-MYC acts as an important regulator of NC cell stemness. In frog, *cmyc* is detected very early in the NB region, the anterior neural fold and the preplacodal ectoderm (Figure 1.3). *C-myc* knockdown results in loss of NC progenitors accompanied by the expansion of the central nervous system marker *sox3*, indicating that c-Myc regulates early ectodermal cell fate decisions between neural and neural crest fates (Bellmeyer et al., 2003). Moreover, c-Myc directly regulates Id3 in the NC, which in turn maintains NC cells in a progenitor state (Light et al., 2005) This is reminiscent of the regulation described above for Hes4 (Nichane et al., 2008a). Together, c-Myc and Id3 control the formation and the maintenance of NC early progenitors. In chick embryos, *c-Myc* is expressed in the pre-migratory NC cells, binds to MIZ1, and regulates the size of the NC cell pool; this promotes the self-renewal and survival of NC progenitors (Kerosuo and Bronner, 2016). In mouse embryos, c-MYC inactivation leads to smaller craniofacial skeleton elements and middle ear and pigmentation defects, further demonstrating the crucial role of c-MYC in mouse neural crest generation (Wei et al., 2007).

1.5   A LONG JOURNEY: NEURAL CREST CELL EMIGRATION FROM THE NEURAL BORDER TOWARDS TARGET ORGANS

The EMT is a complex process in which epithelial cells lose their apical–basal polarity, modify their cell–cell adhesion properties, and convert to a mesenchymal-like phenotype characterized by high motility and invasive behavior (Nieto, 2011). The coordinated action of the NC specifiers promotes NC EMT. Thus, after induction and specification at the NB, the NC cells undergo EMT and migrate throughout the embryo along defined routes to reach their final locations and sites of differentiation. At a molecular level, the EMT includes the reorganization of cytoskeletal proteins, the loss of cell–cell junction proteins and polarity complexes, and an increased expression of metalloproteases (MMPs) (Newgreen and Gibbins, 1982; Nieto, 2011). By secreting those enzymes, NCC remodel the extracellular matrix as they migrate. Interestingly, the EMT events occurring during embryonic morphogenesis, particularly during NC development, share similar regulatory molecular mechanisms with the EMT processes involved in cancer progression and metastasis (Nieto, 2011). NC EMT is thus an excellent model for understanding the principles of EMT in non-malignant conditions.

Since the NC EMT occurs during early embryonic development, vertebrate models that develop externally, such as the frog, chick, and zebrafish, were largely used to study this complex phenomenon. Cranial NC delaminate before neural tube closure in mouse and frog models (Nichols, 1981; Sadaghiani and Thiebaud, 1987), while EMT in chicken embryos begins during the fusion of the neural folds (Theveneau et al., 2007). In addition to transcriptional regulation by NC specifier genes, several signaling pathways are known to control NCC EMT and migration. For example, the activation of WNT and BMP signals is important for trunk NC delamination by coordinating NCC exit from the dorsal neural tube to the cell cycle at the moment of the G1/S transition (Burstyn-Cohen et al., 2004). Additionally, during paraxial mesoderm segmentation, the somites control BMP signaling levels in the dorsal neural tube, which triggers NC EMT in coordination with the morphogenesis of the surrounding tissues (Sela-Donenfeld, and Kalcheim, 2000; Sela-Donenfeld and Kalcheim, 1999). In the following part, we briefly summarize the role of cadherins and extracellular matrix proteins in NC EMT and the signaling cues controlling NC migration.

1.5.1   ROLE OF CADHERINS IN EMT

The expression of cadherins is crucial for cell–cell interactions during NCC migration (Nakagawa and Takeichi, 1995). There are two types of cadherins: type I cadherins, which include CDH1 (E-cadh) and CDH2 (N-cadh); and type II cadherins, including Cadherin 6/7/11. In many contexts, type I cadherins are associated with stable cell assemblies, while type II correlate with low adhesiveness and highly motile cells. NC specifiers SNAI2, FOXD3, SOX9, SOX10, and TWIST1 modulate the expression of *cadherin* genes, thereby controlling NC EMT. Moreover, the membrane localization of cadherins is controlled by complex posttranslational modifications to fine-tune cell–cell interactions. The molecular cascade controlling NC EMT and migration is summarized in Figure 1.5. In frog embryos, *cadherin1* is downregulated at the

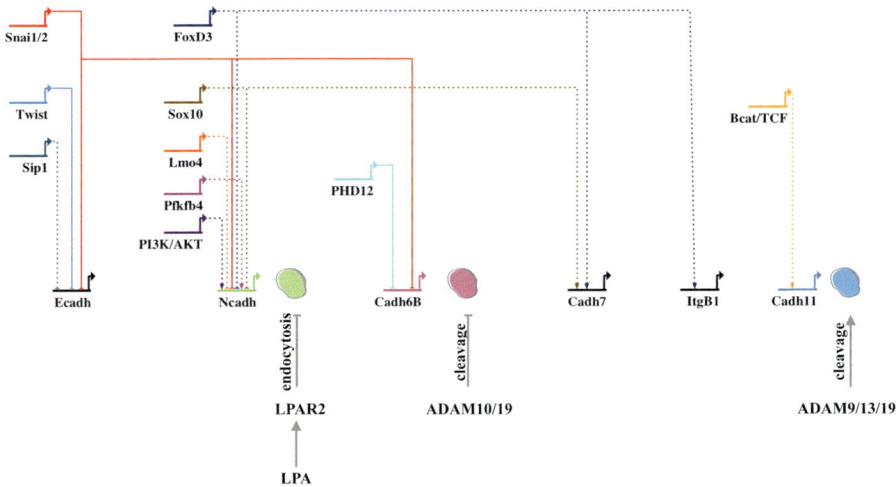

FIGURE 1.5   Main factors involved in neural crest epithelial to mesenchymal transition. This network summarizes the transcriptional activity of NC specifiers and their partners on cadherins and β integrin during NC EMT. It also includes the posttranslational activity of metalloproteinases on cadherins (Bahm et al., 2017; Barriga et al., 2013; Cheung et al., 2005; Fairchild et al., 2014; Ferronha et al., 2013; Figueiredo et al., 2017; Kuriyama et al., 2014; Li et al., 2018; McCusker et al., 2009; Rogers et al., 2013; Schiffmacher et al., 2014; Strobl-Mazzulla and Bronner, 2012; Taneyhill et al., 2007). A solid line indicates a direct regulation, a dotted line an indirect regulation.

start of NC EMT, while *cadherin2* is activated (Theveneau et al., 2010). Later on, *cadherin11* is activated during NCC migration. In chick embryos, a cadherin switch from CDH2 to CDH6B occurs during NC EMT, followed by the upregulation of *cadherin7/11* during NCC migration (Nakagawa and Takeichi, 1995).

1.5.2   EXTRACELLULAR MATRIX PROTEINS (ECM)

Multiple transplantation experiments have illustrated the pivotal role of the extracellular matrix during NCC migration. Both *in vivo* and *in vitro* studies have shown the importance of many ECM proteins such as collagen, fibronectin, and laminin during NCC migration, adhesion, and dispersion (Duband and Thiery, 1987; Newgreen and Thiery, 1980; Perris and Perissinotto, 2000; Rovasio et al., 1983). Interestingly, some repulsive ECM molecules are restricted to the areas non-permissive for migration, thus blocking NCC invasion, while permissive ECM molecules present along the migratory trajectories allow NCC motility (for review see Perris 1997). For example, collagen IX, chondroitin-6-sulfate, and aggrecan are inhibitory/non-permissive for NCC migration (Perris and Perissinotto, 2000). More recently, the proteoglycan versican was also described as an inhibitory signal for NCC migration in frog. Versican is expressed in tissues adjacent to the streams of cranial NC and helps to create physical boundaries to limit the dispersion of NCC during their migration (Szabo et al., 2016). In contrast, Tenascin C is secreted by vagal NCC and is permissive for

NC migration. Tenascin C is particularly important for gut colonization by enteric NCC (Akbareian et al., 2013).

1.5.3 SIGNALING CUES AND CELL–CELL INTERACTIONS CONTROLLING NCC MIGRATION

NC migration is characterized by a complex set of events including intercellular communications between the NC cells themselves and with their neighboring tissues. The main mechanisms are described briefly below.

1.5.3.1 NC Cells Co-Attract During Collective Migration

In *Xenopus laevis*, NC cells attract each other by secreting chemoattractant molecules such as the complement factor C3a, which is coexpressed with its receptor C3aR in NC cells (Carmona-Fontaine et al., 2011). The complement pathway induces a brief cell–cell contact mediated by cdh2/N-cadherin, activating RAC1, resulting in cytoskeleton repolarization (Carmona-Fontaine et al., 2011; Theveneau and Mayor, 2010). Coattraction increases the cohesion between NC cells and prevents the disruption of the cluster of cells during their collective migration.

1.5.3.2 Contact Inhibition of Locomotion (CIL)

Contact inhibition of locomotion is observed when a cell stops migrating as it contacts another cell. This phenomenon has been described *in vivo* using *Xenopus laevis* and zebrafish NC cells (Carmona-Fontaine et al., 2008). When two NC cells meet, they stop moving, collapse their protrusions, and repolarize to move away from each other. In frog, contact inhibition of locomotion is regulated by the non-canonical WNT/PCP pathway and cdh2/N-cadherin (Carmona-Fontaine et al., 2008; Theveneau and Mayor, 2010). Contact inhibition of locomotion alone promotes NC cell dispersion, but together, contact inhibition of locomotion and coattraction control collective NC migration.

1.5.3.3 Chemotaxis

Directional migration of neural crest cells relies on not only on co-attraction and contact inhibition of locomotion but is also controlled by chemoattractants secreted by neighboring tissue, such as the developing cranial placodes. Placode cells attract NCC by secreting the ligand SDF1, which interacts with its receptor CXCR4 expressed on NCC, thus creating an attractive gradient guiding NCC migration towards the placodes (David et al., 2002; Theveneau et al., 2013, 2010). When the NCC meet the placode cells, the placode cells are repelled by CIL shifting the SDF1 source more ventrally, further followed by the NCC (Theveneau et al., 2013). Moreover, SDF1/CXCR4 signaling is also important for cardiac NCC migration; SDF1 secreted by the ectoderm attracts cardiac NCC expressing CXCR4. Alterations of SDF1 signaling leads to a variety of heart anomalies (Escot et al., 2013). More recently, TBX1 expressed in the pharyngeal endoderm and lateral ectoderm was shown to act upstream SDF1 signaling to control the guidance of cranial NCC towards the pharyngeal arches (Escot et al., 2016).

1.5.3.4 Repulsive Signals

In order to avoid NCC invasion into non-targeted tissues and to define the shape of NCC migratory streams, several repulsive signals, such as EPHRINS/EPH, plexins/neuropilins, and SLIT/ROBO signaling, are active between the NCC and their neighboring cells. For example, frog cranial NCC migrating into the second branchial arch express *ephrinB2*, while the NCC of the branchial arches 3 and 4 express *EphA4* and *EphB1* (Helbling et al., 1998; Smith et al., 1997). These repulsive signals are important for separating NC populations into several groups that will migrate into distinct routes. Moreover, in chick cranial and trunk, NCC segregation is triggered by plexin/neuropilin repulsive signals (Gammill et al., 2007, 2006; Osborne et al., 2005). Finally, SLIT/ROBO repulsive signaling delimits the ventral migration of trunk NCC (Jia et al., 2005) and is important for guiding the vagal NCC but not the trunk NC towards the developing gut (De Bellard et al., 2003; Zuhdi et al., 2015). The mesentery expresses SLIT1/2/3 ligands and repulses trunk NCC that express ROBO1/2, therefore preventing their entry into the gut. In contrast, vagal NCC do not express these receptors and therefore are able to enter the gut (De Bellard et al., 2003; Zuhdi et al., 2015).

In sum, NCC migration is a tightly regulated process, orchestrated by a balance between multiple guidance cues from the surrounding cells or from the NCC themselves, acting along the entire trajectory of migration. These signals trigger intracellular response mechanisms that reorganize the cell cytoskeleton, alter the parameters of motility and modulate cell migration. These regulations control the stereotyped organization of NC-derived structures, either in the head, shaping the facial and pharyngeal streams, or in the trunk, with somite segmentation dictating the metameric pattern of NC-derived nerves and ganglia.

Although they were first described long ago (Detwiler, 1933), these interactions have only started to be understood in their cellular and molecular details. During the last decade, detailed studies have helped to better understand the different modes of migration adopted by cranial and trunk NCC. In the head, cranial NCC migrate collectively as a group of cells, whereas at the trunk level, NCC tend to delaminate from the NT as individual cells (Li et al., 2019; Richardson et al., 2016). During collective migration, frog cranial NC intermingle, and their interactions dictate cell polarity (Carmona-Fontaine et al., 2008). Moreover, a supracellular actomyosin ring located at the rear of the cell cluster—but not in the front, thus forming a polarized contractile rear—directs collective migration (Shellard et al., 2018).

Similarly to frog cranial NCC, chick and zebrafish cranial NCC show a constant replacement of the leading cells, not only by the cells present immediately behind them but also by cells that were initially located at the back of the migrating cell group. This observation demonstrates that cranial NC collective cell migration can be accomplished in the absence of specialized leader cells. In contrast to cranial NC, chick and fish trunk NC migration involve leader cells at the front of the group of migrating cells and directing migration, and follower cells (Richardson et al., 2016). The leader cells are larger in size and are polarized towards direction of migration. The laser ablation of a leader cell arrests ventral advance of the stream of cells. The follower cells, behind the ablated cell, remain motile but are unable to restore trunk NC migration (Richardson et al., 2016). In contrast, the laser ablation of a follower cell does not affect trunk NC migration (Richardson et al., 2016). All these studies show that trunk but not cranial NC migration requires the presence of defined leader cells.

In conclusion, the delamination, epithelial to mesenchymal transition, and migration of the neural crest cells are complex and carefully orchestrated processes. The molecular dissection of these mechanisms has moved forward tremendously in recent years. The advent of microfluidics technologies to adapt channels to cell size and manipulate and image cell behavior *in vitro* opens avenues in the analysis of the cellular parameters of migration (Szabó and Mayor, 2018). Moreover, the progress of single cell imaging *in ovo* to follow cell migration and identify and select leader cells allows their transcriptomes and specific properties (Morrison et al., 2017) to be studied. Such new techniques applied to neural crest migration create exciting opportunities to solve the many questions that remain.

1.6   CONCLUSIONS, PERSPECTIVES, AND OPEN QUESTIONS

This chapter attempts to summarize the main discoveries in the field of neural crest development. Since the initial description of NC, over 12,000 articles (more than 500 a year in the past few years) have addressed one of the multifaceted aspects of NC formation. As we are unable to fully acknowledge all the studies contributing to the data mentioned in this review, we apologize to authors we have not cited as we did our best to describe both classical experiments and the latest molecular and cellular approaches. This vast collection of studies has generated the current view of NC development, as regulated by a complex gene regulatory network highly conserved across vertebrates.

In addition to fundamental aspects of developmental biology, the understanding of NC biology is essential for proposing a cellular and molecular basis for congenital malformations and cancers of NC origin. Defects in NC induction, EMT, migration, or differentiation cause major human congenital disorders. Neurocristopathies arise when NCC fail to survive upon EMT, such as in Treacher-Collins syndrome; after defective NCC migration, such as in Hirschsprung syndrome; or after altered postmigration development and differentiation, such as in multiple craniofacial malformations including Branchio-Oculo-Facial syndrome or Pierre Robin sequence (Bolande, 1997, 1974; Keyte and Hutson, 2012; Mundt and Bates, 2010; Noack Watt et al., 2016; Trainor and Richtsmeier, 2015; van Limborgh et al., 1983). While human genetics studies have identified many genes affected in neural crest-linked pathologies, in the vast majority of cases, the molecular cause of neurocristopathies remains to be discovered. This task is complicated by the fact that, as mentioned above, most genes involved in the NC-GRN also play important roles during the development of other organs, creating pleiotropic phenotypes. Moreover, although the main players in the NC regulatory network are globally conserved in all vertebrates studied so far, each animal model may present some species-specific differences for a particular gene or regulation (Sauka-Spengler et al., 2007; Thomas et al., 2008). For example, redundancy after gene duplication or the use of a specific regulatory mechanism allow zebrafish or mouse neurulas to induce NC efficiently despite mutations in genes which are essential in other species, such as *tfap2a*, *snai1/2*, or *pax3* (Li and Cornell,

2007; Murray and Gridley, 2006; Relaix et al., 2004). Furthermore, the proposition that the NC-GRN is structured as a hierarchical series of regulations, elaborated in 2004, has been extremely efficient for positioning key factors in the network, but also has to be modulated with the addition of more complex mechanisms, such as the reiterated use of some factors at multiple steps of the network and the introduction of feedforward and feedback regulatory loops (De Crozé et al., 2011; Meulemans and Bronner-Fraser, 2004; Plouhinec et al., 2014). Finally, in addition to the transcriptional regulations which have focused most of the attention so far, the discovery of other regulatory strategies is emerging, including epigenetic regulations, novel signaling pathways, and posttranslational regulations (Bajpai et al., 2010; Figueiredo et al., 2017; Lander et al., 2013; Rao and LaBonne, 2018; Sanlaville et al., 2006; Sittewelle and Monsoro-Burq, 2018; Strobl-Mazzulla et al., 2010; Zalc et al., 2015).

In addition to congenital malformations, the reactivation of mechanisms involved in NC development is a key element of tumor progression in cancers formed by NC derivatives, as well as in other types of tumor. Either the ectopic expression of transcription factors such as SOX10 or of EMT regulators such as TWIST1 leads to tumor initiation and reactivation of EMT, leading to metastatic NC-derived cancer such as melanoma, neuroblastoma, and pheochromocytoma (Brabletz et al., 2018; Kaufman et al., 2016).

From an evolutionary point of view, the NC is a remarkable structure found only in vertebrates. Together, NC cells and ectodermal placode cells are major elements of head morphogenesis, with key implications for the adaptation and behavior of vertebrate species. Interestingly, the regulatory mechanisms patterning the neural border, upstream of NC induction, are conserved between chordates and vertebrates. This implies that specific novel mechanisms must have been "invented" in vertebrates, allowing the emergence of neural crest cells. Evo-devo approaches have identified several potential scenarios and strategies by comparing the mechanisms described in this chapter for vertebrates to the situation observed in other phyla, as described in the following chapters.

In conclusion, in vertebrates from lamprey to human, the classical studies have identified the main characteristics of the neural crest cells and explored the vast palette of their fates, while the molecular embryology era has discovered genes and signals for the main developmental steps of each fate. In the 21st century, 150 years after the discovery of the neural crest, a daunting challenge remains: to acquire an integrated understanding of these multiple cellular and molecular events and link this knowledge to the exploration of neurocristopathies on one hand, and to the discovery of the mechanisms of neural crest evolution on the other.

ACKNOWLEDGMENTS AND FUNDING

The authors are very grateful to K. Liu, P. Pla, and M. Sittewelle for their proofreading of the manuscript, and to Q. Thuillier and S. Dodier for their technical help in Figure 1.3 preparation. This work was supported by funding from Université Paris Sud/Paris Saclay, Centre National de la Recherche Scientifique (CNRS), Agence Nationale pour la Recherche (ANR Programme Blanc CrestNetMetabo: ANR-15-CE13-0012-01-CRESTNETMETABO), Fondation pour la Recherche

Médicale (FRM, Programme Equipes Labellisées DEQ20150331733), and Institut Universitaire de France to AHMB. M.A. is a Ph.D. fellow funded by Fondation pour la Recherche Médicale (DEQ20150331733).

REFERENCES

Adameyko, I., Lallemend, F., Aquino, J.B., Pereira, J.A., Topilko, P., Müller, T., Fritz, N., Beljajeva, A., Mochii, M., Liste, I., Usoskin, D., Suter, U., Birchmeier, C., Ernfors, P., 2009. Schwann cell precursors from nerve innervation are a cellular origin of melanocytes in skin. *Cell* 139(2), 366–379. doi: 10.1016/j.cell.2009.07.049

Agüero, T.H., Fernández, J.P., López, G.A.V., Tríbulo, C., Aybar, M.J., 2012. Indian hedgehog signaling is required for proper formation, maintenance and migration of Xenopus neural crest. *Dev. Biol.* 364(2), 99–113. doi: 10.1016/j.ydbio.2012.01.020

Akbareian, S.E., Nagy, N., Steiger, C.E., Mably, J.D., Miller, S.A., Hotta, R., Molnar, D., Goldstein, A.M., 2013. Enteric neural crest-derived cells promote their migration by modifying their microenvironment through tenascin-C production. *Dev. Biol.* 382(2), 446–456. doi: 10.1016/j.ydbio.2013.08.006

Alkobtawi, M., Ray, H., Barriga, E.H., Moreno, M., Kerney, R., Monsoro-Burq, A.-H., Saint-Jeannet, J.-P., Mayor, R., 2018. Characterization of Pax3 and Sox10 transgenic Xenopus laevis embryos as tools to study neural crest development. *Dev. Biol.* doi: 10.1016/j.ydbio.2018.02.020

Aoki, Y., Saint-Germain, N., Gyda, M., Magner-Fink, E., Lee, Y.-H., Credidio, C., Saint-Jeannet, J.-P., 2003. Sox10 regulates the development of neural crest-derived melanocytes in Xenopus. *Dev. Biol.* 259(1), 19–33. doi: 10.1016/S0012-1606(03)00161-1

Aoto, K., Sandell, L.L., Butler Tjaden, N.E., Yuen, K.C., Watt, K.E.N., Black, B.L., Durnin, M., Trainor, P.A., 2015. Mef2c-F10N enhancer driven β-galactosidase (LacZ) and Cre recombinase mice facilitate analyses of gene function and lineage fate in neural crest cells. *Dev. Biol.* 402(1), 3–16. doi: 10.1016/j.ydbio.2015.02.022

Arduini, B.L., Bosse, K.M., Henion, P.D., 2009. Genetic ablation of neural crest cell diversification. *Dev. Camb. Engl.* 136(12), 1987–1994. doi: 10.1242/dev.033209

Aruga, J., Yokota, N., Hashimoto, M., Furuichi, T., Fukuda, M., Mikoshiba, K., 1994. A novel zinc finger protein, zic, is involved in neurogenesis, especially in the cell lineage of cerebellar granule cells. *J. Neurochem.* 63(5), 1880–1890.

Bae, C.-J., Park, B.-Y., Lee, Y.-H., Tobias, J.W., Hong, C.-S., Saint-Jeannet, J.-P., 2014. Identification of Pax3 and Zic1 targets in the developing neural crest. *Dev. Biol.* 386(2), 473–483. doi: 10.1016/j.ydbio.2013.12.011

Baggiolini, A., Varum, S., Mateos, J.M., Bettosini, D., John, N., Bonalli, M., Ziegler, U., Dimou, L., Clevers, H., Furrer, R., Sommer, L., 2015. Premigratory and migratory neural crest cells are multipotent in vivo. *Cell Stem Cell* 16(3), 314–322. doi: 10.1016/j.stem.2015.02.017

Bahm, I., Barriga, E.H., Frolov, A., Theveneau, E., Frankel, P., Mayor, R., 2017. PDGF controls contact inhibition of locomotion by regulating N-cadherin during neural crest migration. *Development* 144(13), 2456–2468. doi: 10.1242/dev.147926

Bajpai, R., Chen, D.A., Rada-Iglesias, A., Zhang, J., Xiong, Y., Helms, J., Chang, C.-P., Zhao, Y., Swigut, T., Wysocka, J., 2010. CHD7 cooperates with PBAF to control multipotent neural crest formation. *Nature* 463(7283), 958–962. doi: 10.1038/nature08733

Bang, A.G., Papalopulu, N., Goulding, M.D., Kintner, C., 1999. Expression of Pax-3 in the lateral neural plate is dependent on a Wnt-mediated signal from posterior nonaxial mesoderm. *Dev. Biol.* 212(2), 366–380. doi: 10.1006/dbio.1999.9319

Baroffio, A., Dupin, E., Le Douarin, N.M., 1988. Clone-forming ability and differentiation potential of migratory neural crest cells. *Proc. Natl. Acad. Sci. U.S.A.* 85(14), 5325–5329.

Baroffio, A., Dupin, E., Le Douarin, N.M., 1991. Common precursors for neural and mesectodermal derivatives in the cephalic neural crest. *Dev. Camb. Engl.* 112(1), 301–305.

Barraud, P., Seferiadis, A.A., Tyson, L.D., Zwart, M.F., Szabo-Rogers, H.L., Ruhrberg, C., Liu, K.J., Baker, C.V.H., 2010. Neural crest origin of olfactory ensheathing glia. *Proc. Natl. Acad. Sci. U.S.A.* 107(49), 21040–21045. doi: 10.1073/pnas.1012248107

Barriga, E.H., Maxwell, P.H., Reyes, A.E., Mayor, R., 2013. The hypoxia factor HIF-1α controls neural crest chemotaxis and epithelial to mesenchymal transition. *J. Cell Biol.* 201(5), 759–776. doi: 10.1083/jcb.201212100

Basch, M.L., Bronner-Fraser, M., Garcia-Castro, M.I., 2006. Specification of the neural crest occurs during gastrulation and requires Pax7. *Nature* 441(7090), 218. doi: 10.1038/nature04684

Bellmeyer, A., Krase, J., Lindgren, J., LaBonne, C., 2003. The protooncogene c-myc is an essential regulator of neural crest formation in Xenopus. *Dev. Cell* 4(6), 827–839.

Belmadani, A., Jung, H., Ren, D., Miller, R.J., 2009. The chemokine SDF-1/CXCL12 regulates the migration of melanocyte progenitors in mouse hair follicles. *Differ. Res. Biol. Divers.* 77(4), 395–411. doi: 10.1016/j.diff.2008.10.015

Betancur, P., Bronner-Fraser, M., Sauka-Spengler, T., 2010. Genomic code for Sox10 activation reveals a key regulatory enhancer for cranial neural crest. *Proc. Natl. Acad. Sci. U.S.A.* 107(8), 3570–3575. doi: 10.1073/pnas.0906596107

Bhattacharyya, S., Kulesa, P.M., Fraser, S.E., 2008. Vital labeling of embryonic cells using fluorescent dyes and proteins. *Methods Cell Biol.* 87, 187–210. doi: 10.1016/S0091-679X(08)00210-0

Blaugrund, E., Pham, T.D., Tennyson, V.M., Lo, L., Sommer, L., Anderson, D.J., Gershon, M.D., 1996. Distinct subpopulations of enteric neuronal progenitors defined by time of development, sympathoadrenal lineage markers and Mash-1-dependence. *Development* 122(1), 309–320.

Bolande, R.P., 1974. The neurocristopathies: A unifying concept of disease arising in neural crest maldevelopment. *Hum. Pathol.* 5(4), 409–429. doi: 10.1016/S0046-8177(74)80021-3

Bolande, R.P., 1997. Neurocristopathy: Its growth and development in 20 years. *Pediatr. Pathol. Lab. Med.* 17(1), 1–25. doi: 10.1080/15513819709168343

Bondurand, N., Natarajan, D., Barlow, A., Thapar, N., Pachnis, V., 2006. Maintenance of mammalian enteric nervous system progenitors by SOX10 and endothelin 3 signalling. *Development* 133(10), 2075–2086. doi: 10.1242/dev.02375

Bonstein, L., Elias, S., Frank, D., 1998. Paraxial-fated mesoderm is required for neural crest induction in Xenopus embryos. *Dev. Biol.* 193(2), 156–168. doi: 10.1006/dbio.1997.8795

Borday, C., Parain, K., Thi Tran, H., Vleminckx, K., Perron, M., Monsoro-Burq, A.H., 2018. An atlas of Wnt activity during embryogenesis in *Xenopus tropicalis*. *PLOS ONE* 13(4), e0193606. doi: 10.1371/journal.pone.0193606

Bouchard, P., Garcia, E., 1987. Influence of testosterone substitution on sperm suppression by LHRH agonists. *Horm. Res.* 28(2–4), 175–180. doi: 10.1159/000180942

Brabletz, T., Kalluri, R., Nieto, M.A., Weinberg, R.A., 2018. EMT in cancer. *Nat. Rev. Cancer* 18(2), 128–134. doi: 10.1038/nrc.2017.118

Bradley, C.K., Norton, C.R., Chen, Y., Han, X., Booth, C.J., Yoon, J.K., Krebs, L.T., Gridley, T., 2013. The snail family gene snai3 is not essential for embryogenesis in mice. *PLOS ONE* 8(6), e65344. doi: 10.1371/journal.pone.0065344

Briggs, J.A., Weinreb, C., Wagner, D.E., Megason, S., Peshkin, L., Kirschner, M.W., Klein, A.M., 2018. The dynamics of gene expression in vertebrate embryogenesis at single-cell resolution. *Science* 360(6392). doi: 10.1126/science.aar5780

Britsch, S., Goerich, D.E., Riethmacher, D., Peirano, R.I., Rossner, M., Nave, K.-A., Birchmeier, C., Wegner, M., 2001. The transcription factor Sox10 is a key regulator of peripheral glial development. *Genes Dev.* 15(1), 66–78. doi: 10.1101/gad.186601

Bronner-Fraser, M., Fraser, S.E., 1988. Cell lineage analysis reveals multipotency of some avian neural crest cells. *Nature* 335(6186), 161–164. doi: 10.1038/335161a0

Brugger, S.M., Merrill, A.E., Torres-Vazquez, J., Wu, N., Ting, M.-C., Cho, J.Y.-M., Dobias, S.L., Yi, S.E., Lyons, K., Bell, J.R., Arora, K., Warrior, R., Maxson, R., 2004. A phylogenetically conserved cis-regulatory module in the Msx2 promoter is sufficient for BMP-dependent transcription in murine and Drosophila embryos. *Dev. Camb. Engl.* 131(20), 5153–5165. doi: 10.1242/dev.01390

Buitrago-Delgado, E., Nordin, K., Rao, A., Geary, L., LaBonne, C., 2015. Neurodevelopment. Shared regulatory programs suggest retention of blastula-stage potential in neural crest cells. *Science* 348(6241), 1332–1335. doi: 10.1126/science.aaa3655

Burns, A.J., Douarin, N.M., 1998. The sacral neural crest contributes neurons and glia to the post-umbilical gut: Spatiotemporal analysis of the development of the enteric nervous system. *Dev. Camb. Engl.* 125(21), 4335–4347.

Burstyn-Cohen, T., Stanleigh, J., Sela-Donenfeld, D., Kalcheim, C., 2004. Canonical Wnt activity regulates trunk neural crest delamination linking BMP/noggin signaling with G1/S transition. *Dev. Camb. Engl.* 131(21), 5327–5339. doi: 10.1242/dev.01424

Byrd, N.A., Meyers, E.N., 2005. Loss of Gbx2 results in neural crest cell patterning and pharyngeal arch artery defects in the mouse embryo. *Dev. Biol.* 284(1), 233–245. doi: 10.1016/j.ydbio.2005.05.023

Calloni, G.W., Glavieux-Pardanaud, C., Le Douarin, N.M., Dupin, E., 2007. Sonic Hedgehog promotes the development of multipotent neural crest progenitors endowed with both mesenchymal and neural potentials. *Proc. Natl. Acad. Sci. U.S.A.* 104(50), 19879–19884. doi: 10.1073/pnas.0708806104

Calloni, G.W., Le Douarin, N.M., Dupin, E., 2009. High frequency of cephalic neural crest cells shows coexistence of neurogenic, melanogenic, and osteogenic differentiation capacities. *Proc. Natl. Acad. Sci. U.S.A.* 106(22), 8947–8952. doi: 10.1073/pnas.0903780106

Carmona-Fontaine, C., Matthews, H.K., Kuriyama, S., Moreno, M., Dunn, G.A., Parsons, M., Stern, C.D., Mayor, R., 2008. Contact inhibition of locomotion in vivo controls neural crest directional migration. *Nature* 456(7224), 957–961. doi: 10.1038/nature07441

Carmona-Fontaine, C., Theveneau, E., Tzekou, A., Tada, M., Woods, M., Page, K.M., Parsons, M., Lambris, J.D., Mayor, R., 2011. Complement fragment C3a controls mutual cell attraction during collective cell migration. *Dev. Cell* 21(6), 1026–1037. doi: 10.1016/j.devcel.2011.10.012

Carney, T.J., Dutton, K.A., Greenhill, E., Delfino-Machín, M., Dufourcq, P., Blader, P., Kelsh, R.N., 2006. A direct role for Sox10 in specification of neural crest–derived sensory neurons. *Development* 133(23), 4619–4630. doi: 10.1242/dev.02668

Cebra-Thomas, J.A., Terrell, A., Branyan, K., Shah, S., Rice, R., Gyi, L., Yin, M., Hu, Y., Mangat, G., Simonet, J., Betters, E., Gilbert, S.F., 2013. Late-emigrating trunk neural crest cells in turtle embryos generate an osteogenic ectomesenchyme in the plastron. *Dev. Dyn.* 242, 1223–1235. doi: 10.1002/dvdy.24018

Cerrizuela, S., Vega-López, G.A., Palacio, M.B., Tríbulo, C., Aybar, M.J., 2018. Gli2 is required for the induction and migration of Xenopus laevis neural crest. *Mech. Dev.* 154, 219–239. doi: 10.1016/j.mod.2018.07.010

Chai, Y., Jiang, X., Ito, Y., Bringas, P., Han, J., Rowitch, D.H., Soriano, P., McMahon, A.P., Sucov, H.M., 2000. Fate of the mammalian cranial neural crest during tooth and mandibular morphogenesis. *Dev. Camb. Engl.* 127(8), 1671–1679.

Chai, Y., Maxson, R.E., 2006. Recent advances in craniofacial morphogenesis. *Dev. Dyn.* 235, 2353–2375. doi: 10.1002/dvdy.20833

Chalazonitis, A., Pham, T.D., Rothman, T.P., DiStefano, P.S., Bothwell, M., Blair-Flynn, J., Tessarollo, L., Gershon, M.D., 2001. Neurotrophin-3 is required for the survival–differentiation of subsets of developing enteric neurons. *J. Neurosci.* 21(15), 5620–5636. doi: 10.1523/JNEUROSCI.21-15-05620.2001

Chang, C., Hemmati-Brivanlou, A., 1998. Cell fate determination in embryonic ectoderm. *J. Neurobiol.* 36(2), 128–151. doi: 10.1002/(SICI)1097-4695(199808)36:2<128::AID-NEU3>3.0.CO;2-3

Chang, C., Hemmati Brivanlou, A., 1998. Neural crest induction by Xwnt7B in Xenopus. *Dev. Biol.* 194(1), 129–134. doi: 10.1006/dbio.1997.8820

Chen, C.-L., Broom, D.C., Liu, Y., de Nooij, J.C., Li, Z., Cen, C., Samad, O.A., Jessell, T.M., Woolf, C.J., Ma, Q., 2006. Runx1 determines nociceptive sensory neuron phenotype and is required for thermal and neuropathic pain. *Neuron* 49(3), 365–377. doi: 10.1016/j.neuron.2005.10.036

Chen, G., Ishan, M., Yang, J., Kishigami, S., Fukuda, T., Scott, G., Ray, M.K., Sun, C., Chen, S.-Y., Komatsu, Y., Mishina, Y., Liu, H.-X., 2017. Specific and spatial labeling of P0-Cre versus Wnt1-Cre in cranial neural crest in early mouse embryos. *Genesis*, 55. doi: 10.1002/dvg.23034

Cheng, Y., Cheung, M., Abu-Elmagd, M.M., Orme, A., Scotting, P.J., 2000. Chick sox10, a transcription factor expressed in both early neural crest cells and central nervous system. *Brain Res. Dev. Brain Res.* 121(2), 233–241.

Cheung, M., Briscoe, J., 2003. Neural crest development is regulated by the transcription factor Sox9. *Development* 130(23), 5681–5693. doi: 10.1242/dev.00808

Cheung, M., Chaboissier, M.-C., Mynett, A., Hirst, E., Schedl, A., Briscoe, J., 2005. The transcriptional control of trunk neural crest induction, survival, and delamination. *Dev. Cell* 8(2), 179–192. doi: 10.1016/j.devcel.2004.12.010

Chibon, P., 1964. [Analysis by the method of nuclear labelling WITH tritiated thymidine of derivatives of the cephalic neural crest in the urodele Pleurodeles waltlii MICHAH]. *C. R. Hebd. Seances Acad. Sci.* 259, 3624–3627.

Chibon, P., 1967. [Nuclear labelling by tritiated thymidine of neural crest derivatives in the amphibian urodele Pleurodeles waltlii Michah]. *J. Embryol. Exp. Morphol.* 18(3), 343–358.

Christen, B., Slack, J.M., 1997. FGF-8 is associated with anteroposterior patterning and limb regeneration in Xenopus. *Dev. Biol.* 192(2), 455–466. doi: 10.1006/dbio.1997.8732

Ciurea, A.V., Toader, C., 2009. Genetics of craniosynostosis: Review of the literature. *J. Med. Life* 2(1), 5–17.

Clouthier, D.E., Hosoda, K., Richardson, J.A., Williams, S.C., Yanagisawa, H., Kuwaki, T., Kumada, M., Hammer, R.E., Yanagisawa, M., 1998. Cranial and cardiac neural crest defects in endothelin-A receptor-deficient mice. *Dev. Camb. Engl.* 125(5), 813–824.

Coelho-Aguiar, J.M., Le Douarin, N.M., Dupin, E., 2013. Environmental factors unveil dormant developmental capacities in multipotent progenitors of the trunk neural crest. *Dev. Biol.* 384(1), 13–25. doi: 10.1016/j.ydbio.2013.09.030

Collazo, A., Bronner-Fraser, M., Fraser, S.E., 1993. Vital dye labelling of Xenopus laevis trunk neural crest reveals multipotency and novel pathways of migration. *Dev. Camb. Engl.* 118(2), 363–376.

Conway, S.J., Henderson, D.J., Copp, A.J., 1997a. Pax3 is required for cardiac neural crest migration in the mouse: Evidence from the splotch (Sp2H) mutant. *Development* 124(2), 505–514.

Conway, S.J., Henderson, D.J., Kirby, M.L., Anderson, R.H., Copp, A.J., 1997b. Development of a lethal congenital heart defect in the splotch (Pax3) mutant mouse. *Cardiovasc. Res.* 36(2), 163–173.

Coppola, E., Rallu, M., Richard, J., Dufour, S., Riethmacher, D., Guillemot, F., Goridis, C., Brunet, J.-F., 2010. Epibranchial ganglia orchestrate the development of the cranial neurogenic crest. *Proc. Natl. Acad. Sci. U.S.A.* 107(5), 2066–2071. doi: 10.1073/pnas.0910213107

Couly, G., Creuzet, S., Bennaceur, S., Vincent, C., Le Douarin, N.M., 2002. Interactions between Hox-negative cephalic neural crest cells and the foregut endoderm in patterning the facial skeleton in the vertebrate head. *Dev. Camb. Engl.* 129(4), 1061–1073.

Couly, G., Grapin-Botton, A., Coltey, P., Ruhin, B., Le Douarin, N.M., 1998. Determination of the identity of the derivatives of the cephalic neural crest: Incompatibility between Hox gene expression and lower jaw development. *Dev. Camb. Engl.* 125(17), 3445–3459.

Couly, G.F., Coltey, P.M., Le Douarin, N.M., 1993. The triple origin of skull in higher vertebrates: A study in quail-chick chimeras. *Dev. Camb. Engl.* 117(2), 409–429.

Coutaud, B., Pilon, N., 2013. Characterization of a novel transgenic mouse line expressing Cre recombinase under the control of the Cdx2 neural specific enhancer. *Genesis* 51, 777–784. doi: 10.1002/dvg.22421

Crump, J.G., Swartz, M.E., Eberhart, J.K., Kimmel, C.B., 2006. Moz-dependent Hox expression controls segment-specific fate maps of skeletal precursors in the face. *Dev. Camb. Engl.* 133(14), 2661–2669. doi: 10.1242/dev.02435

Curran, K., Raible, D.W., Lister, J.A., 2009. Foxd3 controls melanophore specification in the zebrafish neural crest by regulation of Mitf. *Dev. Biol.* 332(2), 408–417. doi: 10.1016/j.ydbio.2009.06.010

D'Amico-Martel, A., 1982. Temporal patterns of neurogenesis in avian cranial sensory and autonomic ganglia. *Am. J. Anat.* 163(4), 351–372. doi: 10.1002/aja.1001630407

D'Amico-Martel, A., Noden, D.M., 1983. Contributions of placodal and neural crest cells to avian cranial peripheral ganglia. *Am. J. Anat.* 166(4), 445–468. doi: 10.1002/aja.1001660406

Danielian, P.S., Muccino, D., Rowitch, D.H., Michael, S.K., McMahon, A.P., 1998. Modification of gene activity in mouse embryos in utero by a tamoxifen-inducible form of Cre recombinase. *Curr. Biol.* 8(24), 1323–1326.

Das, A., Crump, J.G., 2012. Bmps and id2a act upstream of Twist1 to restrict ectomesenchyme potential of the cranial neural crest. *PLOS Genet.* 8(5), e1002710. doi: 10.1371/journal.pgen.1002710

Dauger, S., Pattyn, A., Lofaso, F., Gaultier, C., Goridis, C., Gallego, J., Brunet, J.-F., 2003. Phox2b controls the development of peripheral chemoreceptors and afferent visceral pathways. *Dev. Camb. Engl.* 130(26), 6635–6642. doi: 10.1242/dev.00866

David, N.B., Sapede, D., Saint-Etienne, L., Thisse, C., Thisse, B., Dambly-Chaudière, C., Rosa, F.M., Ghysen, A., 2002. Molecular basis of cell migration in the fish lateral line: Role of the chemokine receptor CXCR4 and of its ligand, SDF1. *Proc. Natl. Acad. Sci. U.S.A.* 99(25), 16297–16302. doi: 10.1073/pnas.252339399

Davidson, D., 1995. The function and evolution of Msx genes: Pointers and paradoxes. *Trends Genet.* 11(10), 405–411.

De Bellard, M.E., Rao, Y., Bronner-Fraser, M., 2003. Dual function of Slit2 in repulsion and enhanced migration of trunk, but not vagal, neural crest cells. *J. Cell Biol.* 162(2), 269–279. doi: 10.1083/jcb.200301041

De Crozé, N., Maczkowiak, F., Monsoro-Burq, A.H., 2011. Reiterative Ap2A activity controls sequential steps in the neural crest gene regulatory network. *Proc. Natl. Acad. Sci. U.S.A.* 108(1), 155–160. doi: 10.1073/pnas.1010740107

De Oliveira Bastos, P.R.H., Gardenal, M., Bogo, D., 2008. The social adjustment of bearers of craniofacial abnormalities and the humanist praxis. *Intl. Arch. Otorhinolaryngol.* 12(2), 280–288

Degenhardt, K.R., Milewski, R.C., Padmanabhan, A., Miller, M., Singh, M.K., Lang, D., Engleka, K.A., Wu, M., Li, J., Zhou, D., Antonucci, N., Li, L., Epstein, J.A., 2010. Distinct enhancers at the Pax3 locus can function redundantly to regulate neural tube and neural crest expressions. *Dev. Biol.* 339(2), 519–527. doi: 10.1016/j.ydbio.2009.12.030

Dennis, J.F., Kurosaka, H., Iulianella, A., Pace, J., Thomas, N., Beckham, S., Williams, T., Trainor, P.A., 2012. Mutations in hedgehog acyltransferase (Hhat) perturb hedgehog signaling, resulting in severe Acrania-holoprosencephaly-agnathia craniofacial defects. *PLOS Genet.* 8(10). doi: 10.1371/journal.pgen.1002927

Depew, M.J., Lufkin, T., Rubenstein, J.L.R., 2002. Specification of jaw subdivisions by Dlx genes. *Science* 298(5592), 381–385. doi: 10.1126/science.1075703

Detwiler, S.R., 1933. Experiments upon the segmentation of spinal nerves in salamander embryos. *Proc. Natl. Acad. Sci. U.S.A.* 19(1), 22–29.

Deville, F., Ziller, C., Le Douarin, N.M., 1994. Developmental potentials of enteric neural crest-derived cells in clonal and mass cultures. *Dev. Biol.* 163(1), 141–151.

Deville, S.-S.-C., Ziller, C., Le Douarin, N., 1992. Developmental potentialities of cells derived from the truncal neural crest in clonal cultures. *Brain Res. Dev. Brain Res.* 66, 1–10. doi: 10.1016/0165-3806(92)90134-i

Dixon, A.F., 1904. Professor Wilhelm His. *J. Anat. Physiol.* 38(4), 503–505.

Donnell, M., Hong, C.-S., Huang, X., Delnicki, R.J., Saint-Jeannet, J.-P., 2006. Functional analysis of Sox8 during neural crest development in Xenopus. *Dev. Camb. Engl.* 133, 3817–3826. doi: 10.1242/dev.02558

Dorsky, R.I., Raible, D.W., Moon, R.T., 2000. Direct regulation of nacre, a zebrafish MITF homolog required for pigment cell formation, by the Wnt pathway. *Genes Dev.* 14(2), 158–162.

Dottori, M., Gross, M.K., Labosky, P., Goulding, M., 2001. The winged-helix transcription factor Foxd3 suppresses interneuron differentiation and promotes neural crest cell fate. *Dev. Camb. Engl.* 128(21), 4127–4138.

Le Douarin, N.M., Kalcheim, C., 1999. *The Neural Crest.* Cambridge, UK: Cambridge University Press.

Le Douarin, N.M., Teillet, M.-A., 1973. The migration of neural crest cells to the wall of the digestive tract in avian embryo. *Development* 30, 31–48.

Dougherty, M., Kamel, G., Shubinets, V., Hickey, G., Grimaldi, M., Liao, E.C., 2012. Embryonic fate map of first pharyngeal arch structures in the sox10: Kaede zebrafish transgenic model. *J. Craniofac. Surg.* 23(5), 1333–1337. doi: 10.1097/SCS.0b013e318260f20b

Duband, J.L., Thiery, J.P., 1987. Distribution of laminin and collagens during avian neural crest development. *Dev. Camb. Engl.* 101(3), 461–478.

Dupin, E., Calloni, G.W., Coelho-Aguiar, J.M., Le Douarin, N.M., 2018. The issue of the multipotency of the neural crest cells. *Dev. Biol.* doi: 10.1016/j.ydbio.2018.03.024

Dupin, E., Calloni, G.W., Le Douarin, N.M., 2010. The cephalic neural crest of amniote vertebrates is composed of a large majority of precursors endowed with neural, melanocytic, chondrogenic and osteogenic potentialities. *Cell Cycle Georget. Tex.* 9(2), 238–249. doi: 10.4161/cc.9.2.10491

Dutton, K.A., Pauliny, A., Lopes, S.S., Elworthy, S., Carney, T.J., Rauch, J., Geisler, R., Haffter, P., Kelsh, R.N., 2001. Zebrafish colourless encodes sox10 and specifies nonectomesenchymal neural crest fates. *Dev. Camb. Engl.* 128(21), 4113–4125.

Dyachuk, V., Furlan, A., Shahidi, M.K., Giovenco, M., Kaukua, N., Konstantinidou, C., Pachnis, V., Memic, F., Marklund, U., Müller, T., Birchmeier, C., Fried, K., Ernfors, P., Adameyko, I., 2014. Neurodevelopment. Parasympathetic neurons originate from nerve-associated peripheral glial progenitors. *Science* 345(6192), 82–87. doi: 10.1126/science.1253281

Dykes, I.M., Tempest, L., Lee, S.-I., Turner, E.E., 2011. Brn3a and Islet1 act epistatically to regulate the gene expression program of sensory differentiation. *J. Neurosci.* 31(27), 9789–9799. doi: 10.1523/JNEUROSCI.0901-11.2011

Elkouby, Y.M., Elias, S., Casey, E.S., Blythe, S.A., Tsabar, N., Klein, P.S., Root, H., Liu, K.J., Frank, D., 2010. Mesodermal Wnt signaling organizes the neural plate via Meis3. *Dev. Camb. Engl.* 137(9), 1531–1541. doi: 10.1242/dev.044750

Elms, P., Siggers, P., Napper, D., Greenfield, A., Arkell, R., 2003. Zic2 is required for neural crest formation and hindbrain patterning during mouse development. *Dev. Biol.* 264(2), 391–406. doi: 10.1016/j.ydbio.2003.09.005

Eng, S.R., Dykes, I.M., Lanier, J., Fedtsova, N., Turner, E.E., 2007. POU-domain factor Brn3a regulates both distinct and common programs of gene expression in the spinal and trigeminal sensory ganglia. *Neural Develop.* 2, 3. doi: 10.1186/1749-8104-2-3

Epstein, J.A., Li, J., Lang, D., Chen, F., Brown, C.B., Jin, F., Lu, M.M., Thomas, M., Liu, E., Wessels, A., Lo, C.W., 2000. Migration of cardiac neural crest cells in Splotch embryos. *Dev. Camb. Engl.* 127(9), 1869–1878.

Escot, S., Blavet, C., Faure, E., Zaffran, S., Duband, J.-L., Fournier-Thibault, C., 2016. Disruption of CXCR4 signaling in pharyngeal neural crest cells causes DiGeorge syndrome-like malformations. *Development* 143(4), 582–588. doi: 10.1242/dev.126573

Escot, S., Blavet, C., Härtle, S., Duband, J.-L., Fournier-Thibault, C., 2013. Misregulation of SDF1-CXCR4 signaling impairs early cardiac neural crest cell migration leading to conotruncal defects. *Circ. Res.* 113(5), 505–516. doi: 10.1161/CIRCRESAHA.113.301333

Espinosa-Medina, I., Jevans, B., Boismoreau, F., Chettouh, Z., Enomoto, H., Müller, T., Birchmeier, C., Burns, A.J., Brunet, J.-F., 2017. Dual origin of enteric neurons in vagal Schwann cell precursors and the sympathetic neural crest. *Proc. Natl. Acad. Sci. U.S.A.* 114(45), 11980–11985. doi: 10.1073/pnas.1710308114

Espinosa-Medina, I., Outin, E., Picard, C.A., Chettouh, Z., Dymecki, S., Consalez, G.G., Coppola, E., Brunet, J.-F., 2014. Neurodevelopment. Parasympathetic ganglia derive from Schwann cell precursors. *Science* 345(6192), 87–90. doi: 10.1126/science.1253286

Essex, L.J., Mayor, R., Sargent, M.G., 1993. Expression of Xenopus snail in mesoderm and prospective neural fold ectoderm. *Dev. Dyn.* 198, 108–122. doi: 10.1002/aja.1001980205

Etchevers, H.C., Vincent, C., Le Douarin, N.M., Couly, G.F., 2001. The cephalic neural crest provides pericytes and smooth muscle cells to all blood vessels of the face and forebrain. *Dev. Camb. Engl.* 128(7), 1059–1068.

Fairchild, C.L., Conway, J.P., Schiffmacher, A.T., Taneyhill, L.A., Gammill, L.S., 2014. FoxD3 regulates cranial neural crest EMT via downregulation of Tetraspanin18 independent of its functions during neural crest formation. *Mech. Dev.* 132, 1–12. doi: 10.1016/j.mod.2014.02.004

Faure, S., de Santa Barbara, P., Roberts, D.J., Whitman, M., 2002. Endogenous patterns of BMP signaling during early chick development. *Dev. Biol.* 244(1), 44–65. doi: 10.1006/dbio.2002.0579

Ferronha, T., Rabadán, M.A., Gil-Guiñon, E., Le Dréau, G., de Torres, C., Martí, E., 2013. LMO4 is an essential cofactor in the Snail2-mediated epithelial-to-mesenchymal transition of neuroblastoma and neural crest cells. *J. Neurosci.* 33(7), 2773–2783. doi: 10.1523/JNEUROSCI.4511-12.2013

Figueiredo, A.L., Maczkowiak, F., Borday, C., Pla, P., Sittewelle, M., Pegoraro, C., Monsoro-Burq, A.H., 2017. PFKFB4 control of AKT signaling is essential for premigratory and migratory neural crest formation. *Dev. Camb. Engl.* 144(22), 4183–4194. doi: 10.1242/dev.157644

Forni, P.E., Taylor-Burds, C., Melvin, V.S., Williams, Trevor, Williams, Taylor, Wray, S., 2011. Neural crest and ectodermal cells intermix in the nasal placode to give rise to GnRH-1 neurons, sensory neurons, and olfactory ensheathing cells. *J. Neurosci.* 31(18), 6915–6927. doi: 10.1523/JNEUROSCI.6087-10.2011

Freter, S., Fleenor, S.J., Freter, R., Liu, K.J., Begbie, J., 2013. Cranial neural crest cells form corridors prefiguring sensory neuroblast migration. *Dev. Camb. Engl.* 140(17), 3595–3600. doi: 10.1242/dev.091033

Freyer, L., Aggarwal, V., Morrow, B.E., 2011. Dual embryonic origin of the mammalian otic vesicle forming the inner ear. *Dev. Camb. Engl.* 138(24), 5403–5414. doi: 10.1242/dev.069849

Furlan, A., Adameyko, I., 2018. Schwann cell precursor: A neural crest cell in disguise? *Dev. Biol.* doi: 10.1016/j.ydbio.2018.02.008

Furlan, A., Dyachuk, V., Kastriti, M.E., Calvo-Enrique, L., Abdo, H., Hadjab, S., Chontorotzea, T., Akkuratova, N., Usoskin, D., Kamenev, D., Petersen, J., Sunadome, K., Memic, F., Marklund, U., Fried, K., Topilko, P., Lallemend, F., Kharchenko, P.V., Ernfors, P., Adameyko, I., 2017. Multipotent peripheral glial cells generate neuroendocrine cells of the adrenal medulla. *Science* 357(6346). doi: 10.1126/science.aal3753

Galibert, M.D., Yavuzer, U., Dexter, T.J., Goding, C.R., 1999. Pax3 and regulation of the melanocyte-specific tyrosinase-related protein-1 promoter. *J. Biol. Chem.* 274(38), 26894–26900.

Gammill, L.S., Gonzalez, C., Bronner-Fraser, M., 2007. Neuropilin 2/semaphorin 3F signaling is essential for cranial neural crest migration and trigeminal ganglion condensation. *Dev. Neurobiol.* 67(1), 47–56. doi: 10.1002/dneu.20326

Gammill, L.S., Gonzalez, C., Gu, C., Bronner-Fraser, M., 2006. Guidance of trunk neural crest migration requires neuropilin 2/semaphorin 3F signaling. *Dev. Camb. Engl.* 133(1), 99–106. doi: 10.1242/dev.02187

Garcia-Castro, M.I., Marcelle, C., Bronner-Fraser, M., 2002. Ectodermal Wnt function as a neural crest inducer. *Science* 297(5582), 848–851.

Garnett, A.T., Square, T.A., Medeiros, D.M., 2012. BMP, Wnt and FGF signals are integrated through evolutionarily conserved enhancers to achieve robust expression of Pax3 and Zic genes at the zebrafish neural plate border. *Development* 139(22), 4220–4231. doi: 10.1242/dev.081497

Gfrerer, L., Dougherty, M., Liao, E.C., 2013. Visualization of craniofacial development in the sox10: Kaede transgenic zebrafish line using time-lapse confocal microscopy. *J. Vis. Exp.*, e50525. doi: 10.3791/50525

Gilmour, D.T., Maischein, H.-M., Nusslein-Volhard, C., 2002. Migration and function of a glial subtype in the vertebrate peripheral nervous system. *Neuron* 34(4), 577–588.

Gitelman, I., 1997. Twist protein in mouse embryogenesis. *Dev. Biol.* 189(2), 205–214. doi: 10.1006/dbio.1997.8614

Goda, M., 2017. Rapid integumental color changes due to novel iridophores in the chameleon sand tilefish Hoplolatilus chlupatyi. *Pigment Cell Melanoma Res.* 30(3), 368–371. doi: 10.1111/pcmr.12581

Goda, M., Ohata, M., Ikoma, H., Fujiyoshi, Y., Sugimoto, M., Fujii, R., 2011. Integumental reddish-violet coloration owing to novel dichromatic chromatophores in the teleost fish, Pseudochromis diadema. *Pigment Cell Melanoma Res.* 24(4), 614–617. doi: 10.1111/j.1755-148X.2011.00861.x

Goda, M., Toyohara, J., Visconti, A., Oshima, M., Fujii, N., R., 1994. The blue coloration of the common surgeonfish, Paracanthurus hepatus-1-morphological features of chromatophores. *Zoolog. Sci.* 11, 527–535.

Green, S.A., Uy, B.R., Bronner, M.E., 2017. Ancient evolutionary origin of vertebrate enteric neurons from trunk-derived neural crest. *Nature* 544(7648), 88–91. doi: 10.1038/nature21679

Groves, A.K., George, K.M., Tissier-Seta, J.P., Engel, J.D., Brunet, J.F., Anderson, D.J., 1995. Differential regulation of transcription factor gene expression and phenotypic markers in developing sympathetic neurons. *Dev. Camb. Engl.* 121(3), 887–901.

Guillemot, F., Lo, L.C., Johnson, J.E., Auerbach, A., Anderson, D.J., Joyner, A.L., 1993. Mammalian achaete-scute homolog 1 is required for the early development of olfactory and autonomic neurons. *Cell* 75(3), 463–476.

Hammerschmidt, M., Nusslein-Volhard, C., 1993. The expression of a zebrafish gene homologous to Drosophila snail suggests a conserved function in invertebrate and vertebrate gastrulation. *Dev. Camb. Engl.* 119(4), 1107–1118.

Hammond, W.S., Yntema, C.L., 1947. Depletions in the thoraco-lumbar sympathetic system following removal of neural crest in the chick. *J. Comp. Neurol.* 86(2), 237–265.

Han, D., Wang, S., Hu, Y., Zhang, Y., Dong, X., Yang, Z., Wang, J., Li, J., Deng, X., 2015. Hyperpigmentation results in aberrant immune development in silky fowl (Gallus gallus domesticus Brisson). *PLOS ONE* 10(6), e0125686. doi: 10.1371/journal.pone.0125686

Helbling, P.M., Tran, C.T., Brandli, A.W., 1998. Requirement for EphA receptor signaling in the segregation of Xenopus third and fourth arch neural crest cells. *Mech. Dev.* 78(1–2), 63–79.

Herbarth, B., Pingault, V., Bondurand, N., Kuhlbrodt, K., Hermans-Borgmeyer, I., Puliti, A., Lemort, N., Goossens, M., Wegner, M., 1998. Mutation of the Sry-related Sox10 gene in Dominant megacolon, a mouse model for human Hirschsprung disease. *Proc. Natl. Acad. Sci. U.S.A.* 95(9), 5161–5165. doi: 10.1073/pnas.95.9.5161

His, W., 1868. *Untersuchungen über die erste Anlage Des Wirbelthierleibes : Die erste Entwickelung Des Hühnchens im Ei*. Leipzig: F.C.W. Vogel.

Hochgreb-Hägele, T., Bronner, M.E., 2013. A novel FoxD3 gene trap line reveals neural crest precursor movement and a role for FoxD3 in their specification. *Dev. Biol.* 374(1), 1–11. doi: 10.1016/j.ydbio.2012.11.035

Hockman, D., Adameyko, I., Kaucka, M., Barraud, P., Otani, T., Hunt, A., Hartwig, A.C., Sock, E., Waithe, D., Franck, M.C.M., Ernfors, P., Ehinger, S., Howard, M.J., Brown, N., Reese, J., Baker, C.V.H., 2018. Striking parallels between carotid body glomus cell and adrenal chromaffin cell development. *Dev. Biol.* 444 Suppl 1, S308–S324. doi: 10.1016/j.ydbio.2018.05.016

Holleville, N., Matéos, S., Bontoux, M., Bollerot, K., Monsoro-Burq, A.-H., 2007. Dlx5 drives Runx2 expression and osteogenic differentiation in developing cranial suture mesenchyme. *Dev. Biol.* 304(2), 860–874. doi: 10.1016/j.ydbio.2007.01.003

Holley, J.A., Yu, R.K., 1987. Localization of glycoconjugates recognized by the HNK-1 antibody in mouse and chick embryos during early neural development. *Dev. Neurosci.* 9(2), 105–119. doi: 10.1159/000111613

Hong, C.-S., Park, B.-Y., Saint-Jeannet, J.-P., 2008. Fgf8a induces neural crest indirectly through the activation of Wnt8 in the paraxial mesoderm. *Dev. Camb. Engl.* 135(23), 3903–3910. doi: 10.1242/dev.026229

Hong, C.-S., Saint-Jeannet, J.-P., 2005. Sox proteins and neural crest development. *Semin. Cell. Dev. Biol.* 16(6), 694–703. doi: 10.1016/j.semcdb.2005.06.005

Hong, C.-S., Saint-Jeannet, J.-P., 2007. The activity of Pax3 and Zic1 regulates three distinct cell fates at the neural plate border. *Mol. Biol. Cell* 18(6), 2192–2202. doi: 10.1091/mbc. E06-11-1047

Horstadius, S., 1950. The mechanics of sea urchin development. *Annee Biol.* 26(8), 381–398.

Howard, M.J., Stanke, M., Schneider, C., Wu, X., Rohrer, H., 2000. The transcription factor dHAND is a downstream effector of BMPs in sympathetic neuron specification. *Development* 127(18), 4073–4081.

Huber, K., Brühl, B., Guillemot, F., Olson, E.N., Ernsberger, U., Unsicker, K., 2002. Development of chromaffin cells depends on MASH1 function. *Dev. Camb. Engl.* 129(20), 4729–4738.

Huber, K., Karch, N., Ernsberger, U., Goridis, C., Unsicker, K., 2005. The role of Phox2B in chromaffin cell development. *Dev. Biol.* 279(2), 501–508. doi: 10.1016/j. ydbio.2005.01.007

Imai, K.S., Hino, K., Yagi, K., Satoh, N., Satou, Y., 2004. Gene expression profiles of transcription factors and signaling molecules in the ascidian embryo: Towards a comprehensive understanding of gene networks. *Dev. Camb. Engl.* 131(16), 4047–4058. doi: 10.1242/dev.01270

Inoue, T., Hatayama, M., Tohmonda, T., Itohara, S., Aruga, J., Mikoshiba, K., 2004. Mouse Zic5 deficiency results in neural tube defects and hypoplasia of cephalic neural crest derivatives. *Dev. Biol.* 270(1), 146–162. doi: 10.1016/j.ydbio.2004.02.017

Inoue, T., Ota, M., Ogawa, M., Mikoshiba, K., Aruga, J., 2007. Zic1 and Zic3 regulate medial forebrain development through expansion of neuronal progenitors. *J. Neurosci.* 27(20), 5461–5473. doi: 10.1523/JNEUROSCI.4046-06.2007

Ishii, M., Han, J., Yen, H.-Y., Sucov, H.M., Chai, Y., Maxson, R.E., 2005. Combined deficiencies of Msx1 and Msx2 cause impaired patterning and survival of the cranial neural crest. *Dev. Camb. Engl.* 132(22), 4937–4950. doi: 10.1242/dev.02072

Janesick, A., Tang, W., Ampig, K., Blumberg, B., 2019. Znf703 is a novel RA target in the neural plate border. *Sci. Rep.* 9(1), 8275. doi: 10.1038/s41598-019-44722-1

Jheon, A.H., Schneider, R.A., 2009. The cells that fill the bill: Neural crest and the evolution of craniofacial development. *J. Dent. Res.* 88(1), 12–21. doi: 10.1177/0022034508327757

Jia, L., Cheng, L., Raper, J., 2005. Slit/Robo signaling is necessary to confine early neural crest cells to the ventral migratory pathway in the trunk. *Dev. Biol.* 282(2), 411–421. doi: 10.1016/j.ydbio.2005.03.021

Jiang, R., Lan, Y., Norton, C.R., Sundberg, J.P., Gridley, T., 1998. The Slug gene is not essential for mesoderm or neural crest development in mice. *Dev. Biol.* 198(2), 277–285.

Jiang, X., Iseki, S., Maxson, R.E., Sucov, H.M., Morriss-Kay, G.M., 2002. Tissue origins and interactions in the mammalian skull vault. *Dev. Biol.* 241(1), 106–116. doi: 10.1006/dbio.2001.0487

Johansson, E., Andersson, L., Örnros, J., Carlsson, T., Ingeson-Carlsson, C., Liang, S., Dahlberg, J., Jansson, S., Parrillo, L., Zoppoli, P., Barila, G.O., Altschuler, D.L., Padula, D., Lickert, H., Fagman, H., Nilsson, M., 2015. Revising the embryonic origin of thyroid C cells in mice and humans. *Dev. Camb. Engl.* 142(20), 3519–3528. doi: 10.1242/dev.126581

Johnson, J.E., Birren, S.J., Anderson, D.J., 1990. Two rat homologues of Drosophila achaete-scute specifically expressed in neuronal precursors. *Nature* 346(6287), 858–861. doi: 10.1038/346858a0

Kague, E., Gallagher, M., Burke, S., Parsons, M., Franz-Odendaal, T., Fisher, S., 2012. Skeletogenic fate of zebrafish cranial and trunk neural crest. *PLOS ONE* 7(11), e47394. doi: 10.1371/journal.pone.0047394

Kameda, Y., 2005. Mash1 is required for glomus cell formation in the mouse carotid body. *Dev. Biol.* 283(1), 128–139. doi: 10.1016/j.ydbio.2005.04.004

Kaufman, C.K., Mosimann, C., Fan, Z.P., Yang, S., Thomas, A.J., Ablain, J., Tan, J.L., Fogley, R.D., van Rooijen, E., Hagedorn, E.J., Ciarlo, C., White, R.M., Matos, D.A., Puller, A.-C., Santoriello, C., Liao, E.C., Young, R.A., Zon, L.I., 2016. A zebrafish melanoma model reveals emergence of neural crest identity during melanoma initiation. *Science* 351(6272), aad2197. doi: 10.1126/science.aad2197

Kawasaki-Nishihara, A., Nishihara, D., Nakamura, H., Yamamoto, H., 2011. ET3/Ednrb2 signaling is critically involved in regulating melanophore migration in Xenopus. *Dev. Dyn.* 240, 1454–1466. doi: 10.1002/dvdy.22649

Kerosuo, L., Bronner, M.E., 2016. cMyc regulates the size of the premigratory neural crest stem cell pool. *Cell Rep.* 17, 2648–2659. doi: 10.1016/j.celrep.2016.11.025

Kerosuo, L., Nie, S., Bajpai, R., Bronner, M.E., 2015. Crestospheres: Long-term maintenance of multipotent, premigratory neural crest stem cells. *Stem Cell Rep.* 5(4), 499–507. doi: 10.1016/j.stemcr.2015.08.017

Keyte, A., Hutson, M.R., 2012. The neural crest in cardiac congenital anomalies. *Differ. Res. Biol. Divers.* 84(1), 25–40. doi: 10.1016/j.diff.2012.04.005

Khudyakov, J., Bronner Fraser, M., 2009. Comprehensive spatiotemporal analysis of early chick neural crest network genes. *Dev. Dyn.* 238(3), 716–723. doi: 10.1002/dvdy.21881

Kim, H.-S., Seo, H., Yang, C., Brunet, J.-F., Kim, K.-S., 1998. Noradrenergic-specific transcription of the dopamine β-hydroxylase gene requires synergy of multiple cis-acting elements including at least two Phox2a-binding sites. *J. Neurosci.* 18(20), 8247–8260. doi: 10.1523/JNEUROSCI.18-20-08247.1998

Kim, J., Lo, L., Dormand, E., Anderson, D.J., 2003. SOX10 maintains multipotency and inhibits neuronal differentiation of neural crest stem cells. *Neuron* 38(1), 17–31.

Kirby, B.B., Takada, N., Latimer, A.J., Shin, J., Carney, T.J., Kelsh, R.N., Appel, B., 2006. In vivo time-lapse imaging shows dynamic oligodendrocyte progenitor behavior during zebrafish development. *Nat. Neurosci.* 9(12), 1506–1511. doi: 10.1038/nn1803

Kirby, M.L., Gale, T.F., Stewart, D.E., 1983. Neural crest cells contribute to normal aortico-pulmonary septation. *Science* 220(4601), 1059–1061.

Kirby, M.L., Hutson, M.R., 2010. Factors controlling cardiac neural crest cell migration. *Cell Adhes. Migr.* 4(4), 609–621.

Köntges, G., Lumsden, A., 1996. Rhombencephalic neural crest segmentation is preserved throughout craniofacial ontogeny. *Dev. Camb. Engl.* 122(10), 3229–3242.

Kos, R., Reedy, M.V., Johnson, R.L., Erickson, C.A., 2001. The winged-helix transcription factor FoxD3 is important for establishing the neural crest lineage and repressing melanogenesis in avian embryos. *Dev. Camb. Engl.* 128(8), 1467–1479.

Kramer, I., Sigrist, M., Nooij, J.C. de, Taniuchi, I., Jessell, T.M., Arber, S., 2006. A role for Runx transcription factor signaling in dorsal root ganglion sensory neuron diversification. *Neuron* 49(3), 379–393. doi: 10.1016/j.neuron.2006.01.008

Krispin, S., Nitzan, E., Kassem, Y., Kalcheim, C., 2010. Evidence for a dynamic spatiotemporal fate map and early fate restrictions of premigratory avian neural crest. *Dev. Camb. Engl.* 137(4), 585–595. doi: 10.1242/dev.041509

Kulesa, P.M., McKinney, M.C., McLennan, R., 2013. Developmental imaging: The avian embryo hatches to the challenge. *Birth Defects Res. C Embryo Today* 99(2), 121–133. doi: 10.1002/bdrc.21036

Kuriyama, S., Theveneau, E., Benedetto, A., Parsons, M., Tanaka, M., Charras, G., Kabla, A., Mayor, R., 2014. In vivo collective cell migration requires an LPAR2-dependent increase in tissue fluidity. *J. Cell Biol.* 206(1), 113–127. doi: 10.1083/jcb.201402093

Kwak, J., Park, O.K., Jung, Y.J., Hwang, B.J., Kwon, S.-H., Kee, Y., 2013. Live image profiling of neural crest lineages in zebrafish transgenic lines. *Mol. Cells* 35(3), 255–260. doi: 10.1007/s10059-013-0001-5

LaBonne, C., Bronner-Fraser, M., 1998. Neural crest induction in Xenopus: Evidence for a two-signal model. *Development* 125(13), 2403–2414.

Lander, R., Nasr, T., Ochoa, S.D., Nordin, K., Prasad, M.S., LaBonne, C., 2013. Interactions between Twist and other core epithelial–mesenchymal transition factors are controlled by GSK3-mediated phosphorylation. *Nat. Commun.* 4, 1542. doi: 10.1038/ncomms2543

Laranjeira, C., Sandgren, K., Kessaris, N., Richardson, W., Potocnik, A., Vanden Berghe, P., Pachnis, V., 2011. Glial cells in the mouse enteric nervous system can undergo neurogenesis in response to injury. *J. Clin. Invest.* 121(9), 3412–3424. doi: 10.1172/JCI58200

Laurent-Gengoux, P., Petit, V., Aktary, Z., Gallagher, S., Tweedy, L., Machesky, L., Larue, L., 2018. Simulation of melanoblast displacements reveals new features of developmental migration. *Dev. Camb. Engl.* 145(12). doi: 10.1242/dev.160200

Le Douarin, N.M., 1975. The neural crest in the neck and other parts of the body. *Birth Defects Orig. Artic. Ser.* 11(7), 19–50.

Le Lievre, C.S., 1978. Participation of neural crest-derived cells in the genesis of the skull in birds. *J. Embryol. Exp. Morphol.* 47, 17–37.

Le Lièvre, C.S., Le Douarin, N.M., 1975. Mesenchymal derivatives of the neural crest: Analysis of chimaeric quail and chick embryos. *J. Embryol. Exp. Morphol.* 34(1), 125–154.

Lecoin, L., Lahav, R., Martin, F.H., Teillet, M.A., Le Douarin, N.M., 1995. Steel and c-kit in the development of avian melanocytes: A study of normally pigmented birds and of the hyperpigmented mutant silky fowl. *Dev. Dyn.* 203, 106–118. doi: 10.1002/aja.1002030111

Lee, K.E., Nam, S., Cho, E., Seong, I., Limb, J.-K., Lee, S., Kim, J., 2008. Identification of direct regulatory targets of the transcription factor Sox10 based on function and conservation. *BMC Genomics* 9, 408. doi: 10.1186/1471-2164-9-408

Lee, R.T.H., Thiery, J.P., Carney, T.J., 2013. Dermal fin rays and scales derive from mesoderm, not neural crest. *Curr. Biol.* 23(9), R336–337. doi: 10.1016/j.cub.2013.02.055

Lee, Y.-H., Aoki, Y., Hong, C.-S., Saint-Germain, N., Credidio, C., Saint-Jeannet, J.-P., 2004. Early requirement of the transcriptional activator Sox9 for neural crest specification in Xenopus. *Dev. Biol.* 275(1), 93–103. doi: 10.1016/j.ydbio.2004.07.036

Lefebvre, V., de Crombrugghe, B., 1998. Toward understanding SOX9 function in chondrocyte differentiation. *Matrix Biol. J. Int. Soc. Matrix Biol.* 16(9), 529–540.

Lekven, A.C., Thorpe, C.J., Waxman, J.S., Moon, R.T., 2001. Zebrafish wnt8 encodes two Wnt8 proteins on a bicistronic transcript and is required for mesoderm and neurectoderm patterning. *Dev. Cell* 1(1), 103–114. doi: 10.1016/S1534-5807(01)00007-7

Levy, C., Khaled, M., Fisher, D.E., 2006. MITF: Master regulator of melanocyte development and melanoma oncogene. *Trends Mol. Med.* 12(9), 406–414. doi: 10.1016/j.molmed.2006.07.008

Lewis, A.E., Vasudevan, H.N., O'Neill, A.K., Soriano, P., Bush, J.O., 2013. The widely used Wnt1-Cre transgene causes developmental phenotypes by ectopic activation of Wnt signaling. *Dev. Biol.* 379(2), 229–234. doi: 10.1016/j.ydbio.2013.04.026

Lewis, J.L., Bonner, J., Modrell, M., Ragland, J.W., Moon, R.T., Dorsky, R.I., Raible, D.W., 2004. Reiterated Wnt signaling during zebrafish neural crest development. *Development* 131(6), 1299–1308. doi: 10.1242/dev.01007

Li, B., Kuriyama, S., Moreno, M., Mayor, R., 2009. The posteriorizing gene Gbx2 is a direct target of Wnt signalling and the earliest factor in neural crest induction. *Dev. Camb. Engl.* 136(19), 3267–3278. doi: 10.1242/dev.036954

Li, J., Chen, F., Epstein, J.A., 2000. Neural crest expression of Cre recombinase directed by the proximal Pax3 promoter in transgenic mice. *Genesis* 26, 162–164.

Li, J., Huang, X., Xu, X., Mayo, J., Bringas, P., Jiang, R., Wang, S., Chai, Y., 2011. SMAD4-mediated WNT signaling controls the fate of cranial neural crest cells during tooth morphogenesis. *Dev. Camb. Engl.* 138(10), 1977–1989. doi: 10.1242/dev.061341

Li, J., Liu, K.C., Jin, F., Lu, M.M., Epstein, J.A., 1999. Transgenic rescue of congenital heart disease and spina bifida in Splotch mice. *Dev. Camb. Engl.* 126(11), 2495–2503.

Li, J., Perfetto, M., Materna, C., Li, R., Thi Tran, H., Vleminckx, K., Duncan, M.K., Wei, S., 2019. A new transgenic reporter line reveals Wnt-dependent Snai2 re-expression and cranial neural crest differentiation in Xenopus. *Sci. Rep.* 9(1), 11191. doi: 10.1038/s41598-019-47665-9

Li, J., Perfetto, M., Neuner, R., Bahudhanapati, H., Christian, L., Mathavan, K., Bridges, L.C., Alfandari, D., Wei, S., 2018. Xenopus ADAM19 regulates Wnt signaling and neural crest specification by stabilizing ADAM13. *Development* 158154. doi: 10.1242/dev.158154

Li, W., Cornell, R.A., 2007. Redundant activities of Tfap2a and Tfap2c are required for neural crest induction and development of other nonneural ectoderm derivatives in zebrafish embryos. *Dev. Biol.* 304(1), 338–354. doi: 10.1016/j.ydbio.2006.12.042

Li, Y., Vieceli, F.M., Gonzalez, W.G., Li, A., Tang, W., Lois, C., Bronner, M.E., 2019. In vivo quantitative imaging provides insights into trunk neural crest migration. *Cell Rep.* 26(6), 1489–1500.e3. doi: 10.1016/j.celrep.2019.01.039

Light, W., Vernon, A.E., Lasorella, A., Iavarone, A., LaBonne, C., 2005. Xenopus Id3 is required downstream of Myc for the formation of multipotent neural crest progenitor cells. *Dev. Camb. Engl.* 132(8), 1831–1841. doi: 10.1242/dev.01734

Lignell, A., Kerosuo, L., Streichan, S.J., Cai, L., Bronner, M.E., 2017. Identification of a neural crest stem cell niche by spatial genomic analysis. *Nat. Commun.* 8(1), 1830. doi: 10.1038/s41467-017-01561-w

Lo, C.W., Cohen, M.F., Huang, G.Y., Lazatin, B.O., Patel, N., Sullivan, R., Pauken, C., Park, S.M., 1997. Cx43 gap junction gene expression and gap junctional communication in mouse neural crest cells. *Dev. Genet.* 20(2), 119–132. doi: 10.1002/(SICI)1520-6408(1997)20:2<119::AID-DVG5>3.0.CO;2-A

Lo, L., Tiveron, M.C., Anderson, D.J., 1998. MASH1 activates expression of the paired homeodomain transcription factor Phox2a, and couples pan-neuronal and subtype-specific components of autonomic neuronal identity. *Development* 125(4), 609–620.

Lukoseviciute, M., Gavriouchkina, D., Williams, R.M., Hochgreb-Hagele, T., Senanayake, U., Chong-Morrison, V., Thongjuea, S., Repapi, E., Mead, A., Sauka-Spengler, T., 2018. From pioneer to repressor: Bimodal foxd3 activity dynamically remodels neural crest regulatory landscape in vivo. *Dev. Cell* 47(5), 608–628.e6. doi: 10.1016/j.devcel.2018.11.009

Lumb, R., Schwarz, Q., 2015. Sympathoadrenal neural crest cells: The known, unknown and forgotten? *Dev. Growth Differ.* 57(2), 146–157. doi: 10.1111/dgd.12189

Luo, T., Lee, Y.-H., Saint-Jeannet, J.-P., Sargent, T.D., 2003. Induction of neural crest in Xenopus by transcription factor AP2alpha. *Proc. Natl. Acad. Sci. U.S.A.* 100(2), 532–537. doi: 10.1073/pnas.0237226100

Macatee, T.L., Hammond, B.P., Arenkiel, B.R., Francis, L., Frank, D.U., Moon, A.M., 2003. Ablation of specific expression domains reveals discrete functions of ectoderm- and endoderm-derived FGF8 during cardiovascular and pharyngeal development. *Dev. Camb. Engl.* 130(25), 6361–6374. doi: 10.1242/dev.00850

Maczkowiak, F., Matéos, S., Wang, E., Roche, D., Harland, R., Monsoro-Burq, A.H., 2010. The Pax3 and Pax7 paralogs cooperate in neural and neural crest patterning using distinct molecular mechanisms, in Xenopus laevis embryos. *Dev. Biol.* 340(2), 381–396. doi: 10.1016/j.ydbio.2010.01.022

Mancilla, A., Mayor, R., 1996. Neural crest formation in Xenopus laevis: Mechanisms of Xslug induction. *Dev. Biol.* 177(2), 580–589. doi: 10.1006/dbio.1996.0187

Marchant, L., Linker, C., Ruiz, P., Guerrero, N., Mayor, R., 1998. The inductive properties of mesoderm suggest that the neural crest cells are specified by a BMP gradient. *Dev. Biol.* 198(2), 319–329. doi: 10.1016/S0012-1606(98)80008-0

Martinez-Morales, P.L., Diez del Corral, R., Olivera-Martínez, I., Quiroga, A.C., Das, R.M., Barbas, J.A., Storey, K.G., Morales, A.V., 2011. FGF and retinoic acid activity gradients control the timing of neural crest cell emigration in the trunk. *J. Cell Biol.* 194(3), 489–503. doi: 10.1083/jcb.201011077

Matsumoto, J., 1965. Studies on fine structure and cytochemical properties of erythrophores in swordtail, Xiphophorus helleri, with special reference to their pigment granules (Pterinosomes). *J. Cell Biol.* 27(3), 493–504.

Matsuoka, T., Ahlberg, P.E., Kessaris, N., Iannarelli, P., Dennehy, U., Richardson, W.D., McMahon, A.P., Koentges, G., 2005a. Neural crest origins of the neck and shoulder. *Nature* 436(7049), 347–355. doi: 10.1038/nature03837

Matsuoka, T., Ahlberg, P.E., Kessaris, N., Iannarelli, P., Dennehy, U., Richardson, W.D., McMahon, A.P., Koentges, G., 2005b. Neural crest origins of the neck and shoulder. *Nature* 436(7049), 347–355. doi: 10.1038/nature03837

Mayor, R., Morgan, R., Sargent, M.G., 1995. Induction of the prospective neural crest of Xenopus. *Dev. Camb. Engl.* 121(3), 767–777.

McCusker, C., Cousin, H., Neuner, R., Alfandari, D., 2009. Extracellular cleavage of cadherin-11 by ADAM metalloproteases is essential for Xenopus cranial neural crest cell migration. *Mol. Biol. Cell* 20(1), 78–89. doi: 10.1091/mbc.e08-05-0535

McKinney, M.C., Fukatsu, K., Morrison, J., McLennan, R., Bronner, M.E., Kulesa, P.M., 2013. Evidence for dynamic rearrangements but lack of fate or position restrictions in premigratory avian trunk neural crest. *Dev. Camb. Engl.* 140(4), 820–830. doi: 10.1242/dev.083725

Medeiros, D.M., 2013. The evolution of the neural crest: New perspectives from lamprey and invertebrate neural crest-like cells. *Wiley Interdiscip. Rev. Dev. Biol.* 2(1), 1–15. doi: 10.1002/wdev.85

Merzdorf, C.S., 2007. Emerging roles for zic genes in early development. *Dev. Dyn.* 236, 922–940. doi: 10.1002/dvdy.21098

Meulemans, D., Bronner-Fraser, M., 2004. Gene-regulatory interactions in neural crest evolution and development. *Dev. Cell* 7(3), 291–299. doi: 10.1016/j.devcel.2004.08.007

Milet, C., Maczkowiak, F., Roche, D.D., Monsoro-Burq, A.H., 2013. Pax3 and Zic1 drive induction and differentiation of multipotent, migratory, and functional neural crest in Xenopus embryos. *Proc. Natl. Acad. Sci. U.S.A.* 110(14), 5528–5533. doi: 10.1073/pnas.1219124110

Milet, C., Monsoro-Burq, A.H., 2012. Neural crest induction at the neural plate border in vertebrates. *Dev. Biol. Neural Crest* 366(1), 22–33. doi: 10.1016/j.ydbio.2012.01.013

Miller, C.T., Yelon, D., Stainier, D.Y.R., Kimmel, C.B., 2003. Two endothelin 1 effectors, hand2 and bapx1, pattern ventral pharyngeal cartilage and the jaw joint. *Dev. Camb. Engl.* 130(7), 1353–1365.

Minchin, J.E.N., Hughes, S.M., 2008. Sequential actions of Pax3 and Pax7 drive xanthophore development in zebrafish neural crest. *Dev. Biol.* 317(2), 508–522. doi: 10.1016/j.ydbio.2008.02.058

Modrell, M.S., Hockman, D., Uy, B., Buckley, D., Sauka-Spengler, T., Bronner, M.E., Baker, C.V.H., 2014. A fate-map for cranial sensory ganglia in the sea lamprey. *Dev. Biol.* 385(2), 405–416. doi: 10.1016/j.ydbio.2013.10.021

Mollaaghababa, R., Pavan, W.J., 2003. The importance of having your SOX on: Role of SOX10 in the development of neural crest-derived melanocytes and glia. *Oncogene* 22(20), 3024–3034. doi: 10.1038/sj.onc.1206442

Mongera, A., Nusslein-Volhard, C., 2013. Scales of fish arise from mesoderm. *Curr. Biol.* 23(9), R338–339. doi: 10.1016/j.cub.2013.02.056

Mongera, A., Singh, A.P., Levesque, M.P., Chen, Y.-Y., Konstantinidis, P., Nüsslein-Volhard, C., 2013. Genetic lineage labeling in zebrafish uncovers novel neural crest contributions to the head, including gill pillar cells. *Dev. Camb. Engl.* 140(4), 916–925. doi: 10.1242/dev.091066

Monsoro-Burq, A.H., 2015. PAX transcription factors in neural crest development. *Semin. Cell Dev. Biol.* 44, 87–96. doi: 10.1016/j.semcdb.2015.09.015

Monsoro-Burq, A.-H., Fletcher, R.B., Harland, R.M., 2003. Neural crest induction by paraxial mesoderm in Xenopus embryos requires FGF signals. *Dev. Camb. Engl.* 130(14), 3111–3124.

Monsoro-Burq, A.-H., Wang, E., Harland, R., 2005. Msx1 and Pax3 cooperate to mediate FGF8 and WNT signals during Xenopus neural crest induction. *Dev. Cell* 8(2), 167–178. doi: 10.1016/j.devcel.2004.12.017

Mori-Akiyama, Y., Akiyama, H., Rowitch, D.H., de Crombrugghe, B., 2003. Sox9 is required for determination of the chondrogenic cell lineage in the cranial neural crest. *Proc. Natl. Acad. Sci. U.S.A.* 100(16), 9360–9365. doi: 10.1073/pnas.1631288100

Morrison, J.A., McLennan, R., Wolfe, L.A., Gogol, M.M., Meier, S., McKinney, M.C., Teddy, J.M., Holmes, L., Semerad, C.L., Box, A.C., Li, H., Hall, K.E., Perera, A.G., Kulesa, P.M., 2017. Singlecell transcriptome analysis of avian neural crest migration reveals signatures of invasion and molecular transitions. *Elife.* 6:e28415. doi: 10.7554/eLife.28415.

Morrison, R.L., 1995. A transmission electron microscopic (TEM) method for determining structural colors reflected by lizard iridophores. *Pigment Cell Res.* 8(1), 28–36.

Motohashi, T., Yamanaka, K., Chiba, K., Miyajima, K., Aoki, H., Hirobe, T., Kunisada, T., 2011. Neural crest cells retain their capability for multipotential differentiation even after lineage-restricted stages. *Dev. Dyn.* 240, 1681–1693. doi: 10.1002/dvdy.22658

Muller, S.M., Stolt, C.C., Terszowski, G., Blum, C., Amagai, T., Kessaris, N., Iannarelli, P., Richardson, W.D., Wegner, M., Rodewald, H.-R., 2008. Neural crest origin of perivascular mesenchyme in the adult thymus. *J. Immunol.* 1950 180(8), 5344–5351.

Mundell, N.A., Labosky, P.A., 2011. Neural crest stem cell multipotency requires Foxd3 to maintain neural potential and repress mesenchymal fates. *Dev. Camb. Engl.* 138(4), 641–652. doi: 10.1242/dev.054718

Mundt, E., Bates, M.D., 2010. Genetics of Hirschsprung disease and anorectal malformations. *Semin. Pediatr. Surg.* 19(2), 107–117. doi: 10.1053/j.sempedsurg.2009.11.015

Murdoch, B., DelConte, C., García-Castro, M.I., 2012. Pax7 lineage contributions to the mammalian neural crest. *PLOS ONE* 7(7), e41089. doi: 10.1371/journal.pone.0041089

Murray, S.A., Gridley, T., 2006. Snail family genes are required for left-right asymmetry determination, but not neural crest formation, in mice. *Proc. Natl. Acad. Sci. U.S.A.* 103(27), 10300–10304. doi: 10.1073/pnas.0602234103

Nagatomo, K.-I., Hashimoto, C., 2007. Xenopus hairy2 functions in neural crest formation by maintaining cells in a mitotic and undifferentiated state. *Dev. Dyn.* 236(6), 1475–1483. doi: 10.1002/dvdy.21152

Nagy, N., Goldstein, A.M., 2006. Endothelin-3 regulates neural crest cell proliferation and differentiation in the hindgut enteric nervous system. *Dev. Biol.* 293(1), 203–217. doi: 10.1016/j.ydbio.2006.01.032

Nakagawa, S., Takeichi, M., 1995. Neural crest cell-cell adhesion controlled by sequential and subpopulation-specific expression of novel cadherins. *Development* 121(5), 1321–1332.

Nakamura, S., Senzaki, K., Yoshikawa, M., Nishimura, M., Inoue, K., Ito, Y., Ozaki, S., Shiga, T., 2008. Dynamic regulation of the expression of neurotrophin receptors by Runx3. *Development* 135(9), 1703–1711. doi: 10.1242/dev.015248

Newgreen, D., Gibbins, I., 1982. Factors controlling the time of onset of the migration of neural crest cells in the fowl embryo. *Cell Tissue Res.* 224(1), 145–160.

Newgreen, D., Thiery, J.P., 1980. Fibronectin in early avian embryos: Synthesis and distribution along the migration pathways of neural crest cells. *Cell Tissue Res.* 211(2), 269–291.

Nguyen, V.H., Schmid, B., Trout, J., Connors, S.A., Ekker, M., Mullins, M.C., 1998. Ventral and lateral regions of the zebrafish gastrula, including the neural crest progenitors, are established by a bmp2b/swirl pathway of genes. *Dev. Biol.* 199(1), 93–110. doi: 10.1006/dbio.1998.8927

Nichane, M., de Crozé, N., Ren, X., Souopgui, J., Monsoro-Burq, A.H., Bellefroid, E.J., 2008a. Hairy2-Id3 interactions play an essential role in Xenopus neural crest progenitor specification. *Dev. Biol.* 322(2), 355–367. doi: 10.1016/j.ydbio.2008.08.003

Nichane, M., Ren, X., Souopgui, J., Bellefroid, E.J., 2008b. Hairy2 functions through both DNA-binding and non DNA-binding mechanisms at the neural plate border in Xenopus. *Dev. Biol.* 322(2), 368–380. doi: 10.1016/j.ydbio.2008.07.026

Nichols, D.H., 1981. Neural crest formation in the head of the mouse embryo as observed using a new histological technique. *Development* 64, 105–120.

Nieto, M.A., 2011. The ins and outs of the epithelial to mesenchymal transition in health and disease. *Annu. Rev. Cell Dev. Biol.* 27, 347–376. doi: 10.1146/annurev-cellbio-092910-154036

Nieto, M.A., 2018. A snail tale and the chicken embryo. *Int. J. Dev. Biol.* 62(1–2), 121–126. doi: 10.1387/ijdb.170301mn

Nieto, M.A., Bennett, M.F., Sargent, M.G., Wilkinson, D.G., 1992. Cloning and developmental expression of Sna, a murine homologue of the Drosophila snail gene. *Dev. Camb. Engl.* 116(1), 227–237.

Nieto, M.A., Sargent, M.G., Wilkinson, D.G., Cooke, J., 1994. Control of cell behavior during vertebrate development by Slug, a zinc finger gene. *Science* 264(5160), 835–839.

Nikitina, N., Tong, L., Bronner, M.E., 2011. Ancestral network module regulating prdm1 expression in the lamprey neural plate border. *Dev. Dyn.* 240, 2265–2271. doi: 10.1002/dvdy.22720

Nikitina, N.V., Bronner-Fraser, M., 2009. Gene regulatory networks that control the specification of neural-crest cells in the lamprey. *Biochim. Biophys. Acta* 1789(4), 274–278. doi: 10.1016/j.bbagrm.2008.03.006

Nishiyama, C., Uesaka, T., Manabe, T., Yonekura, Y., Nagasawa, T., Newgreen, D.F., Young, H.M., Enomoto, H., 2012. Trans-mesenteric neural crest cells are the principal source of the colonic enteric nervous system. *Nat. Neurosci.* 15(9), 1211–1218. doi: 10.1038/nn.3184

Niss, K., Leutz, A., 1998. Expression of the homeobox gene GBX2 during chicken development. *Mech. Dev.* 76(1–2), 151–155.

Nitzan, E., Krispin, S., Pfaltzgraff, E.R., Klar, A., Labosky, P.A., Kalcheim, C., 2013. A dynamic code of dorsal neural tube genes regulates the segregation between neurogenic and melanogenic neural crest cells. *Dev. Camb. Engl.* 140(11), 2269–2279. doi: 10.1242/dev.093294

Noack Watt, K.E., Achilleos, A., Neben, C.L., Merrill, A.E., Trainor, P.A., 2016. The roles of RNA polymerase I and III subunits Polr1c and Polr1d in craniofacial development and in zebrafish models of Treacher Collins syndrome. *PLOS Genet.* 12(7), e1006187. doi: 10.1371/journal.pgen.1006187

Noden, D.M., 1975. An analysis of migratory behavior of avian cephalic neural crest cells. *Dev. Biol.* 42(1), 106–130.

Noden, D.M., 1983. The role of the neural crest in patterning of avian cranial skeletal, connective, and muscle tissues. *Dev. Biol.* 96(1), 144–165. doi: 10.1016/0012-1606(83)90318-4

Ogura, H., Aruga, J., Mikoshiba, K., 2001. Behavioral abnormalities of Zic1 and Zic2 mutant mice: Implications as models for human neurological disorders. *Behav. Genet.* 31(3), 317–324.

Olden, T., Akhtar, T., Beckman, S.A., Wallace, K.N., 2008. Differentiation of the zebrafish enteric nervous system and intestinal smooth muscle. *Genesis* 46(9), 484–498. doi: 10.1002/dvg.20429

Osborne, N.J., Begbie, J., Chilton, J.K., Schmidt, H., Eickholt, B.J., 2005. Semaphorin/neuropilin signaling influences the positioning of migratory neural crest cells within the hindbrain region of the chick. *Dev. Dyn.* 232, 939–949. doi: 10.1002/dvdy.20258

Pardal, R., Ortega-Sáenz, P., Durán, R., López-Barneo, J., 2007. Glia-like stem cells sustain physiologic neurogenesis in the adult mammalian carotid body. *Cell* 131(2), 364–377. doi: 10.1016/j.cell.2007.07.043

Parr, L.A., 2011. The evolution of face processing in primates. *Philos. Trans. R. Soc. Lond. B* 366(1571), 1764–1777. doi: 10.1098/rstb.2010.0358

Patthey, C., Edlund, T., Gunhaga, L., 2009. Wnt-regulated temporal control of BMP exposure directs the choice between neural plate border and epidermal fate. *Dev. Camb. Engl.* 136(1), 73–83. doi: 10.1242/dev.025890

Pattyn, A., Morin, X., Cremer, H., Goridis, C., Brunet, J.F., 1999. The homeobox gene Phox2b is essential for the development of autonomic neural crest derivatives. *Nature* 399(6734), 366–370. doi: 10.1038/20700

Peirano, R.I., Goerich, D.E., Riethmacher, D., Wegner, M., 2000. Protein zero gene expression is regulated by the glial transcription factor Sox10. *Mol. Cell. Biol.* 20(9), 3198–3209.

Perez, S.E., Rebelo, S., Anderson, D.J., 1999. Early specification of sensory neuron fate revealed by expression and function of neurogenins in the chick embryo. *Development* 126(8), 1715–1728.

Perris, R., 1997. The extracellular matrix in neural crest-cell migration. *Trends Neurosci.* 20(1), 23–31. doi: 10.1016/S0166-2236(96)10063-1

Perris, R., Perissinotto, D., 2000. Role of the extracellular matrix during neural crest cell migration. *Mech. Dev.* 95(1–2), 3–21.

Petersen, J., Adameyko, I., 2017. Nerve-associated neural crest: Peripheral glial cells generate multiple fates in the body. *Curr. Opin. Genet. Dev.* 45, 10–14. doi: 10.1016/j.gde.2017.02.006

Phillips, B.T., Kwon, H.-J., Melton, C., Houghtaling, P., Fritz, A., Riley, B.B., 2006. Zebrafish msxB, msxC and msxE function together to refine the neural-nonneural border and regulate cranial placodes and neural crest development. *Dev. Biol.* 294(2), 376–390. doi: 10.1016/j.ydbio.2006.03.001

Pietri, T., Eder, O., Blanche, M., Thiery, J.P., Dufour, S., 2003. The human tissue plasminogen activator-Cre mouse: A new tool for targeting specifically neural crest cells and their derivatives in vivo. *Dev. Biol.* 259(1), 176–187.

Pla, P., Alberti, C., Solov'eva, O., Pasdar, M., Kunisada, T., Larue, L., 2005. Ednrb2 orients cell migration towards the dorsolateral neural crest pathway and promotes melanocyte differentiation. *Pigment Cell Res.* 18(3), 181–187. doi: 10.1111/j.1600-0749.2005.00230.x

Plouhinec, J.-L., Medina-Ruiz, S., Borday, C., Bernard, E., Vert, J.-P., Eisen, M.B., Harland, R.M., Monsoro-Burq, A.H., 2017. A molecular atlas of the developing ectoderm defines neural, neural crest, placode, and nonneural progenitor identity in vertebrates. *PLOS Biol.* 15(10), e2004045. doi: 10.1371/journal.pbio.2004045

Plouhinec, J.-L., Roche, D.D., Pegoraro, C., Figueiredo, A.-L., Maczkowiak, F., Brunet, L.J., Milet, C., Vert, J.-P., Pollet, N., Harland, R.M., Monsoro-Burq, A.H., 2014. Pax3 and Zic1 trigger the early neural crest gene regulatory network by the direct activation of multiple key neural crest specifiers. *Dev. Biol.* 386(2), 461–472. doi: 10.1016/j.ydbio.2013.12.010

Plouhinec, J.-L., Zakin, L., Moriyama, Y., De Robertis, E.M., 2013. Chordin forms a self-organizing morphogen gradient in the extracellular space between ectoderm and mesoderm in the Xenopus embryo. *Proc. Natl. Acad. Sci. U.S.A.* 110(51), 20372–20379. doi: 10.1073/pnas.1319745110

Polak, J.M., Pearse, A.G.E., Lièvre, C.L., Fontaine, J., Douarin, N.M.L., 1974. Immunocytochemical confirmation of the neural crest origin of avian calcitonin-producing cells. *Histochemistry* 40(3), 209–214. doi: 10.1007/BF00501955

Prasad, M.S., Uribe-Querol, E., Marquez, J., Vadasz, S., Yardley, N., Shelar, P.B., Charney, R.M., Garcia-Castro, M.I., 2019. Blastula stage specification of avian neural crest. *bioRxiv* 705731. doi: 10.1101/705731

Rada-Iglesias, A., Bajpai, R., Prescott, S., Brugmann, S., Swigut, T., Wysocka, J., 2012. Epigenomic annotation of enhancers predicts transcriptional regulators of human neural crest. *Cell Stem Cell* 11(5), 633–648. doi: 10.1016/j.stem.2012.07.006

Raible, D.W., Eisen, J.S., 1994. Restriction of neural crest cell fate in the trunk of the embryonic zebrafish. *Dev. Camb. Engl.* 120(3), 495–503.

Rao, A., LaBonne, C., 2018. Histone deacetylase activity plays an essential role in establishing and maintaining the vertebrate neural crest. *Dev. Camb. Engl.* doi: 10.1242/dev.163386

Raven, C.P., 1931. Zur Entwicklung der Ganglienleiste. I. Die Kinematik der Ganglienleistenentwicklung bei den Urodelen. *Wilhelm Roux Arch. Entwicklungsmechanik Org.* 125(2–3), 210–292. doi: 10.1007/BF00576356

Relaix, F., Rocancourt, D., Mansouri, A., Buckingham, M., 2004. Divergent functions of murine Pax3 and Pax7 in limb muscle development. *Genes Dev.* 18(9), 1088–1105. doi: 10.1101/gad.301004

Relaix, F., Rocancourt, D., Mansouri, A., Buckingham, M., 2005. A Pax3/Pax7-dependent population of skeletal muscle progenitor cells. *Nature* 435(7044), 948–953. doi: 10.1038/nature03594

Rhinn, M., Lun, K., Amores, A., Yan, Y.-L., Postlethwait, J.H., Brand, M., 2003. Cloning, expression and relationship of zebrafish gbx1 and gbx2 genes to Fgf signaling. *Mech. Dev.* 120(8), 919–936. doi: 10.1016/S0925-4773(03)00135-7

Rice, R., Cebra-Thomas, J., Haugas, M., Partanen, J., Rice, D.P.C., Gilbert, S.F., 2017. Melanoblast development coincides with the late emerging cells from the dorsal neural tube in turtle Trachemys scripta. *Sci. Rep.* 7(1), 12063. doi: 10.1038/s41598-017-12352-0

Richardson, J., Gauert, A., Briones Montecinos, L., Fanlo, L., Alhashem, Z.M., Assar, R., Marti, E., Kabla, A., Härtel, S., Linker, C., 2016. Leader cells define directionality of trunk, but not cranial, neural crest cell migration. *Cell Rep.* 15(9), 2076–2088. doi: 10.1016/j.celrep.2016.04.067

Rodrigues, F.S.L.M., Doughton, G., Yang, B., Kelsh, R.N., 2012. A novel transgenic line using the Cre-lox system to allow permanent lineage-labeling of the zebrafish neural crest. *Genesis* 50(10), 750–757. doi: 10.1002/dvg.22033

Roellig, D., Tan-Cabugao, J., Esaian, S., Bronner, M.E., 2017. Dynamic transcriptional signature and cell fate analysis reveals plasticity of individual neural plate border cells. *eLife* 6. doi: 10.7554/eLife.21620

Rogers, C.D., Saxena, A., Bronner, M.E., 2013. Sip1 mediates an E-cadherin-to-N-cadherin switch during cranial neural crest EMT. *J. Cell Biol.* 203(5), 835–847. doi: 10.1083/jcb.201305050

Rothstein, M., Simoes-Costa, M., 2020. Heterodimerization of TFAP2 pioneer factors drives epigenomic remodeling during neural crest specification. *Genome Res.* 30(1), 35–48. doi: 10.1101/gr. 249680.119.

Rovasio, R.A., Delouvee, A., Yamada, K.M., Timpl, R., Thiery, J.P., 1983. Neural crest cell migration: Requirements for exogenous fibronectin and high cell density. *J. Cell Biol.* 96(2), 462–473.

Sadaghiani, B., Thiebaud, C.H., 1987. Neural crest development in the Xenopus laevis embryo, studied by interspecific transplantation and scanning electron microscopy. *Dev. Biol.* 124(1), 91–110. doi: 10.1016/0012-1606(87)90463-5

Saint-Jeannet, J.P., He, X., Varmus, H.E., Dawid, I.B., 1997. Regulation of dorsal fate in the neuraxis by Wnt-1 and Wnt-3a. *Proc. Natl. Acad. Sci. U.S.A.* 94(25), 13713–13718.

Saint-Jeannet, J.-P., He, X., Varmus, H.E., Dawid, I.B., 1997. Regulation of dorsal fate in the neuraxis by Wnt-1 and Wnt-3a. *Proc. Natl. Acad. Sci. U.S.A.* 94(25), 13713–13718. doi: 10.1073/pnas.94.25.13713

Sanlaville, D., Etchevers, H.C., Gonzales, M., Martinovic, J., Clément-Ziza, M., Delezoide, A.-L., Aubry, M.-C., Pelet, A., Chemouny, S., Cruaud, C., Audollent, S., Esculpavit, C., Goudefroye, G., Ozilou, C., Fredouille, C., Joye, N., Morichon-Delvallez, N., Dumez, Y., Weissenbach, J., Munnich, A., Amiel, J., Encha-Razavi, F., Lyonnet, S., Vekemans, M., Attié-Bitach, T., 2006. Phenotypic spectrum of CHARGE syndrome in fetuses with CHD7 truncating mutations correlates with expression during human development. *J. Med. Genet.* 43(3), 211–217. doi: 10.1136/jmg.2005.036160

Santagati, F., Minoux, M., Ren, S.-Y., Rijli, F.M., 2005. Temporal requirement of Hoxa2 in cranial neural crest skeletal morphogenesis. *Dev. Camb. Engl.* 132(22), 4927–4936. doi: 10.1242/dev.02078

Santiago, A., Erickson, C.A., 2002. Ephrin-B ligands play a dual role in the control of neural crest cell migration. *Dev. Camb. Engl.* 129(15), 3621–3632.

Sasai, N., Mizuseki, K., Sasai, Y., 2001. Requirement of FoxD3-class signaling for neural crest determination in Xenopus. *Development* 128(13), 2525–2536.

Sato, T., Sasai, N., Sasai, Y., 2005. Neural crest determination by co-activation of Pax3 and Zic1 genes in Xenopus ectoderm. *Development* 132(10), 2355–2363. doi: 10.1242/dev.01823

Sauka-Spengler, T., Meulemans, D., Jones, M., Bronner-Fraser, M., 2007. Ancient evolutionary origin of the neural crest gene regulatory network. *Dev. Cell* 13(3), 405–420. doi: 10.1016/j.devcel.2007.08.005

Saunders, L.M., Mishra, A.K., Aman, A.J., Lewis, V.M., Toomey, M.B., Packer, J.S., Qiu, X., McFaline-Figueroa, J.L., Corbo, J.C., Trapnell, C., Parichy, D.M., 2019. Thyroid hormone regulates distinct paths to maturation in pigment cell lineages. *eLife* 8. doi: 10.7554/eLife.45181

Scerbo, P., Monsoro-Burq, A.H., 2020. The vertebratespecific VENTX/NANOG gene empowers neural crest with ectomesenchyme potential. *Science Advances.* 6(18). doi: 10.1126/sciadv.aaz1469

Schiffmacher, A.T., Padmanabhan, R., Jhingory, S., Taneyhill, L.A., 2014. Cadherin-6B is proteolytically processed during epithelial-to-mesenchymal transitions of the cranial neural crest. *Mol. Biol. Cell* 25(1), 41–54. doi: 10.1091/mbc.E13-08-0459

Schille, C., Bayerlová, M., Bleckmann, A., Schambony, A., 2016. Ror2 signaling is required for local upregulation of GDF6 and activation of BMP signaling at the neural plate border. *Dev. Camb. Engl.* 143(17), 3182–3194. doi: 10.1242/dev.135426

Schlosser, G., Ahrens, K., 2004. Molecular anatomy of placode development in Xenopus laevis. *Dev. Biol.* 271(2), 439–466. doi: 10.1016/j.ydbio.2004.04.013

Schmid, B., Fürthauer, M., Connors, S.A., Trout, J., Thisse, B., Thisse, C., Mullins, M.C., 2000. Equivalent genetic roles for bmp7/snailhouse and bmp2b/swirl in dorsoventral pattern formation. *Dev. Camb. Engl.* 127(5), 957–967.

Schmidt, C., McGonnell, I.M., Allen, S., Otto, A., Patel, K., 2007. Wnt6 controls amniote neural crest induction through the noncanonical signaling pathway. *Dev. Dyn.* 236(9), 2502–2511. doi: 10.1002/dvdy.21260

Schneider, C., Wicht, H., Enderich, J., Wegner, M., Rohrer, H., 1999. Bone morphogenetic proteins are required in vivo for the generation of sympathetic neurons. *Neuron* 24(4), 861–870.

Schorle, H., Meier, P., Buchert, M., Jaenisch, R., Mitchell, P.J., 1996. Transcription factor AP-2 essential for cranial closure and craniofacial development. *Nature* 381(6579), 235–238. doi: 10.1038/381235a0

Schumacher, J.A., Hashiguchi, M., Nguyen, V.H., Mullins, M.C., 2011. An intermediate level of BMP signaling directly specifies cranial neural crest progenitor cells in zebrafish. *PLOS ONE* 6(11), e27403. doi: 10.1371/journal.pone.0027403

Seberg, H.E., Van Otterloo, E., Cornell, R.A., 2017. Beyond MITF: Multiple transcription factors directly regulate the cellular phenotype in melanocytes and melanoma. *Pigment Cell Melanoma Res.* 30(5), 454–466. doi: 10.1111/pcmr.12611

Sela-Donenfeld, D., Kalcheim, C., 1999. Regulation of the onset of neural crest migration by coordinated activity of BMP4 and Noggin in the dorsal neural tube. *Dev. Camb. Engl.* 126(21), 4749–4762.

Sela-Donenfeld, D., Kalcheim, C., 2000. Inhibition of noggin expression in the dorsal neural tube by somitogenesis: A mechanism for coordinating the timing of neural crest emigration. *Dev. Camb. Engl.* 127(22), 4845–4854.

Selleck, M.A., Bronner-Fraser, M., 1995. Origins of the avian neural crest: The role of neural plate-epidermal interactions. *Dev. Camb. Engl.* 121(2), 525–538.

Serbedzija, G.N., Bronner-Fraser, M., Fraser, S.E., 1994. Developmental potential of trunk neural crest cells in the mouse. *Dev. Camb. Engl.* 120(7), 1709–1718.

Serbedzija, G.N., Fraser, S.E., Bronner-Fraser, M., 1990. Pathways of trunk neural crest cell migration in the mouse embryo as revealed by vital dye labelling. *Dev. Camb. Engl.* 108(4), 605–612.

Shellard, A., Szabó, A., Trepat, X., Mayor, R., 2018. Supracellular contraction at the rear of neural crest cell groups drives collective chemotaxis. *Science* 362(6412), 339–343. doi: 10.1126/science.aau3301

Sieber-Blum, M., Cohen, A.M., 1980. Clonal analysis of quail neural crest cells: They are pluripotent and differentiate in vitro in the absence of noncrest cells. *Dev. Biol.* 80(1), 96–106.

Simoes-Costa, M., Bronner, M.E., 2015. Establishing neural crest identity: A gene regulatory recipe. *Dev. Camb. Engl.* 142(2), 242–257. doi: 10.1242/dev.105445

Simoes-Costa, M., Bronner, M.E., 2016. Reprogramming of avian neural crest axial identity and cell fate. *Science* 352(6293), 1570–1573. doi: 10.1126/science.aaf2729

Simões-Costa, M., Stone, M., Bronner, M.E., 2015. Axud1 integrates Wnt signaling and transcriptional inputs to drive neural crest formation. *Dev. Cell* 34(5), 544–554. doi: 10.1016/j.devcel.2015.06.024

Simon, J.D., Peles, D., Wakamatsu, K., Ito, S., 2009. Current challenges in understanding melanogenesis: Bridging chemistry, biological control, morphology, and function. *Pigment Cell Melanoma Res.* 22(5), 563–579. doi: 10.1111/j.1755-148X.2009.00610.x

Singh, A.P., Dinwiddie, A., Mahalwar, P., Schach, U., Linker, C., Irion, U., Nüsslein-Volhard, C., 2016. Pigment cell progenitors in zebrafish remain multipotent through metamorphosis. *Dev. Cell* 38(3), 316–330. doi: 10.1016/j.devcel.2016.06.020

Singh, N., Gupta, M., Trivedi, C.M., Singh, M.K., Li, L., Epstein, J.A., 2013. Murine craniofacial development requires Hdac3-mediated repression of Msx gene expression. *Dev. Biol.* 377(2), 333–344. doi: 10.1016/j.ydbio.2013.03.008

Sittewelle, M., Monsoro-Burq, A.H., 2018. AKT signaling displays multifaceted functions in neural crest development. *Dev. Biol.* doi: 10.1016/j.ydbio.2018.05.023

Smith, A., Robinson, V., Patel, K., Wilkinson, D.G., 1997. The EphA4 and EphB1 receptor tyrosine kinases and ephrin-B2 ligand regulate targeted migration of branchial neural crest cells. *Curr. Biol.* 7(8), 561–570.

Smith, D.E., Franco del Amo, F., Gridley, T., 1992. Isolation of Sna, a mouse gene homologous to the Drosophila genes snail and escargot: Its expression pattern suggests multiple roles during postimplantation development. *Dev. Camb. Engl.* 116(4), 1033–1039.

Sobkow, L., Epperlein, H.-H., Herklotz, S., Straube, W.L., Tanaka, E.M., 2006. A germline GFP transgenic axolotl and its use to track cell fate: Dual origin of the fin mesenchyme during development and the fate of blood cells during regeneration. *Dev. Biol.* 290(2), 386–397. doi: 10.1016/j.ydbio.2005.11.037

Soldatov, R., Kaucka, M., Kastriti, M.E., Petersen, J., Chontorotzea, T., Englmaier, L., Akkuratova, N., Yang, Y., Häring, M., Dyachuk, V., Bock, C., Farlik, M., Piacentino, M.L., Boismoreau, F., Hilscher, M.M., Yokota, C., Qian, X., Nilsson, M., Bronner,

M.E., Croci, L., Hsiao, W.-Y., Guertin, D.A., Brunet, J.-F., Consalez, G.G., Ernfors, P., Fried, K., Kharchenko, P.V., Adameyko, I., 2019. Spatiotemporal structure of cell fate decisions in murine neural crest. *Science* 364(6444). doi: 10.1126/science.aas9536

Sommer, L., Shah, N., Rao, M., Anderson, D.J., 1995. The cellular function of MASH1 in autonomic neurogenesis. *Neuron* 15(6), 1245–1258.

Soo, K., O'Rourke, M.P., Khoo, P.-L., Steiner, K.A., Wong, N., Behringer, R.R., Tam, P.P.L., 2002. Twist function is required for the morphogenesis of the cephalic neural tube and the differentiation of the cranial neural crest cells in the mouse embryo. *Dev. Biol.* 247(2), 251–270. doi: 10.1006/dbio.2002.0699

Southard-Smith, E.M., Kos, L., Pavan, W.J., 1998. SOX10 mutation disrupts neural crest development in Dom Hirschsprung mouse model. *Nat. Genet.* 18(1), 60–64. doi: 10.1038/ng0198-60

Spokony, R.F., Aoki, Y., Saint-Germain, N., Magner-Fink, E., Saint-Jeannet, J.-P., 2002. The transcription factor Sox9 is required for cranial neural crest development in Xenopus. *Development* 129(2), 421–432.

Stanke, M., Junghans, D., Geissen, M., Goridis, C., Ernsberger, U., Rohrer, H., 1999. The Phox2 homeodomain proteins are sufficient to promote the development of sympathetic neurons. *Development* 126(18), 4087–4094.

Steventon, B., Araya, C., Linker, C., Kuriyama, S., Mayor, R., 2009. Differential requirements of BMP and Wnt signalling during gastrulation and neurulation define two steps in neural crest induction. *Dev. Camb. Engl.* 136(5), 771–779. doi: 10.1242/dev.029017

Stewart, R.A., Arduini, B.L., Berghmans, S., George, R.E., Kanki, J.P., Henion, P.D., Look, A.T., 2006. Zebrafish foxd3 is selectively required for neural crest specification, migration and survival. *Dev. Biol.* 292(1), 174–188. doi: 10.1016/j.ydbio.2005.12.035

Stolfi, A., Ryan, K., Meinertzhagen, I.A., Christiaen, L., 2015. Migratory neuronal progenitors arise from the neural plate borders in tunicates. *Nature* 527(7578), 371–374. doi: 10.1038/nature15758

Stone, L.S., 1929. Experiments showing the role of migrating neural crest (mesectoderm) in the formation of head skeleton and loose connective tissue in Rana palustris. *Wilhelm Roux Arch. Entwicklungsmechanik Org.* 118(1), 40–77. doi: 10.1007/BF02108871

Streit, A., 2002. Extensive cell movements accompany formation of the otic placode. *Dev. Biol.* 249(2), 237–254.

Strobl-Mazzulla, P.H., Bronner, M.E., 2012. A PHD12–Snail2 repressive complex epigenetically mediates neural crest epithelial-to-mesenchymal transition. *J. Cell Biol.* 198(6), 999–1010. doi: 10.1083/jcb.201203098

Strobl-Mazzulla, P.H., Sauka-Spengler, T., Bronner-Fraser, M., 2010. Histone demethylase JmjD2A regulates neural crest specification. *Dev. Cell* 19(3), 460–468. doi: 10.1016/j.devcel.2010.08.009

Sun Rhodes, L.S., Merzdorf, C.S., 2006. The zic1 gene is expressed in chick somites but not in migratory neural crest. *Gene Expr. Patterns* 6(5), 539–545. doi: 10.1016/j.modgep.2005.10.006

Sun, Y., Dykes, I.M., Liang, X., Eng, S.R., Evans, S.M., Turner, E.E., 2008. A central role for Islet1 in sensory neuron development linking sensory and spinal gene regulatory programs. *Nat. Neurosci.* 11(11), 1283–1293. doi: 10.1038/nn.2209

Suzuki, A., Ueno, N., Hemmati-Brivanlou, A., 1997. Xenopus msx1 mediates epidermal induction and neural inhibition by BMP4. *Dev. Camb. Engl.* 124(16), 3037–3044.

Suzuki, H.R., Padanilam, B.J., Vitale, E., Ramirez, F., Solursh, M., 1991. Repeating developmental expression of G-Hox 7, a novel homeobox-containing gene in the chicken. *Dev. Biol.* 148(1), 375–388.

Szabó, A., Mayor, R., 2018. Mechanisms of neural crest migration. *Annu. Rev. Genet.* 52, 43–63. doi: 10.1146/annurev-genet-120417-031559

Szabo, A., Melchionda, M., Nastasi, G., Woods, M.L., Campo, S., Perris, R., Mayor, R., 2016. In vivo confinement promotes collective migration of neural crest cells. *J. Cell Biol.* 213(5), 543–555. doi: 10.1083/jcb.201602083

Takahashi, Y., Bontoux, M., Le Douarin, N.M., 1991. Epithelio--mesenchymal interactions are critical for Quox 7 expression and membrane bone differentiation in the neural crest derived mandibular mesenchyme. *EMBO J.* 10(9), 2387–2393.

Talbot, J.C., Johnson, S.L., Kimmel, C.B., 2010. Hand2 and Dlx genes specify dorsal, intermediate and ventral domains within zebrafish pharyngeal arches. *Dev. Camb. Engl.* 137(15), 2507–2517. doi: 10.1242/dev.049700

Taneyhill, L.A., Coles, E.G., Bronner-Fraser, M., 2007. Snail2 directly represses cadherin6B during epithelial-to-mesenchymal transitions of the neural crest. *Development* 134(8), 1481–1490. doi: 10.1242/dev.02834

Teng, L., Mundell, N.A., Frist, A.Y., Wang, Q., Labosky, P.A., 2008. Requirement for Foxd3 in maintenance of neural crest progenitors. *Dev. Camb. Engl.* 135, 1615–1624. doi: 10.1242/dev.012179

Theveneau, E., Duband, J.-L., Altabef, M., 2007. Ets-1 confers cranial features on neural crest delamination. *PLOS ONE* 2(11), e1142. doi: 10.1371/journal.pone.0001142

Theveneau, E., Marchant, L., Kuriyama, S., Gull, M., Moepps, B., Parsons, M., Mayor, R., 2010. Collective chemotaxis requires contact-dependent cell polarity. *Dev. Cell* 19(1), 39–53. doi: 10.1016/j.devcel.2010.06.012

Theveneau, E., Mayor, R., 2010. Integrating chemotaxis and contact-inhibition during collective cell migration: Small GTPases at work. *Small GTPases* 1(2), 113–117. doi: 10.4161/sgtp.1.2.13673

Theveneau, E., Steventon, B., Scarpa, E., Garcia, S., Trepat, X., Streit, A., Mayor, R., 2013. Chase-and-run between adjacent cell populations promotes directional collective migration. *Nat. Cell Biol.* 15(7), 763–772. doi: 10.1038/ncb2772

Thisse, C., Thisse, B., Postlethwait, J.H., 1995. Expression of snail2, a second member of the zebrafish snail family, in cephalic mesendoderm and presumptive neural crest of wild-type and spadetail mutant embryos. *Dev. Biol.* 172(1), 86–99. doi: 10.1006/dbio.1995.0007

Thomas, S., Thomas, M., Wincker, P., Babarit, C., Xu, P., Speer, M.C., Munnich, A., Lyonnet, S., Vekemans, M., Etchevers, H.C., 2008. Human neural crest cells display molecular and phenotypic hallmarks of stem cells. *Hum. Mol. Genet.* 17(21), 3411–3425. doi: 10.1093/hmg/ddn235

Thompson, H., Blentic, A., Watson, S., Begbie, J., Graham, A., 2010. The formation of the superior and jugular ganglia: Insights into the generation of sensory neurons by the neural crest. *Dev. Dyn.* 239, 439–445. doi: 10.1002/dvdy.22179

Thompson, H., Tucker, A.S., 2013. Dual origin of the epithelium of the mammalian middle ear. *Science* 339(6126), 1453–1456. doi: 10.1126/science.1232862

Tibboel, D., Meijers, J.H., Klück, P., van der Kamp, A.W., ten Kate, F.W., van Haperen-Heuts, I.C., Molenaar, J.C., 1987. Monoclonal antibodies for diagnosis and research in enteric nervous system pathology. A review. *Dev. Neurosci.* 9(3), 133–143. doi: 10.1159/000111617

Tien, C.-L., Jones, A., Wang, H., Gerigk, M., Nozell, S., Chang, C., 2015. Snail2/Slug cooperates with Polycomb repressive complex 2 (PRC2) to regulate neural crest development. *Dev. Camb. Engl.* 142(4), 722–731. doi: 10.1242/dev.111997

Tonks, I.D., Nurcombe, V., Paterson, C., Zournazi, A., Prather, C., Mould, A.W., Kay, G.F., 2003. Tyrosinase-Cre mice for tissue-specific gene ablation in neural crest and neuroepithelial-derived tissues. *Genesis* 37(3), 131–138. doi: 10.1002/gene.10242

Tour, E., Pillemer, G., Gruenbaum, Y., Fainsod, A., 2002. Gbx2 interacts with Otx2 and patterns the anterior–posterior axis during gastrulation in Xenopus. *Mech. Dev.* 112(1–2), 141–151. doi: 10.1016/S0925-4773(01)00653-0

Trainor, P.A., Richtsmeier, J.T., 2015. Facing up to the challenges of advancing Craniofacial Research. *Am. J. Med. Genet. A.* 167(7), 1451–1454. doi: 10.1002/ajmg.a.37065

Tribulo, C., Aybar, M.J., Nguyen, V.H., Mullins, M.C., Mayor, R., 2003. Regulation of Msx genes by a Bmp gradient is essential for neural crest specification. *Dev. Camb. Engl.* 130(26), 6441–6452. doi: 10.1242/dev.00878

Tsarovina, K., Pattyn, A., Stubbusch, J., Müller, F., Wees, J. van der, Schneider, C., Brunet, J.-F., Rohrer, H., 2004. Essential role of Gata transcription factors in sympathetic neuron development. *Development* 131(19), 4775–4786. doi: 10.1242/dev.01370

Tucker, G.C., Aoyama, H., Lipinski, M., Tursz, T., Thiery, J.P., 1984. Identical reactivity of monoclonal antibodies HNK-1 and NC-1: Conservation in vertebrates on cells derived from the neural primordium and on some leukocytes. *Cell Differ.* 14(3), 223–230.

Tucker, G.C., Delarue, M., Zada, S., Boucaut, J.C., Thiery, J.P., 1988. Expression of the HNK-1/NC-1 epitope in early vertebrate neurogenesis. *Cell Tissue Res.* 251(2), 457–465.

Uesaka, T., Enomoto, H., 2010. Neural precursor death is central to the pathogenesis of intestinal aganglionosis in ret hypomorphic mice. *J. Neurosci.* 30(15), 5211–5218. doi: 10.1523/JNEUROSCI.6244-09.2010

Uesaka, T., Nagashimada, M., Enomoto, H., 2015. Neuronal differentiation in Schwann cell lineage underlies postnatal neurogenesis in the enteric nervous system. *J. Neurosci.* 35(27), 9879–9888. doi: 10.1523/JNEUROSCI.1239-15.2015

Vallin, J., Thuret, R., Giacomello, E., Faraldo, M.M., Thiery, J.P., Broders, F., 2001. Cloning and characterization of three Xenopus slug promoters reveal direct regulation by Lef/beta-catenin signaling. *J. Biol. Chem.* 276(32), 30350–30358. doi: 10.1074/jbc.M103167200

van Limborgh, J., Song, Lieuw Kie, Been, S.H., W., 1983. Cleft lip and palate due to deficiency of mesencephalic neural crest cells. *Cleft Palate J.* 20(3), 251–259.

Villanueva, S., Glavic, A., Ruiz, P., Mayor, R., 2002. Posteriorization by FGF, Wnt, and retinoic acid is required for neural crest induction. *Dev. Biol.* 241(2), 289–301. doi: 10.1006/dbio.2001.0485

Vincent, C., Bontoux, M., Le Douarin, N.M., Pieau, C., Monsoro-Burq, A.-H., 2003. Msx genes are expressed in the carapacial ridge of turtle shell: A study of the European pond turtle, *Emys orbicularis. Dev. Genes Evol.* 213(9), 464–469. doi: 10.1007/s00427-003-0347-3

Vincent, M., Duband, J.-L., Thiery, J.P., 1983. A cell surface determinant expressed early on migrating avian neural crest cells. *Brain Res.* 285(2), 235–238.

Wada, N., Javidan, Y., Nelson, S., Carney, T.J., Kelsh, R.N., Schilling, T.F., 2005. Hedgehog signaling is required for cranial neural crest morphogenesis and chondrogenesis at the midline in the zebrafish skull. *Dev. Camb. Engl.* 132(17), 3977–3988. doi: 10.1242/dev.01943

Wan, P., Hu, Y., He, L., 2011. Regulation of melanocyte pivotal transcription factor MITF by some other transcription factors. *Mol. Cell. Biochem.* 354(1–2), 241–246. doi: 10.1007/s11010-011-0823-4

Wang, S.-K., Komatsu, Y., Mishina, Y., 2011. Potential contribution of neural crest cells to dental enamel formation. *Biochem. Biophys. Res. Commun.* 415(1), 114–119. doi: 10.1016/j.bbrc.2011.10.026

Wassarman, K.M., Lewandoski, M., Campbell, K., Joyner, A.L., Rubenstein, J.L., Martinez, S., Martin, G.R., 1997. Specification of the anterior hindbrain and establishment of a normal mid/hindbrain organizer is dependent on Gbx2 gene function. *Development* 124(15), 2923–2934.

Wei, K., Chen, J., Akrami, K., Galbraith, G.C., Lopez, I.A., Chen, F., 2007. Neural crest cell deficiency of c-myc causes skull and hearing defects. *Genesis* 45(6), 382–390. doi: 10.1002/dvg.20304

Weston, J.A., 1963. A radioautographic analysis of the migration and localization of trunk neural crest cells in the chick. *Dev. Biol.* 6, 279–310.

Weston, J.A., 1970. The migration and differentiation of neural crest cells. *Adv. Morphog.* 8, 41–114.

Wijchers, P.J.E.C., Burbach, J.P.H., Smidt, M.P., 2006. In control of biology: Of mice, men and foxes. *Biochem. J.* 397(2), 233–246. doi: 10.1042/BJ20060387

Williams, A.L., Bohnsack, B.L., 2017. Multi-photon time lapse imaging to visualize development in real-time: Visualization of migrating neural crest cells in zebrafish embryos. *J. Vis. Exp.* doi: 10.3791/56214

Williams, R.M., Candido-Ferreira, I., Repapi, E., Gavriouchkina, D., Senanayake, U., Ling, I.T.C., Telenius, J., Taylor, S., Hughes, J., Sauka-Spengler, T., 2019. Reconstruction of the global neural crest gene regulatory network in vivo. *Dev Cell.* 51(2), 255–276. doi: 10.1016/j.devcel.2019.10.003.

Wilson, S.I., Rydström, A., Trimborn, T., Willert, K., Nusse, R., Jessell, T.M., Edlund, T., 2001. The status of Wnt signalling regulates neural and epidermal fates in the chick embryo. *Nature* 411(6835), 325–330. doi: 10.1038/35077115

Wynn, M.L., Rupp, P., Trainor, P.A., Schnell, S., Kulesa, P.M., 2013. Follow-the-leader cell migration requires biased cell-cell contact and local microenvironmental signals. *Phys. Biol.* 10(3), 035003. doi: 10.1088/1478-3975/10/3/035003

Yamauchi, Y., Abe, K., Mantani, A., Hitoshi, Y., Suzuki, M., Osuzu, F., Kuratani, S., Yamamura, K., 1999. A novel transgenic technique that allows specific marking of the neural crest cell lineage in mice. *Dev. Biol.* 212(1), 191–203. doi: 10.1006/dbio.1999.9323

Yan, H., Bergner, A.J., Enomoto, H., Milbrandt, J., Newgreen, D.F., Young, H.M., 2004. Neural cells in the esophagus respond to glial cell line-derived neurotrophic factor and neurturin, and are RET-dependent. *Dev. Biol.* 272(1), 118–133. doi: 10.1016/j.ydbio.2004.04.025

Yntema, C.L., Hammond, W.S., 1954. The origin of intrinsic ganglia of trunk viscera from vagal neural crest in the chick embryo. *J. Comp. Neurol.* 101(2), 515–541. doi: 10.1002/cne.901010212

Young, H.M., Bergner, A.J., Anderson, R.B., Enomoto, H., Milbrandt, J., Newgreen, D.F., Whitington, P.M., 2004. Dynamics of neural crest-derived cell migration in the embryonic mouse gut. *Dev. Biol.* 270(2), 455–473. doi: 10.1016/j.ydbio.2004.03.015

Zalc, A., Rattenbach, R., Auradé, F., Cadot, B., Relaix, F., 2015. Pax3 and Pax7 play essential safeguard functions against environmental stress-induced birth defects. *Dev. Cell* 33(1), 56–66. doi: 10.1016/j.devcel.2015.02.006

Ziegler, I., 2003. The pteridine pathway in zebrafish: Regulation and specification during the determination of neural crest cell-fate. *Pigment Cell Res.* 16(3), 172–182.

Zou, J., Beermann, F., Wang, J., Kawakami, K., Wei, X., 2006. The Fugu tyrp1 promoter directs specific GFP expression in zebrafish: Tools to study the RPE and the neural crest-derived melanophores. *Pigment Cell Res.* 19(6), 615–627. doi: 10.1111/j.1600-0749.2006.00349.x

Zuhdi, N., Ortega, B., Giovannone, D., Ra, H., Reyes, M., Asención, V., McNicoll, I., Ma, L., de Bellard, M.E., 2015. Slit molecules prevent entrance of trunk neural crest cells in developing gut. *Int. J. Dev. Neurosci.* 41, 8–16. doi: 10.1016/j.ijdevneu.2014.12.003

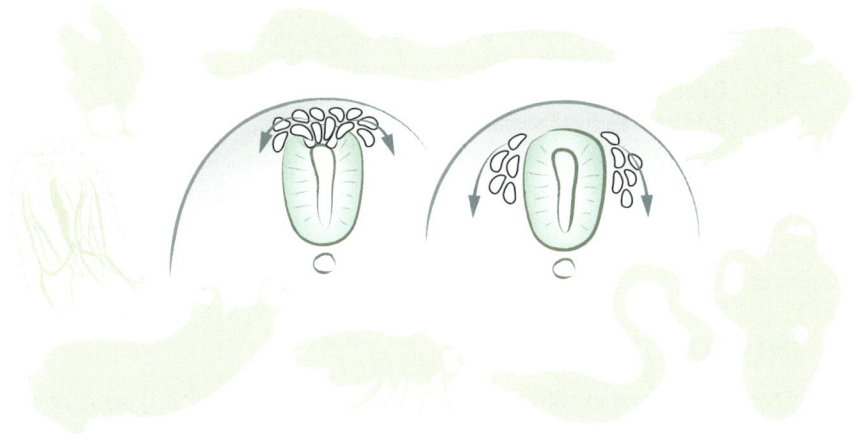

2 The Evolution of Cellular EMT and Migration

Joshua R. York, Kevin Zehnder,
and David W. McCauley

CONTENTS

2.1 INTRODUCTION

During early vertebrate evolution, the acquisition of a novel cell type—the neural crest—was seminal in establishing much of the vertebrate body plan and provided the substrate for diversification of vertebrate morphology (Gans and Northcutt, 1983). Neural crest cells form in the dorsalmost part of the embryonic neural tube, from which they detach and then migrate throughout much of the embryo, where they give rise to many tissues and organ systems (Le Douarin and Kalcheim, 1999) (Figures 2.1, 2.2 show a generic model for neural crest migration in vertebrate embryos). For example, they comprise the core of the vertebrate head, including much of the cartilage, bone, and muscle of the skull and jaws, as well as contributing to paired sensory organ systems (Trainor, 2013). Importantly, many of these structures fundamentally distinguish the body plan of vertebrates from that of their closest relatives, the invertebrate protochordates (Gans and Northcutt, 1983; Northcutt and Gans, 1983). Thus, neural crest cells are a prime example of how a key developmental and evolutionary innovation can drive the origin and diversification of a major animal clade.

The neural crest forms in all vertebrate embryos in a highly stereotyped manner at both molecular and cellular levels. It is established in a region of intermediate bone morphogenetic protein (BMP) expression between the medial neural plate and lateral epidermal ectoderm, referred to as the neural plate border (Groves and LaBonne,

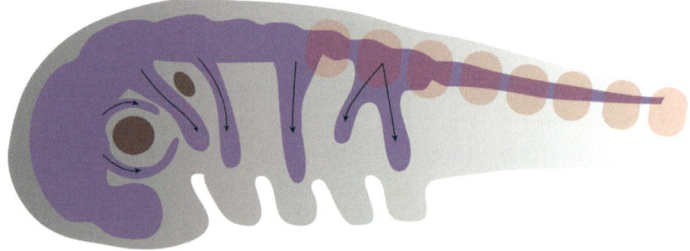

FIGURE 2.1 Neural crest migration in a generalized vertebrate embryo. A lateral view shows migratory neural crest streams that originate in the midbrain and hindbrain of the central nervous system (purple) descending ventrally (arrows) into the head and pharyngeal arches. Large and small brown ovals indicate eye and otic vesicle, respectively. Somites are indicated by pink rectangles. Anterior is left, dorsal is up.

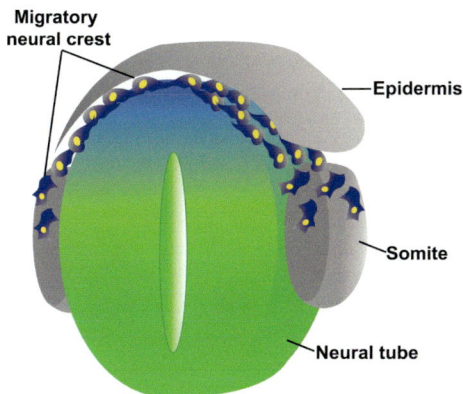

**Migratory
neural crest**

—**Epidermis**

—**Somite**

—**Neural tube**

FIGURE 2.2   EMT, delamination and early migration of neural crest from the dorsal neural tube in a generalized vertebrate embryo. Premigratory neural crest cells (light blue shading) are specified in the dorsal-most aspect of the neural tube. Soon after specification, these cells undergo EMT, delaminate from the underlying neural epithelium and then begin to migrate (dark blue) laterally and ventrally, and in doing so invade surrounding tissues.

2014; Milet and Monsoro-Burq, 2012). Within this border region, intercellular inductive signaling by *BMP*, *WNT*, *FGF*, and *Delta-Notch* activates a core set of genes known as neural plate border specifiers (e.g., *Zic1*, *Dlx5*, *Msx1/2*, *Pax3/7*) (Pla and Monsoro-Burq, 2018). These, in turn, carve out a specific embryonic territory that confers competence on cells within the neural plate border to respond to downstream signals that will eventually specify their fate as bona fide neural crest (Betancur et al., 2010; Monsoro-Burq et al., 2005). Such neural crest specifiers are activated by expression of neural plate border specifiers, combined with reiterated expression of neural crest inducers. These specifiers include members of the *SoxE* family, *Tfap2α*, *Id*, *Snail1/Snail2*, *Myc*, *Twist*, *Ets*, and a host of others that are directly responsible for establishing neural crest identity (Martik and Bronner, 2017; Rogers and Nie, 2018).

One of the hallmarks of neural crest cells is their ability to undergo a dramatic change in cell shape and molecular architecture, resulting in what is known as an epithelial-to-mesenchymal transition (EMT), a feature that confers on neural crest cells an ability to embark on long-distance migrations to specific locations throughout the vertebrate embryo (Bronner, 2012; Cheung et al., 2005b; Cordero et al., 2011; Duband et al., 1995; Gouignard et al., 2018; Hutchins and Bronner, 2018; Kalcheim, 2018; Kerosuo and Bronner-Fraser, 2012; Kulesa et al., 2010; Morales et al., 2005; Savagner, 2010) (Figure 2.3 shows a simple model for molecular and cellular features of neural crest EMT). Although all metazoans have cells that undergo EMT and migrate at some point during development (Fritzenwanker et al., 2004b; Hay, 1995; Hay, 2005; Kee et al., 2007), neural crest cells in vertebrates are unique in several respects. For example, the neural crest possesses a unique "molecular anatomy" that is defined by combinatorial expression of a unique suite of transcription factors and signaling molecules (Green et al., 2015; Nikitina et al., 2008; Sauka-Spengler and Bronner-Fraser, 2008a). In addition, the extent to which neural crest cells migrate as a proliferative, stem cell–like population throughout the embryo, while managing to

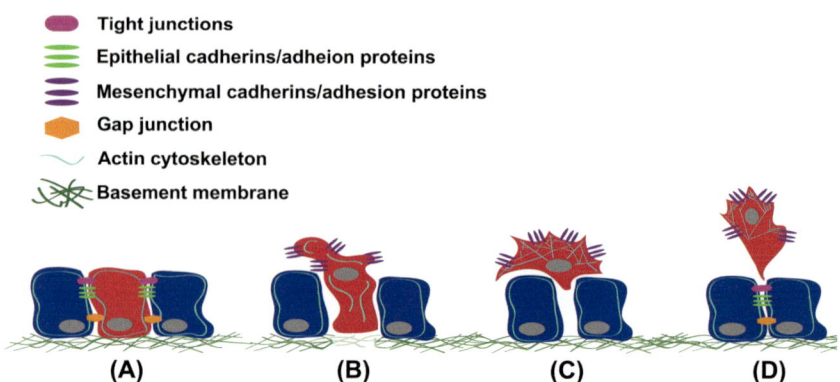

FIGURE 2.3 Model for canonical cellular EMT and delamination program to initiate neural crest migration. (a) Shortly after being specified in the dorsal neural tube, premigratory neural crest cells (red) activate a genetic program that directs changes at the cell surface of intercellular adhesion and junction proteins which allows neural crest cells to break free from neighboring neuroepithelial cells (purple cells) and the underlying basement membrane (b). Concomitant with changes in cell surface proteins is breakdown and reorganization of the actin cytoskeleton to establish a leading edge (c) and begin directed migration (d).

avoid succumbing to cell death and finding their often distant target regions in the developing vertebrate embryo, is unique among metazoan cell types (Cordero et al., 2011; Kulesa et al., 2004; Trainor, 2013).

Given the central role of EMT and migration during neural crest development, an outstanding question in vertebrate evolutionary-developmental biology is how these cells acquired this ability, and what the patterns and mechanics were like when they appeared in the first vertebrates over 500 million years ago (Muñoz and Trainor, 2015). Currently, we understand very little about the evolutionary history of the core features of the EMT and migration module of the neural crest GRN and therefore have no context for how it was assembled and/or coopted from molecular circuitry that most likely originated from simpler precursors among invertebrates. Addressing these issues is critical to understanding both neural crest evolution specifically and vertebrate evolution more broadly because it would identify the cellular and molecular control mechanisms that allowed the spread of multipotent progenitor cells throughout the vertebrate embryo and contributed to the developmental origin of novel morphological features.

In this chapter, we review the molecular, cellular, and genetic mechanisms controlling EMT and migration of neural crest cells across vertebrates in an attempt to highlight key features that can be generalized for a range of vertebrate clades. We also compare and contrast similar mechanisms mediating EMT and cell migration in diverse metazoans in order to better understand the phylogenetic context in which neural crest cells acquired the ability to break free from the neural tube and migrate. Finally, using a comparative approach, we lay out a scenario for the origin of EMT and migratory properties in vertebrate neural crest cells, and speculate on the assemblage and topology of the EMT-migration module of the neural crest GRN and how this module has been rewired across almost 500 million years of vertebrate evolution.

2.2 NEURAL CREST EMT AND MIGRATION IN VERTEBRATES

There is a large volume of literature describing the molecular, cellular, and genetic underpinnings of neural crest EMT and migration in vertebrates. Many of these studies have been conducted in established model systems (e.g., mouse, zebrafish, chick, frog), especially those that have genetic lines available and/or are relatively easy to obtain and manipulate in laboratory settings (Barriga et al., 2013; Carl et al., 1999; Coles et al., 2007; Kubota and Ito, 2000; Linker et al., 2000; Morales et al., 2005; Scarpa et al., 2015; Tucker, 2004; Vallin et al., 1998; Vannier et al., 2013; Willems et al., 2015). However, although this provides an extensive catalogue of the transcription factors, signaling cascades, and downstream effectors involved in neural crest EMT and migration, there is very little context for how these studies can shed light on the evolutionary origin of neural crest EMT, how these factors were integrated into the neural crest EMT-migration module of the ancestral GRN, and how elaboration and modification of this module helped shape the vertebrate body plan. Below, we first describe the developmental genetics of EMT, delamination, and migration across jawed (gnathostome) vertebrates, with an emphasis on features that are shared across multiple gnathostome taxa and hence are likely to be evolutionarily conserved. We then compare and contrast these mechanisms operating in gnathostomes with what we currently know from jawless (agnathan) vertebrates to gain insight into what may be ancient features of vertebrate neural crest EMT versus those that are elaborations on the ancestral EMT program.

2.2.1 GNATHOSTOME NEURAL CREST EMT AND MIGRATION

2.2.1.1 Intercellular Signaling Pathways in Neural Crest EMT

Similar to induction of the neural crest at the neural plate border, the initiation of neural crest EMT occurs by receptor–ligand interactions and the subsequent activation of intracellular signaling pathways in the dorsal neural tube. These pathways, which include *TGFβ/BMP* (Sela-Donenfeld and Kalcheim, 1999), *FGF*, *PDGF*, and *WNT* signaling (Burstyn-Cohen et al., 2004; De Calisto et al., 2005), are highly conserved across metazoans and have been "plugged in" to diverse GRNs (Davidson and Erwin, 2006). During induction in the dorsal neural tube, presumptive neural crest cells often express the receptors for these signaling molecules (e.g., *FGFR2/3*, *PDGFR-α*) on their cell surface and activate intracellular signaling events once bound to their corresponding ligands, which emanate from surrounding tissues (Huang and Saint-Jeannet, 2004; Knecht and Bonner-Fraser, 2002; Nikitina et al., 2008). The result is signal transduction to the nucleus (e.g., by β-catenin and Smads) and subsequent regulation of gene expression, often by activating expression of key transcription factors that will repress epithelial fate and promote the conversion of premigratory neural crest to mesenchyme (Nikitina et al., 2008) (described below). For example, canonical WNT signaling, together with the neural plate border specifiers *Pax3* and *Zic1*, coordinately activate expression of *Snail1* (Sato et al., 2005), a key transcriptional regulator of neural crest EMT. Similarly, *BMP4* signaling stimulates transcriptional activity that leads to an increased number of premigratory neural crest cells undergoing EMT (Shoval et al., 2007).

2.2.1.2 Transcriptional Control of Neural Crest EMT

As described above, early intercellular inductive signaling converges on the regulation of expression of several transcription factors that are directly responsible for initiating neural crest EMT, delamination, and early migration (Nikitina et al., 2008; Thiery and Sleeman, 2006). Many of these transcription factors also control specification of bona fide neural crest, including SoxE (Cheung et al., 2005b; McKeown et al., 2005), FoxD (Fairchild et al., 2014; Stewart et al., 2006), *Snail1/Snail2* (Bolos et al., 2003; Manzanares et al., 2001; Nieto et al., 1994), Twist (Lander et al., 2011; Yang et al., 2004; Yang and Wu, 2008), Sip1 (Comijn et al., 2001; Vandewalle et al., 2005), Zeb1 (Sánchez-Tilló et al., 2010; Vannier et al., 2013), LMO4 (Ochoa and Labonne, 2011; Ochoa et al., 2012), E12/E47 (Moreno-Bueno et al., 2006; Perez-Moreno et al., 2001; Zheng and Kang, 2014), among many others. The transcriptional control of neural crest EMT involves the coordination of both gene activation and repression programs, which collectively inhibit expression of proepithelial gene batteries and upregulate those promoting a mesenchymal state (Coles et al., 2007; Liu and Jessell, 1998; Perez-Alcala et al., 2004; Pla et al., 2001).

In gnathostomes, the HMG box transcription factor sub-family, SoxE, consists of three paralogues, Sox8, Sox9, and Sox10 (Cheung and Briscoe, 2003; Heeg-truesdell and Labonne, 2004; Kim et al., 2003; Lee et al., 2016). Each of these genes is involved in some capacity early in neural crest specification, although their exact functions may vary and be redundant among taxa (Heeg-truesdell and Labonne, 2004). However, after specification, SoxE factors are also important transcriptional regulators of neural crest EMT and migration. In the trunk neural crest of chick embryos, for example, Sox9, together with activity of Snail1 and/or Snail2 (formerly identified as Snail and Slug respectively) proteins, is sufficient to induce EMT and delamination, and there is evidence that *Sox9* promotes *Snail1/Snail2* expression for neural crest EMT in a *BMP*-dependent manner (Sakai et al., 2006). Similarly, *Sox10* is expressed in delaminating neural crest, and its upregulation is concomitant with decreased expression of proepithelial genes (Cheung et al., 2005a). In addition, forced expression of *Sox10* alone is sufficient not only to specify more neural crest, but also to promote ectopic migration of cells from the entire dorsal–ventral axis of the neural tube (McKeown et al., 2005).

The forkhead box transcription factor FoxD3 is another critical upstream specifier of neural crest identity that is also required later for trunk neural crest EMT and migration in both fish and birds. For example, FoxD3 appears to function primarily in controlling differential expression of intercellular adhesion proteins, a prerequisite for neural crest EMT in gnathostomes (Cheung et al., 2005a; Dottori et al., 2001) (described below). Current evidence suggests that FoxD3 represses expression of certain epithelial cadherins and other epithelial gene batteries, whereas it may promote expression of intercellular adhesion proteins that facilitate neural crest delamination and migration (Fairchild et al., 2014; Fairchild and Gammill, 2013).

The zinc finger transcription factors Snail1 and Snail2 occupy key nodes in the vertebrate neural crest GRN and play important roles early in neural crest specification (Hemavathy et al., 2000; Nieto, 2002). Both *Snail1* and *Snail2* control the onset of neural crest EMT, as functional perturbation of either gene may inhibit delamination of premigratory crest from the neural tube, whereas overexpression promotes ectopic migration (del Barrio and Nieto, 2002). There is also strong evidence from both embryos and cell culture that Snail1/Snail2 factors are direct transcriptional repressors of epithelial genes, including type I and type II cadherins (Bolos et al., 2003; Guaita et al., 2002; Taneyhill et al., 2007). The mechanism by which this occurs—binding of Snail proteins to consensus E-box (CANNTG) elements in the target gene promoter—is thought to downregulate transcription (Nieto, 2002). Although transcriptional repression may be directly related to Snail promoter occupancy, there is evidence that Snail1/Snail2 recruit other proteins that inhibit transcription, such as histone deacetylases (Peinado et al., 2004a) and/or other pro-EMT transcription factors such as LMO4 (Ferronha et al., 2013), Sox9 (Cheung and Briscoe, 2003; Liu et al., 2013) and LIM homeodomain proteins (Langer et al., 2008). Interestingly, despite widespread evolutionary conservation of *Snail1/Snail2* in neural crest development, there have been changes in which the specific *Snail* gene is recruited for neural crest EMT across gnathostome lineages (Locascio et al., 2002). For example, in anamniotes, *Snail1*—rather than *Snail2*—is the primary regulator of neural crest EMT, and this is likely to be the ancestral condition in vertebrates. Near the origin of amniotes, however, *Snail1* activity in the neural crest was swapped for *Snail2*, and there was yet again an apparent secondary reversion back to *Snail1* activity in the neural crest with the evolution of mammals. These data have provided evidence that significant shuffling of *Snail1/Snail2* activity has occurred during the evolution of the neural crest EMT module, although the significance of this remains unknown (Locascio et al., 2002).

In addition to the SoxE, FoxD, and Snail families, the transcription factors Twist, Sip, and Zeb have emerged as key players in neural crest EMT in gnathostomes (Hopwood et al., 1989; Lander et al., 2011; Vandewalle et al., 2005). Similar to Snail1/Snail2, the mechanism of action by these proteins is thought to be direct transcriptional repression of genes that promote an epithelial state, often achieved by coordinated activity with corepressors acting on the promoter of the target gene (Connerney et al., 2006; Lehmann et al., 2016). On the other hand, there may be context-dependent roles for transcriptional activation, as nuclear localization of Twist results in upregulation of N-cadherin and migration in cell culture (Alexander et al., 2006). In chick embryos, *Sip1* is thought to promote neural crest delamination since loss of *Sip1* function impairs delamination and EMT in premigratory crest (Rogers et al., 2013). Finally, although less well-studied, there is growing evidence for the importance of other transcription factors in neural crest EMT, including Ets-1 in cranial neural crest migration in chick (Tahtakran and Selleck, 2003; Théveneau et al., 2007), LMO4 (Ochoa et al., 2012), HIF1α (Barriga et al., 2013) in neural crest EMT in *Xenopus*, and E12/E47 during EMT in cultured cells (Perez-Moreno et al., 2001).

2.2.1.3 Intercellular Adhesion Proteins

Mesenchymal cells such as migratory neural crest are distinguished from epithelia by their behavior, overall morphology, and gene expression profiles (Duband et al., 1995; Hay, 1995; Hay, 2005; Savagner, 2001; Thiery, 2003). Mesenchymal cells have unique markers that characterize their affinities for other cells and provide a reliable means of distinguishing them from epithelia (Fendrich et al., 2009; Kalluri and Weinberg, 2009; Mani et al., 2008). The classical cadherins have long been recognized as diagnostic of epithelia versus mesenchyme and are divided into type I (E-cadherin, N-cadherin, R-cadherin, P-cadherin) and type II (Cadherin-6–15) (Kemler, 1992; Peinado et al., 2004b; Tepass et al., 2000) subgroups. Although other intercellular adhesion proteins have been implicated in neural crest EMT (e.g., occludins, claudins, connexins; Jourdeuil and Taneyhill, 2018), we focus here on the role of classical cadherins. For detailed discussions of other intercellular adhesion proteins involved in neural crest EMT, please refer to Trainor (Trainor, 2013).

Cadherins are transmembrane proteins that promote the ability of cells to adhere together (Tepass et al., 2000). In the classic model of EMT, premigratory neural crest cells undergo cadherin "switching", a process by which they downregulate type I cadherins (e.g., E- or N-cadherin) and upregulate type II cadherins (Cadherin 6b, Cadherin 7, Cadherin 11) (Cousin, 2017; Nakagawa and Takeichi, 1998; Pla et al., 2001; Wheelock et al., 2008). This outcome is achieved through direct repression of type I cadherin genes (and other pro-epithelial gene batteries) by transcription factors such as Snail1/Snail2, Twist, Sip1, Zeb1, and E12/E47, followed by activation of type II cadherins. It was thought that cadherin switching was required for neural crest EMT and migration because the binding affinity of type I cadherins is much greater than that of type II cadherins, and consequently type I cadherins restrict cell movement (Katsamba et al., 2009). However, it is now thought that the concept of a singular type I–type II cadherin switch oversimplifies the complex process of neural crest EMT. In both frog and chick embryos, for example, the type I "pro-epithelial" cadherins such as N-Cadherin and E-Cadherin are expressed in and may even be required for cranial neural crest migration, whereas type II "pro-migration" cadherins, such as Cadherin 6b in the chick midbrain, are repressed in order to allow neural crest cells to migrate (Huang et al., 2016; Rogers et al., 2013; Taneyhill et al., 2007). What these findings suggest is that regulation of type I/type II cadherins in the neural crest is likely to involve subtle yet complex shifts in gene expression—relative to persistent type I cadherin expression in the rest of the neural tube proper—that promote the individuation of the neural crest from the rest of the embryonic neural tube.

2.2.1.4   Reorganization of the Cytoskeleton

The marked shift in intercellular adhesion proteins during neural crest EMT is accompanied by an equally important series of changes to the cellular cytoskeleton (Duband et al., 1995; Hill et al., 2008). Although there are numerous ways in which cytoskeletal changes promote EMT, the most thoroughly characterized of these involve a fundamental reorganization of the structural properties of actin filaments to establish cell polarity with a leading edge that allows directed migration (Genuth et al., 2018; Savagner, 2001). Cytoskeletal reorganization is regulated in large part by the Rho family of small GTPases, a subfamily of the Ras superfamily of small G-protein signaling molecules (Clay and Halloran, 2011; Sadok and Marshall, 2014). The Rho subfamily includes RhoA/B/C, and each Rho protein has distinct developmental functions in cell polarity and migration. Among the Rho A/B/C group, RhoB has figured prominently in studies of neural crest EMT and migration (Liu and Jessell, 1998). Through interactions with the Rho associated kinase (ROCK), Rho GTPases are responsible for regulating the spatial-temporal assembly of actin microfilaments, as well as their contractility (Lai et al., 2005). These functions are most obvious in Rho-mediated organization of stress fibers and actin filaments in filapodia and lamellipodia of migratory cells (Nobes and Hall, 1995). At the transcriptional level, the HMG box transcription factor SoxD is known to regulate expression of RhoB; loss of SoxD-mediated RhoB expression prevents premigratory neural crest from exiting the neural tube (Perez-Alcala et al., 2004). BMP and WNT signaling also help reorganize the cytoskeleton to enhance neural crest migration by promoting RhoB expression in premigratory and migratory crest, and almost certainly do so by regulating expression of transcription factors that directly promote neural crest EMT and migration (Burstyn-Cohen et al., 2004; Sela-Donenfeld and Kalcheim, 1999; Taneyhill and Bronner-Fraser, 2005). Taken together, these findings demonstrate that cytoskeletal mobilization and reorganization—mediated in large part by Rho GTPases—are instrumental to neural crest EMT and migration in gnathostomes (Merchant et al., 2018).

2.2.1.5 Breakdown of the Basal Lamina and Early Migration

After neurulation, the basal lamina—a specialized extracellular matrix of fibrous protein—forms on the basal surface of the neural tube. The basal lamina has many different functions, including inhibition of EMT and cell migration to maintain structural integrity of the neuroepithelium (Tyler, 2003b). Neural crest cells, in order to emigrate from the neural tube, must overcome the barriers imposed by the neural tube basal lamina (Erickson, 1987). In some cases, the default state is that the dorsal neural tube delays production of a basal lamina until all neural crest cells have migrated, rendering this problem obsolete (Martins-Green and Erickson, 1987). In other cases, however, premigratory neural crest cells undergoing EMT must actively break down and degrade the basal lamina to fully emigrate from the neural tube (Kerosuo and Bronner-Fraser, 2012).

Neural crest cells express several different transcription factors and proteases that help to break down and reorganize the basal lamina and alter the structural properties of the surrounding ECM (Hutchins and Bronner, 2019). Together, these processes create a favorable environment for migration and are necessary for a proper EMT. The two most widely studied of these proteins are ADAMs (A Disintegrin And Metalloprotease) and MMPs (Matrix MetalloProteases) (Neuner et al., 2009) (Alfandari et al., 2001; Alfandari and Taneyhill, 2018; Cai et al., 2000). Both proteases cleave cell surface proteins and signaling molecules, with the resulting fragments serving to drive the EMT program by transcriptional regulation. *Xenopus* ADAM13 is required for migration of cranial neural crest (Alfandari et al., 2001; McCusker et al., 2009), and ADAM19 is expressed prominently in the neural crest (Neuner et al., 2009). *MMP2* and *MMP14* are expressed in chick and frog migratory neural crest, respectively (Anderson, 2003; Cai et al., 2000; Duong and Erickson, 2004; Garmon et al., 2018). These proteases are thought to degrade and remodel ECM proteins surrounding neural crest cells, such as fibronectin, in order to create an ECM pathway favorable to neural crest migration (Thiery and Sleeman, 2006; Trainor, 2013). In chick, for example, transcriptional regulation of ADAMs and MMPs is mediated by Ets1, leading to breakdown of the basal lamina and early neural crest migration (Théveneau et al., 2007).

2.2.2   Cyclostome Neural Crest EMT and Migration

The cyclostomes (lampreys and hagfish) are the only surviving relics of an ancient and ecologically dominant group of jawless fish (agnathans) from the Paleozoic. Because they are the sister group to gnathostomes (Heimberg et al., 2010), comparative studies of cyclostome biology have strong potential to offer insights into the genetic and morphological innovations likely present in the vertebrate ancestor (McCauley et al., 2015; York et al., 2019a). Given the relative ease of obtaining, culturing, and manipulating embryos (McCauley et al., 2015), as well as the availability of an annotated genome (Smith et al., 2013; Smith et al., 2018) and modern molecular-genetic tools (Parker et al., 2019; Parker et al., 2014; Square et al., 2015; York et al., 2018; York et al., 2017; York et al., 2019b; Yuan et al., 2018; Zu et al., 2016), lampreys have emerged as the leading cyclostome model system for studying the origin of vertebrate traits (McCauley et al., 2015; York et al., 2019a). Similar to gnathostomes, lampreys have neural crest cells that contribute to the head skeleton (Cattell et al., 2011; Jandzik et al., 2015; Lakiza et al., 2011; McCauley and Bronner-Fraser, 2006; Square et al., 2016), pigment (Lakiza et al., 2011; McCauley and Bronner-Fraser, 2003) and cranial sensory and enteric neurons and glia (Green et al., 2017; Modrell et al., 2014). Early cell labeling and gene expression studies showed that many defining properties of neural crest cells are conserved in lamprey (McCauley and Bronner-Fraser, 2003). Similar to gnathostomes, lamprey neural crest cells form in the dorsal neural tube, delaminate, and then migrate in three primary streams in the head. Some of these streams segregate further, for example, around the eye, similar to that in gnathostomes (McCauley and Bronner-Fraser, 2003; Szabó et al., 2019). Moreover, gene expression and functional analysis has shown that much of the neural crest GRN in agnathans is very similar overall to that of gnathostomes, suggesting that the molecular features of vertebrate neural crest cells may be conserved to the base of vertebrates (Sauka-Spengler et al., 2007). Notably, however, very little work has been done on specific modules within the broader neural crest GRN, including the regulatory circuit that controls defining features of the neural crest—EMT, delamination, and cell migration.

2.2.2.1   Intercellular Signaling and Transcriptional Control

In gnathostomes, *BMP2* and *BMP4* are expressed in the dorsal neural tube and are crucial for activating early transcriptional regulators of neural crest EMT, such as *Snail1/Snail2* (Nikitina et al., 2008; Raible and Ragland, 2005). By contrast, despite having three *BMP2/4* paralogues (*BMP2/4a, BMP2/4b, BMP2/4c*), lamprey never expresses any of these genes during the onset of neural crest EMT and migration, localizing instead to the early neurula, neural plate, neural plate border, endoderm, and postmigratory crest (McCauley and Bronner-Fraser, 2004; Sauka-Spengler et al., 2007). Expression of the intercellular inducer *WNT8* occurs in both the neural plate border and dorsal neural tube, suggesting a possible *WNT*-mediated signaling role during lamprey neural crest development that is conserved with gnathostomes (Sauka-Spengler and Bronner-Fraser, 2008b). However, functional analysis is needed to tease apart inductive signaling of the neural crest in general from a possible direct role for *WNTs* and other intercellular signaling molecules during lamprey neural crest EMT.

The transcriptional control of lamprey neural crest EMT and migration shows both evolutionary conservation as well as important differences compared to gnathostomes. Lamprey has three SoxE group transcription factors (SoxE1, SoxE2, SoxE3), and phylogenetic analysis shows that SoxE2 and SoxE3 are likely to be homologous to gnathostome Sox10 and Sox9, respectively (Lee et al., 2016). Both *SoxE1* and *SoxE2* are expressed in neural crest cells undergoing EMT and migration from the neural tube (Lakiza et al., 2011). However, functional knockdown of either gene results in nearly complete loss of premigratory neural crest rather than an arrest of EMT and migration, making it unclear whether or not SoxE factors actually regulate EMT independent of an earlier role in neural crest specification (Lakiza et al., 2011; McCauley and Bronner-Fraser, 2006). Although lamprey does not have a strict paralog to gnathostome *FoxD3*, a related gene known as *FoxD-A* is expressed in lamprey premigratory and migratory neural crest, and functional perturbation of *FoxD-A* results in loss of migratory neural crest and neural crest derivatives (Sauka-Spengler et al., 2007). In contrast to *SoxE* and *FoxD*, other lamprey *Twist* (Betancur et al., 2010; Hopwood et al., 1989) and *Ets* (Betancur et al., 2009) genes are not expressed in premigratory neural crest, despite both of these transcription factors being important for promoting cranial neural crest delamination and EMT in gnathostomes. Rather, these factors in lamprey are both expressed in postmigratory cranial neural crest–derived cartilage and embryonic vasculature, respectively (Sauka-Spengler et al., 2007).

Preliminary studies on lamprey *Snail*, a "master" regulator of neural crest EMT and migration in gnathostomes, suggested that this factor was not expressed in lamprey premigratory or migratory neural crest cells (Rahimi et al., 2009). However, it has recently been shown that *Snail* is indeed critical for early neural crest development in lamprey, as *Snail* CRISPants show defects in neural crest migration, fail to express a type II cadherin (*CadIIA*) in premigratory and migratory neural crest, and lose neural crest derivatives such as cartilage and cranial sensory neurons (York et al., 2017). It was also shown that lamprey has homologues of both *Sip1* and *Zeb1* that are expressed in domains overlapping with *Snail*, *CadIIA*, and *Pax3/7* in the dorsal neural tube during the onset of neural crest migration, raising the possibility that Sip and Zeb transcription factors are evolutionarily conserved regulators of neural crest EMT across vertebrates (York et al., 2017).

2.2.2.2 Intercellular Adhesion Proteins

In contrast to gnathostomes, lampreys have a relatively simple genomic complement of classical cadherin adhesion proteins, with only a single representative of the type I (*CadIA*) and type II (*CadIIA*) cadherins, similar to the condition in invertebrate chordates (Gallin, 1998; Hulpiau and Van Roy, 2009; York et al., 2017). Interestingly, *CadIA* is never expressed anywhere in the neural tube during the early stages of neural crest EMT/delamination, whereas the promesenchymal *CadIIA* localizes to early premigratory and migratory cranial neural crest (York et al., 2017). This stands in stark contrast to the situation in gnathostomes in which type I cadherins are first expressed in the dorsal neuroepithelium and then are gradually replaced by promesenchymal type II cadherins to facilitate migration (*cadherin switching*, described above). This suggests that neural crest cells in early vertebrates may not have required modulation of cadherin intercellular adhesion proteins to facilitate neural crest migration, and that cadherin-switching–mediated neural crest EMT and migration may be a gnathostome novelty (York et al., 2017).

2.2.2.3 Cytoskeletal Reorganization and Breakdown of Basal Lamina

Compared to detailed analysis both *in vivo* and *in vitro* of the downstream mechanics of cellular EMT and migration in gnathostomes (cytoskeletal changes, ECM remodeling), almost nothing is known regarding the operation of these processes in agnathans. For example, it is unclear what homologues of ECM remodeling proteins (e.g., *MMPs*, *ADAMs*) or cytoskeletal regulators (*Rho* GTPases) are present in agnathan genomes, much less where these genes may be expressed and what their functional relationships are to neural crest EMT, delamination, and migration.

2.2.3 Summary

Taken together, studies of neural crest development in lamprey suggest that the topology of the EMT/migration-specific module in agnathans shows evolutionarily conserved features (e.g., expression of *Snail*, *SoxE*, and *FoxD* in premigratory and migratory crest). But there are also features in lamprey quite different from that of gnathostomes. For example, a lack of expression of key transcriptional regulators of neural crest EMT such as *Twist* and *Ets*, as well as a lack of apparent switching between intercellular adhesion proteins such as classical cadherins in agnathans, may indicate that the EMT of ancestral vertebrate neural crest cells may have involved alternative means to initiate migration from the neural tube. These differences raise the possibility that the molecular-genetic and cellular mechanisms of neural crest EMT in gnathostomes may have diverged from the ancestral condition in early vertebrates if lamprey neural crest regulatory mechanisms are taken to be grossly similar to that of ancestral vertebrates. However, it is clear that a more detailed analysis of the functional roles of the neural crest EMT module during agnathan development is needed in order to clarify what the ancestral state of the vertebrate EMT module may have been and how this module has been altered over the course of vertebrate evolution.

2.3 ANCIENT ORIGIN OF CELLULAR EMT AND AN EMT GENE REGULATORY NETWORK

Vertebrate neural crest cells are defined in part by their ability to undergo a coordinated EMT and initiate migration. However, this feature alone does not distinguish neural crest cells from other cell types. Indeed, cells undergoing EMT have been described for almost all metazoan clades, and an ability to undergo EMT may in fact be a defining feature of metazoans (Figure 2.4 summarizes diverse cellular EMTs in a wide range of metazoan groups). The origin of EMT and mesenchyme, in conjunction with a complex extracellular matrix (ECM) (Ereskovsky et al., 2013; Hynes, 2012), most likely facilitated the integration and communication of large groups of cells, thereby allowing the establishment of discrete structures and, eventually, organs and organ systems. The interface of epithelium and ECM facilitated the ability of individual cells

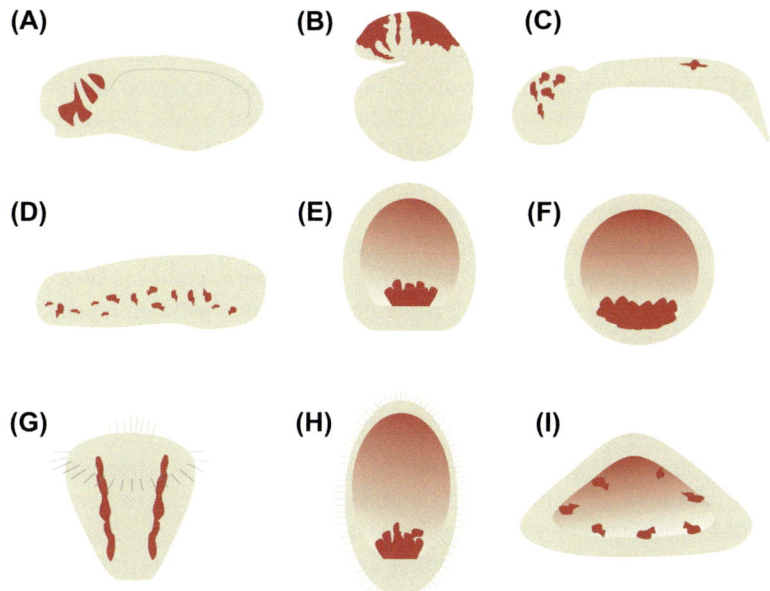

FIGURE 2.4 Examples of cellular EMTs in diverse metazoan embryos including: cranial neural crest in amphibians (a, *Xenopus*) and agnathans (b, lamprey); neural crest-like cells in tunicates (c, cranial melanocytes left, bipolar tail neurons, right); epidermal sensory neurons in amphioxus (d); primary mesenchyme in sea urchins (e); visceral mesoderm in *Drosophila* (f); neuroblasts in trochophore larvae of annelids (g); interstitial mesenchyme in planula larvae of Cnidarians (h); and migratory mesenchyme in sponges (i). Solid red shading indicates cells undergoing EMT in each embryo. Panels a-d, g show whole mount cartoons of the embryo, whereas e, f, h, i represent cross-sections showing the interior of the embryo.

or groups of cells to undergo morphogenetic movements and establish basic structural properties within animal embryos. However, it was the ability of a subset of cells to undergo EMT during metazoan embryonic development that was largely responsible for generating the varied and complex developmental morphologies across groups as diverse as sponges and humans. By deploying EMTs and migrating into new regions of the embryo, these cell types could encounter new cellular environments and intercellular signaling cues, thereby potentiating the evolution of novel cell types and embryonic structures. Below, we review several examples of EMTs across invertebrate metazoans with a focus on the molecular, cellular, and genetic similarities to vertebrate neural crest cells in order to provide context for a program for cellular EMT that could have been incorporated into the ancestral neural crest GRN.

2.3.1 Non-Bilaterian Metazoans

2.3.1.1 Sponges

Sponges are made up of a variety of loosely arranged cell types that appear to lack belt-like adhesion proteins and a basal lamina (Boute et al., 1996; Tyler, 2003a).

Nonetheless, sponge embryos perform gastrulation-like movements, and there are several points during sponge development in which individual cells or groups of cells undergo EMT-type movements involving ingression, delamination, and invagination (Ereskovsky, 2010; Ereskovsky et al., 2010; Nakanishi et al., 2014). Interestingly, there are no classical cadherin homologues in sponge genomes, making it unclear if the EMT mechanisms operating during sponge development are like those operating in other animals. This is accentuated by the fact that genomic analysis of the *Amphimedon queenslandica* genome failed to uncover many key transcriptional regulators of EMT and migration, including *Twist*, *Snail*, and *Forkhead* (Nakanishi et al., 2014; Srivastava et al., 2010). Nonetheless, the downstream mechanics of cell polarity and migration (*Rho*, *Rac*, *CDC42*) are likely conserved, as these features appear to predate metazoans (Boureux et al., 2007).

2.3.1.2   Placozoans

Placozoans are small, amoeba-like animals composed of two simple epithelial layers, each with a small number of specialized cell types (Schierwater, 2005; Smith et al., 2014; Srivastava et al., 2008). The exact placement of placozoans within the metazoan tree of life is still hotly debated, and the solution to this problem has important implications for our understanding of early animal evolution (Miller and Ball, 2008). Sequencing of the *Trichoplax adherens* genome revealed a rich repertoire of transcription factors and signaling pathways also found in the genomes of many other animals that control EMT and delamination, including members of the *Wnt*, *TGFβ*, *Sox*, *bHLH*, *Fox*, and *Snail* families (DuBuc et al., 2019; Ryan and Chiodin, 2015; Srivastava et al., 2008). Placozoans have no obvious ECM or basal lamina, but they do have intercellular protein junctions, including cadherins, as well as proteins that control the actin cytoskeleton (Smith et al., 2014; Srivastava et al., 2008). Thus, although there are no clear-cut examples of *bona fide* cellular EMTs in placozoans, current evidence suggests that they at least have the molecular machinery requisite for EMT-type cell behaviors.

2.3.1.3   Cnidarians

The origin of cnidarians was a milestone in animal evolution because they were possibly the first lineage of animals to have true epithelial cells (ectoderm, gastrodermis) and an ECM. The evolution of an interface between epithelia and ECM was important because it eventually facilitated the detachment and migration into the ECM of individual cells and even whole groups of cells. Overall, cnidarian development primarily involves epithelial morphogenesis (Magie et al., 2007; Magie and Martindale, 2008). However, changes in cellular morphology reminiscent of EMT are observed during gastrulation, where invaginating cells of the gastrodermis express *Forkhead* and *Snail* (Fritzenwanker et al., 2004a). Presumably, these transcription factors function to apically constrict invaginating cells, but there is also evidence that individual cells can ingress and migrate shortly after invagination (Byrum, 2001; Kraus and Technau, 2006; Magie and Martindale, 2008). Additional support for this can be found in recent work where overexpression of *Snail* genes and disruption of cell polarity can induce EMT-like cell behaviors in cnidarian embryos (Salinas-Saavedra et al., 2018). Cnidarians therefore provide evidence of a simple

GRN in early eumetazoans involving *Snail* and *Forkhead* to control morphogenesis and cell ingression.

2.3.1.4 Ctenophores

As with many of the other nonbilaterian metazoans, ctenophores have been moved back and forth within the metazoan phylogeny. Traditionally, they were viewed as being close allies of cnidarians based on shared characters such as diploblasty, but recent molecular phylogenies controversially place ctenophores as sister to all other metazoans (Dunn et al., 2008; Hejnol et al., 2009; Moroz, 2015; Nielsen, 2012). Regardless of their true phylogenetic position, there is evidence from lineage tracing experiments in *Mnemiopsis leidyia* that ctenophore embryos have migratory/mesen-chymal cells (Martindale and Henry, 1999). Some of these cells may be of endodermal origin; others are described as stellate mesenchymal cells within the mesoglea and seem capable of movement (Martindale and Henry, 1999). Interestingly, analysis of ctenophore genomes suggests that many of the familiar regulators of EMT and delami-nation, including *Twist* and *Snail*, may not be present (Ryan et al., 2013). Thus, if cel-lular EMTs do occur during ctenophore development, it is possible that the underlying molecular mechanisms may be quite different from those in other metazoans.

2.3.2 BILATERIANS

Bilaterians—which are divided into the ecdysozoans, lophotrochozoans, and deu-terostomes—include all other metazoans above sponge and cnidarian-grade organi-zation (Aguinaldo et al., 1997; Halanych et al., 1995). With the origin of bilaterians, we see a dramatic increase in body-plan complexity commensurate with the evolu-tion of novel cellular interactions. These interactions were in turn likely facilitated by diverse cell types that could take advantage of an EMT program promoting cell migration.

2.3.2.1 Ecdysozoans

2.3.2.1.1 Insects

In the fruit fly *Drosophila*, cadherin switching from *DE*-Cadherin to *DN*-Cadherin is observed during gastrulation, as the mesoderm invaginates and is eventually inter-nalized (Hemavathy et al., 1997; Oda et al., 1998). This process involves deployment of both Twist and Snail protein activity to directly repress transcription of epithe-lial genes, thereby facilitating apical constriction and invagination. The internalized mesodermal cells eventually migrate throughout the embryo to generate body-wall musculature (Wheelock et al., 2008). Migration of mesoderm requires Rho GTPase activity, as loss of the Rho guanine nucleotide exchange factor, Pebble, results in maintenance of epithelial traits and failure of mesodermal cells to create protrusions (Smallhorn et al., 2004). The exact *cis*-regulatory relationships remain elusive, but it seems that both Twist and Snail proteins intersect with the gene regulatory program that activates Rho expression in delaminating mesoderm (Leptin, 1999), providing a striking example of how EMTs in invertebrates deploy almost identical basic gene regulatory programs to initiate EMT and cell migration.

2.3.2.1.2   Nematodes

In the nematode, *C. elegans*, many of the cellular adhesion proteins that mediate neural crest EMT and migration in vertebrates, such as cadherins (Hill et al., 2001), similarly regulate tissue morphogenesis (Costa et al., 1998) and cell migration (Montell, 1999). There is also evidence that cells of endodermal fate ingress individually (Leptin, 2005). Yet, it is unknown if EMT-mediated cell migration during ingression involves the same regulatory control as that which occurs in other ecdysozoans. For example, although nematodes contain a *Snail* homologue (CES-1), this protein functions primarily to repress proapoptotic genes during cell fate determination (Metzstein and Horvitz, 1999; Reece-Hoyes et al., 2009; Thellmann et al., 2003). Snail family members across bilaterians are known to inhibit apoptosis during cell migration in numerous contexts (Barrallo-Gimeno and Nieto, 2005; Vega et al., 2004), but it is not clear if maintenance of cell survival is the sole function of *CES-1*, or if it is also critical for the initiation of EMT and migration in ingressing endoderm. *Zag-1*, a zinc finger homeodomain transcription factor, is homologous to vertebrate *Zeb1* and is expressed in and required for proper neural development in *C. elegans* (Clark and Chiu, 2003). Loss of *Zag-1* results in failure of neural progenitors to migrate and leads to loss of neural differentiation (Clark and Chiu, 2003), but similar to *CES-1*, it is not clear if the lack of neural migration results from loss of regulatory mechanisms similar to those in the vertebrate neural crest.

2.3.2.2   Lophotrochozoans

2.3.2.2.1   Annelids and Molluscs

Relatively little is known about the operation of EMT processes in lophotrochozoans apart from gene expression analysis. In polychaete annelids, one of two *Snail* paralogues is expressed in neuroectodermal derivatives, including migrating neuroblasts and maturing neurons in the central nervous system (Dill et al., 2007). In contrast to *Drosophila*, *Capitella sp. Twist* is not expressed in migrating cells during gastrulation, but this may be due to functional compensation by *Snail* (Dill et al., 2007). Contrary to most other invertebrates, *Snail* genes in the gastropod mollusk, *Patella vulgata*, are never expressed in involuting mesoderm, and instead are expressed mostly in sensory neurons and in several parts of the early larva in which ectodermal clefts or folds are forming (Lespinet et al., 2002). This raises the possibility that *Snail* may drive apical constriction and EMT-like movements similar to that of ventral furrow formation in *Drosophila*.

2.3.2.2.2   Acoels

The acoelomates (also known as the *acoels* or *acoelomorpha*) are worm-like creatures that have become increasingly important for understanding the origin of bilaterian traits (Philippe et al., 2011). Planarians in particular are an excellent model for studying cell migration and EMT as they are capable of regenerating many of their organs, which inevitably involves production of mesenchyme. Nevertheless, we currently lack insight into the molecular-genetic control of how these processes operate. Descriptions of embryogenesis in one planarian, *S. polychroa*, suggest that early developmental events involving EMT (e.g., gastrulation) coincide with expression of key EMT regulatory factors such as *Twist* and *Snail* (Martín-Durán et al., 2010). More extensive gene expression and functional analyses are required to form a comparative framework for studying the evolution of EMT mechanisms in these organisms.

2.3.2.3   Deuterostomes

Although less diverse than their lophotrochozoan and ecdysozoan relatives, the deuterostomes—the third main division of bilaterians—include the vertebrates and their closest extant relatives: the echinoderms, hemichordates, and invertebrate chordates. It is within the deuterostomes that we see the appearance of cells within the central nervous system that begin to take on cellular and molecular features of EMT that are strikingly similar to that of vertebrate neural crest.

2.3.2.3.1   *Echinoderms*

During gastrulation in sea urchins, a group of cells known as the *primary mesenchyme* undergoes EMT and detaches from epithelial cells of the vegetal plate (Saunders and McClay, 2014; Shook and Keller, 2003). As individual cells of the primary mesenchyme begin to ingress, they endocytose a proepithelial cadherin protein (Cad-1) (Miller and McClay, 1997; Wu et al., 2007). This is mediated by *Twist* and *Snail* activity, as functional perturbation of these genes results in failure of primary mesenchyme ingression and loss of mesenchyme-derived skeleton (Wu and McClay, 2007; Wu et al., 2008).

2.3.2.3.2   *Hemichordates*

Hemichordates have become increasingly important models for studying the evolutionary-developmental biology of deuterostome features such as the organization of the central nervous system, pharynx, and mesoderm (Gillis et al., 2012; Green et al., 2013a; Lowe et al., 2003). However, there is a paucity of detailed studies on hemichordate EMT or cell migration. Gene expression analyses show that key EMT regulators such as *Snail* and *Forkhead* are expressed in early mesoderm (Green et al., 2013b), but the functions, if any, of these genes within the context of EMT/cell migration are unknown. Similarly, although the hemichordate *Ptychodera flava* has a type I (*PfCad1*) and type II (*PfCad2*) classical cadherin (Oda and Takeichi, 2011), it is unknown how they influence EMT and cell migration.

2.3.2.3.3   *Invertebrate Chordates—Amphioxus and Tunicates*

The embryonic development of the invertebrate chordates—amphioxus and tunicates—has occupied a central place in the study of vertebrate origins for nearly 150 years (Dohrn, 1875; Gee, 1996; Lacalli, 2010; Laubichler and Maienschein, 2007; Wada, 2001). These animals are the closest extant relatives of vertebrates and share many embryological and genomic features with vertebrates (Putnam et al., 2008). Despite these similarities, however, it had been assumed throughout much of the history of neural crest research that invertebrate chordates do not have migratory neural crest cells, or even likely homologues of the vertebrate neural crest (Gans and Northcutt, 1983). However, over the past two decades, analysis of gene expression and function, as well as cell lineage tracing, has questioned this thinking, and in doing so has provided new and exciting insights into the evolutionary origin of the migratory properties of the neural crest.

The cephalochordates, represented by amphioxus, are invertebrate chordates that appear to have retained many of the ancestral morphological and genomic traits of the last common chordate ancestor (Holland et al., 2004; Putnam et al., 2008). Amphioxus development relies almost exclusively upon morphogenetic, sheet-like cellular movements rather than individual or collective cell migration. For example, somite development in amphioxus does not appear to involve cellular EMT as occurs in vertebrates (Mansfield et al., 2015). On the other hand, there are a few cases of migratory cells that may undergo EMT, such as sensory neurons that derive from the ventral epidermal ectoderm and migrate a short distance before reinserting into the epidermis (Kaltenbach et al., 2009). Similarly, gene expression analysis revealed that *Distal-less* transcripts (homologous to vertebrate *Dlx* genes) localize in dorsal epidermal cells that move as a sheet toward the dorsal midline during neurulation, similar to early *Dlx* expression in the open neural plate of vertebrate embryos during establishment of the neural plate border (Holland et al., 1996). These cells have lamellipodial extensions and appear to detach from the underlying neuroepithelium, leading the authors to speculate that these dorsal epidermal cells may be homologous with vertebrate neural crest. Yet, unlike the neural crest, these cells never completely delaminate, nor do they migrate away from their site of origin, and they never form any cell type other than epidermis. Thus, although gene expression patterns have been informative for outlining the ancestral framework for the neural crest GRN, there have been no key cell types identified in cephalochordates having the cellular and molecular properties that offer a compelling link to migratory neural crest.

It has been suggested that the lack of migratory neural crest in amphioxus is a consequence of their lacking much of the neural crest specifier and EMT program in the dorsal neural tube, lending support to the notion that these parts of the neural crest GRN were coopted from other cell types (e.g., mesoderm, endoderm) to the neural tube early in the origin of the vertebrates (Hall, 2008; Meulemans and Bronner-Fraser, 2005). Although the notion of gene cooption has figured prominently in studies of neural crest evolution, there is one interesting example that runs counter to this trend. The transcription factor Snail—a key regulator of neural crest EMT and migration in gnathostome vertebrates—is expressed in the amphioxus neural tube (Langeland et al., 1998). In gnathostomes, forced expression of *Snail1 or Snail2* results in ectopic neural crest EMT and migration, suggesting that these factors are sufficient to induce cells to emigrate from the neural tube (Hemavathy et al., 2000) (del Barrio and Nieto, 2002; Guaita et al., 2002). Yet, amphioxus lacks migratory neural crest, raising questions as to why *Snail*-positive cells in the neural tube are unable to migrate. Although much of the neural crest regulatory apparatus controlling induction, neural border specification and cell differentiation are conserved among invertebrate chordates, amphioxus lacks expression of many of the downstream effectors that are largely responsible for downregulating epithelial state and promoting mesenchymal state in the neural tube (Yu et al., 2008). For example, only a single *RhoA/B/C* gene is present in the amphioxus genome in contrast to individual *RhoA*, *RhoB*, and *RhoC* genes in vertebrate genomes (Boureux et al., 2007), which implies that gene duplication may have been an important step in promoting the migration of "proto-neural crest cells" from the neural tube. In addition, the downstream effectors of migration, such as the classical cadherins, may be highly derived and not function in the same context as vertebrate classical cadherins that promote neural crest migration. For example, the classical cadherin proteins in *Branchiostoma belcheri* (Bb1C and Bb2C) are structurally similar to vertebrate E- and N-cadherins, respectively, but they have swapped functions because *Bb1C* is expressed in the neural tube and somites, whereas Bb2C is expressed in the epidermis—expression patterns exactly opposite to that of *E-Cadherin* and *N-Cadherin* in vertebrates (Oda et al., 2004; Oda et al., 2002). Amphioxus cadherins also lack extracellular domains, yet are capable of holding cells together, suggesting that the fundamental mechanisms of cadherin-mediated adhesion in the amphioxus neural tube are substantively different from that in any other chordate (Oda et al., 2004).

Molecular phylogenetic analysis places the tunicates as the sister group to vertebrates, and therefore makes them the best model system for studying the evolution of migratory neural crest that appeared after the divergence of vertebrates from invertebrate chordates (Delsuc et al., 2006). As described above, it had often been assumed that migratory neural crest cells first appeared in early vertebrates, with little or no vestiges of a rudimentary neural crest among invertebrate chordates (Gans and Northcutt, 1983). However, vital dye (DiI) labeling experiments in the mangrove tunicate (*Ecteinascidia turbinata*) were the first to reveal that cells originating near the dorsal neural tube of the tadpole larva were not only capable of migrating, but could also differentiate into pigment cells, a cellular derivative of vertebrate neural crest (Jeffery et al., 2004). Moreover, these cells expressed HNK-1 protein and *ZicA* mRNA, similar to that of migratory neural crest in vertebrates (Jeffery et al., 2004). These results were later confirmed in other ascidian species, suggesting that tunicates possessed so-called migratory "neural crest-like cells" (NCLCs) (Jeffery, 2006; Jeffery et al., 2004). Subsequent studies revealed that NCLCs originate near the neural plate border and express several markers of neural crest cells, including homologues of *Twist, Myc, FoxD*, type II *Cadherin* and *Rho A/B/C* GTPases (Jeffery et al., 2008). Although these cells are similar to bona fide neural crest, there are some important differences. For example, NCLCs do not exit immediately from the neural folds, but rather remain stationary for a prolonged period to proliferate prior to delamination. This is unlike the case in many vertebrates in which neural crest cells either migrate before or immediately after neural tube closure (Jeffery, 2006; Jeffery et al., 2008). Also, the NCLCs of ascidians originate from the *a*7.6 trunk lateral cell lineage, which is not within the neural plate border and is also distinct from other types of NCLCs reported in different tunicate species (Jeffery et al., 2008) (described below).

There have been recent reports of other types of NCLCs in tunicates that are completely different from those described in the *a*7.6 cell lineage. In *Ciona intestinalis*, pigment-cell precursors from the *a*9.49 lineage in the head contribute to cranial sensory structures such as the otolith and ocellus and are regulated by a conserved *Wnt7-FoxD* axis that operates in neural crest–derived melanocytes in vertebrates (Abitua et al., 2012). Although they normally delaminate and migrate only a short distance within the neural tube, cells from the *a*9.49 lineage can be induced to migrate out of the neural tube as mesenchymal cells upon forced expression of *Twist* (Abitua et al., 2012). This suggests that NCLCs with minimal migratory potential could have been directed to undergo a full EMT and migrate extensively throughout a chordate embryo simply by cooption of a single pro-migration transcriptional regulator such as Twist.

Yet another distinct NCLC has been recently identified in *Ciona*, a cell lineage that contributes to the bipolar tail neuron (BTN) (Stolfi et al., 2015). BTN precursors form within a *Snail-Pax3/7-Msx*-positive neural plate border region, delaminate, and migrate (Stolfi et al., 2015). Eventually, these cells differentiate into neurons that express *Neurogenin*, which, the authors argue, is similar to development of neural crest–derived dorsal root ganglia in vertebrates. Interestingly, these cells also appear to downregulate Cadherin.b, a classical cadherin homologue found in epithelial cells of the neural tube, and forced expression of a proepithelial protocadherin in the neural tube prevented BTN precursor migration. Taken together, these results provide the first evidence that migratory NCLCs in invertebrate chordates control differential expression of intercellular adhesion molecules to facilitate migration, as occurs during EMT in gnathostome vertebrate neural crest cells (Stolfi et al., 2015).

2.4 EVOLUTIONARY EMERGENCE OF NEURAL CREST EMT AND MIGRATION

Taking into account the use of a wide array of transcription factors and signaling molecules in both vertebrate neural crest cells and similar migratory cell types among invertebrates, it becomes clear that there is more than one way to execute cellular EMT during embryonic development. From our comparative analysis across diverse metazoans, we propose that a similar set of genes and cellular responses govern EMT and migration in several different developmental contexts and cell types (Figure 2.5 summarizes the most conserved features of cellular EMT across metazoan clades). These can range from ingression of neuroblasts and primary mesenchyme in insects and echinoderms to invagination and delamination of sensory cells and mesenchyme in cnidarians, nematodes, and molluscs. We propose that this common regulatory process sets into motion a molecular chain of events that regulates diverse types of cellular EMT and has underlain the repeated evolution of EMTs in different metazoan cell types.

An important theme in evolutionary developmental biology is deep homology—the concept that similar developmental processes and embryonic structures may arise independently in different lineages by using common genes or even entire regulatory networks (Shubin et al., 2009). Classic examples of deep homology include the shared molecular circuitry controlling development of nonhomologous limbs and eyes across bilaterians (Shubin et al., 2009). Thus, although these structures, or the cell types that comprise them, are not homologous *sensu stricto*, the regulatory interactions of the genes that control their development are and have been deployed over and over again to accomplish the same developmental goal. Likewise, we propose that there is a deeply homologous metazoan EMT regulatory network for cell delamination and migration (Figure 2.5). Although there are almost certainly clade-specific

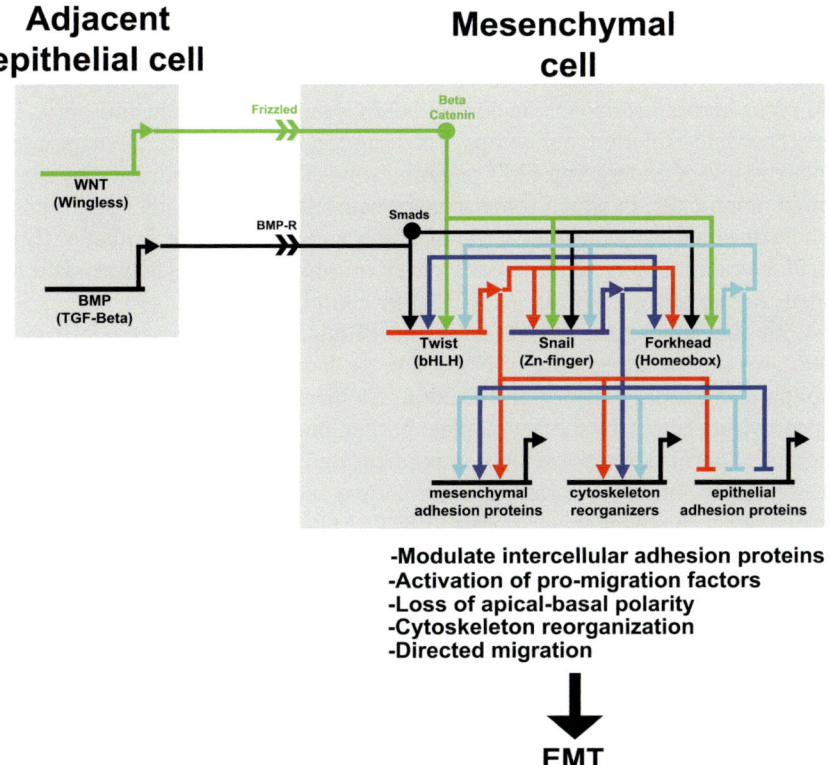

FIGURE 2.5 Hypothetical gene regulatory network (GRN) showing conserved elements governing EMT during embryonic development across metazoans.

features that have been lost from or superimposed onto this core network, we suggest that a highly conserved EMT program is a recurring motif in EMTs that have evolved throughout metazoan evolution and likely formed the basis for the evolution of a neural crest EMT module in NCLCs in invertebrate chordates and in *bona fide* neural crest in vertebrates.

Some of the most ancient components of this EMT network may include intercellular signaling pathways, such as those from TGFβ (e.g., BMPs) and canonical Wingless (WNT) signaling pathways (Figure 2.5). These signaling systems show little variation across metazoans and rely on conserved intracellular effectors to control expression of target genes (Davidson and Erwin, 2006). During neural crest EMT, these signaling pathways converge on the activation of one or more transcription factors, including members of the Snail, Twist, and Forkhead families, which are also activated for EMTs in diverse metazoan groups. Once expressed, these transcriptional regulators directly bind and repress gene batteries responsible for maintaining epithelial fate, such as certain cadherin intercellular adhesion proteins (Figure 2.5). These same factors then either directly or indirectly turn on cytoskeletal regulators that reorganize the cytoskeleton and prepare cells to detach and migrate.

Observations that migratory cells of various developmental and phylogenetic origins all activate this core EMT-migration program provide compelling evidence that it has been inserted *in toto* into diverse gene regulatory programs that define many cell types across metazoans, including NCLCs and vertebrate neural crest. During early chordate evolution, our comparisons suggest that there were perhaps a few cell types capable of undergoing EMT and migrating. This is approximated by the condition in amphioxus, in which some sensory neurons and a few other cell types move a short distance away from their site of origin and then become epithelial (Lu et al., 2012). Notably, however, these cells never originate from the neural plate border, suggesting no obvious affinity to NCLCs or neural crest.

With the appearance of Olfactoreans (Tunicates+Vertebrates) (Delsuc et al., 2006), we see the appearance of NCLCs in the head and trunk of chordate embryos (Abitua et al., 2012; Jeffery et al., 2008; Stolfi et al., 2015). Some of these cells appear within or nearby the neural plate border, undergo EMT, delaminate, and then migrate to form derivatives such as pigment and sensory neurons. Thus, NCLCs are strikingly similar to vertebrate neural crest and likely used the same or similar molecular mechanisms to accomplish EMT (Abitua et al., 2012; Stolfi et al., 2015). For example, it is likely that NCLCs in ascidians undergo EMT and migrate by modulating intercellular adhesion proteins such as cadherins (Stolfi et al., 2015), and these cells originate nearby or within the neural border and express transcription factors that promote EMT, including *Snail* and *Pax3/7* (Stolfi et al., 2015).

Although it remains unclear whether NCLCs are homologous to *bona fide* neural crest, these cells do not form major structures or organ systems as occurs in vertebrates, but rather form isolated cells or small cell populations such as sensory neurons or other sensory cell types (Abitua et al., 2012; Stolfi et al., 2015). It seems likely then that the EMT module in NCLCs was either coopted multiple times by individual cells in the neural plate border or, more likely, by one or a few progenitor cell types that could divide and spread throughout the head and trunk. What is unknown is if early neural crest cells first migrated collectively in stream-like populations in the head, or as individual cells more akin to that observed in the trunk (Shellard et al., 2018; Theveneau and Mayor, 2011, 2012a,b). On this note, it is worth pointing out that ectopic migration of NCLCs in tunicate embryos seems to involve migration of individual mesenchymal cells, rather than a mass of collectively migrating cells (Abitua et al., 2012). If this result reflects the ancestral condition, then it would imply that NCLCs and possibly neural crest in early vertebrates migrated as a population of loosely arranged individual cells, with the collective, sheet-like migration pattern evolving later. Another important difference between NCLCs and vertebrate neural crest is that NCLCs are not multipotent, forming only single cell types (Stolfi et al., 2015). Thus, although invertebrate chordate NCLCs would have established much of the core neural crest GRN that operates in vertebrate neural crest, including the EMT/migration module, this program was not linked to multipotency. The acquisition of a multipotency program would have likely endowed migratory NCLCs with the ability to generate the diverse set of cellular derivatives that distinguish *bona fide* vertebrate neural crest (York et al., 2017).

During the origin of early vertebrates, the EMT module likely consisted of a very simple network, similar to that operating in invertebrate NCLCs and nonchordate deuterostomes. This is bolstered by analysis of the transcription factor repertoire of neural crest EMT/migration in agnathans, which suggests that many "key" signal transduction pathways and transcriptional regulators of neural crest EMT (e.g., *Twist*, *Ets*, *BMPs*), as well as cellular mechanisms (e.g., cadherin switching) may be dispensable for neural crest EMT in jawless vertebrates (Sauka-Spengler et al., 2007; York et al., 2017). Thus, early vertebrates—represented by extant agnathans—may offer key insights into the stepwise assembly of the EMT module of the neural crest GRN operating in higher (gnathostome) vertebrates.

Based on the apparent differences in their EMT modules, there was likely a large scale "rewiring" of the EMT module after the divergence of agnathans and gnathostomes. Presumably, this would have occurred by *cis*-regulatory evolution that brought novel transcriptional regulators and intercellular signaling pathways such as *Twist*, *Ets*, *BMP2/4*, and others into the dorsal neural tube, which would have been facilitated by and integrated with additional lineage-specific gene duplications (Donoghue and Purnell, 2005; Ohno, 1970; Wada and Makabe, 2006). Although the significance of such changes to the EMT module is not clear, one possibility is that these additional genes would have endowed migratory neural crest cells with functional redundancy to ensure a properly timed and coordinated EMT. This can be seen in extant gnathostomes in which one of the seminal events of neural crest EMT—transcriptional repression of epithelial gene batteries—often involves coordinate repression by numerous proteins, including Snail, Twist, SoxE, Sip1, Zeb1, and many others (Thiery and Sleeman, 2006).

2.5 CONCLUSIONS

Of the many molecular and cellular features that define vertebrate neural crest cells, an ability to undergo EMT, delaminate, and migrate from the embryonic neural tube is one of their hallmark traits, yet the evolutionary origin of the neural crest EMT program has remained obscure. By comparing the molecular, genetic, and cellular features of EMT in vertebrate neural crest cells with similar mechanisms in diverse invertebrate cell types, we identify a core conserved set of genes and cellular mechanisms that may constitute an ancient regulatory program that served as the basis for the independent origin of cellular EMTs during animal evolution. This network was likely deployed during the evolution of neural crest cells in early vertebrates and has been elaborated upon significantly with the divergence of agnathans and gnathostomes.

REFERENCES

Abitua, P.B., Wagner, E., Navarrete, I.A., Levine, M., 2012. Identification of a rudimentary neural crest in a non-vertebrate chordate. *Nature* 492(7427), 104–107.

Aguinaldo, A.M.A., Turbeville, J.M., Linford, L.S., Rivera, M.C., Garey, J.R., Raff, R.A., Lake, J.A., 1997. Evidence for a clade of nematodes, arthropods and other moulting animals. *Nature* 387(6632), 489–493.

Alexander, N.R., Tran, N.L., Rekapally, H., Summers, C.E., Glackin, C., Heimark, R.L., 2006. N-cadherin gene expression in prostate carcinoma is modulated by integrin-dependent nuclear translocation of Twist1. *Cancer Res.* 66(7), 3365–3369.

Alfandari, D., Cousin, H., Gaultier, A., Smith, K., White, J.M., Darribère, T., DeSimone, D.W., 2001. Xenopus ADAM 13 is a metalloprotease required for cranial neural crest-cell migration. *Curr. Biol.* 11(12), 918–930.

Alfandari, D., Taneyhill, L.A., 2018. Cut loose and run: The complex role of ADAM proteases during neural crest cell development. *Genesis* 56(6–7), e23095.

Anderson, R.B., 2003. Matrix metalloproteinase-2 is involved in the migration and network formation of enteric neural crest-derived cells. *Int. J. Dev. Biol.* 54(1), 63–69.

Barrallo-Gimeno, A., Nieto, M.A., 2005. The Snail genes as inducers of cell movement and survival: Implications in development and cancer. *Development* 132(14), 3151–3161.

Barriga, E.H., Maxwell, P.H., Reyes, A.E., Mayor, R., 2013. The hypoxia factor HIF-1 alpha controls neural crest chemotaxis and epithelial to mesenchymal transition. *J. Cell Biol.* 201(5), 759–776.

Betancur, P., Bronner-Fraser, M., Sauka-Spengler, T., 2010. Assembling neural crest regulatory circuits into a gene regulatory network. *Annu. Rev. Cell. Dev. Biol.* 26, 581–603.

Betancur, P., Sauka-Spengler, T., Bronner-Fraser, M., 2009. c-Myb, Ets- 1 and Sox9 directly activate a Sox10 core enhancer in delaminating cranial neural crest. *Proc. Natl. Acad. Sci. U.S.A.* 331(2), 438–438.

Bolos, V., Peinado, H., Perez-Moreno, M.A., Fraga, M.F., Esteller, M., Cano, A., 2003. The transcription factor Slug represses E-cadherin expression and induces epithelial to mesenchymal transitions: A comparison with Snail and E47 repressors. *J. Cell Sci.* 116(3), 499–511.

Boureux, A., Vignal, E., Faure, S., Fort, P., 2007. Evolution of the Rho family of ras-like GTPases in eukaryotes. *Mol. Biol. Evol.* 24(1), 203–216.

Boute, N., Exposito, J.-Y., Boury-Esnault, N., Vacelet, J., Noro, N., Miyazaki, K., Yoshizato, K., Garrone, R., 1996. Type IV collagen in sponges, the missing link in basement membrane ubiquity. *Biol. Cell* 88(1–2), 37–44.

Bronner, M., 2012. Formation and migration of neural crest cells in the vertebrate embryo. *Histochem. Cell Biol.* 138(2), 179–186.

Burstyn-Cohen, T., Stanleigh, J., Sela-Donenfeld, D., Kalcheim, C., 2004. Canonical Wnt activity regulates trunk neural crest delamination linking BMP/noggin signaling with G1/S transition. *Development* 131(21), 5327–5339.

Byrum, C.A., 2001. An analysis of hydrozoan gastrulation by unipolar ingression. *Dev. Biol.* 240(2), 627–640.

Cai, D.H., Vollberg, T.M., Hahn-Dantona, E., Quigley, J.P., Brauer, P.R., 2000. MMP-2 expression during early avian cardiac and neural crest morphogenesis. *Anat. Rec.* 259(2), 168–179.

Carl, T.F., Dufton, C., Hanken, J., Klymkowsky, M.W., 1999. Inhibition of neural crest migration in Xenopus using antisense slug RNA. *Dev. Biol.* 213(1), 101–115.

Cattell, M., Lai, S., Cerny, R., Medeiros, D.M., 2011. A new mechanistic scenario for the origin and evolution of vertebrate cartilage. *PLOS ONE* 6(7), 13.

Cheung, M., Briscoe, J., 2003. Neural crest development is regulated by the transcription factor Sox9. *Development* 130(23), 5681–5693.

Cheung, M., Chaboissier, M.-C., Mynett, A., Hirst, E., Schedl, A., Briscoe, J., 2005a. The transcriptional control of trunk neural crest induction, survival, and delamination. *Dev. Cell* 8(2), 179–192.

Clark, S.G., Chiu, C., 2003. *C. elegans* ZAG-1, a Zn-finger-homeodomain protein, regulates axonal development and neuronal differentiation. *Development* 130(16), 3781–3794.

Clay, M.R., Halloran, M.C., 2011. Regulation of cell adhesions and motility during initiation of neural crest migration. *Curr. Opin. Neurobiol.* 21(1), 17–22.

Coles, E.G., Taneyhill, L.A., Bronner-Fraser, M., 2007. A critical role for Cadherin6B in regulating avian neural crest emigration. *Dev. Biol.* 312(2), 533–544.

Comijn, J., Berx, G., Vermassen, P., Verschueren, K., van Grunsven, L., Bruyneel, E., Mareel, M., Huylebroeck, D., van Roy, F., 2001. The two-handed E box binding zinc finger protein SIP1 downregulates E-cadherin and induces invasion. *Mol. Cell* 7(6), 1267–1278.

Connerney, J., Andreeva, V., Leshem, Y., Muentener, C., Mercado, M.A., Spicer, D.B., 2006. Twist1 dimer selection regulates cranial suture patterning and fusion. *Dev. Dyn.* 235(5), 1334–1346.

Cordero, D.R., Brugmann, S., Chu, Y.N., Bajpai, R., Jame, M., Helms, J.A., 2011. Cranial neural crest cells on the move: Their roles in craniofacial development. *Am. J. Med. Genet. A* 155A(2), 270–279.

Costa, M., Raich, W., Agbunag, C., Leung, B., Hardin, J., Priess, J.R., 1998. A putative catenin–cadherin system mediates morphogenesis of the *Caenorhabditis elegans* embryo. *J. Cell Biol.* 141(1), 297–308.

Cousin, H., 2017. Cadherins function during the collective cell migration of Xenopus cranial neural crest cells: Revisiting the role of E-cadherin. *Mech. Dev.* 148, 79–88.

Davidson, E.H., Erwin, D.H., 2006. Gene regulatory networks and the evolution of animal body plans. *Science* 311(5762), 796–800.

De Calisto, J., Araya, C., Marchant, L., Riaz, C.F., Mayor, R., 2005. Essential role of non-canonical Wnt signalling in neural crest migration. *Development* 132(11), 2587–2597.

del Barrio, M.G., Nieto, M.A., 2002. Overexpression of Snail family members highlights their ability to promote chick neural crest formation. *Development* 129(7), 1583–1593.

Delsuc, F., Brinkmann, H., Chourrout, D., Philippe, H., 2006. Tunicates and not cephalochordates are the closest living relatives of vertebrates. *Nature* 439(7079), 965–968.

Dill, K.K., Thamm, K., Seaver, E.C., 2007. Characterization of twist and snail gene expression during mesoderm and nervous system development in the polychaete annelid *Capitella* sp I. *Dev. Genes Evol.* 217(6), 435–447.

Dohrn, A., 1875. *Der Ursprung der Wirbelthiere und das Princip Des Functionswechsels: Genealogische Skizzen.* W. Engelmann, Leipzig.

Donoghue, P.C.J., Purnell, M.A., 2005. Genome duplication, extinction and vertebrate evolution. *Trends Ecol. Evol. (Amst.)* 20(6), 312–319.

Dottori, M., Gross, M.K., Labosky, P., Goulding, M., 2001. The winged-helix transcription factor Foxd3 suppresses interneuron differentiation and promotes neural crest cell fate. *Development* 128(21), 4127–4138.

Duband, J.-L., Monier, F., Delannet, M., Newgreen, D., 1995. Epithelium-mesenchyme transition during neural crest development. *Acta Anat.* 154(1), 63–78.

DuBuc, T.Q., Ryan, J.F., Martindale, M.Q., 2019. "Dorsal–Ventral" genes are part of an ancient axial patterning system: Evidence from *Trichoplax adhaerens* (Placozoa). *Mol. Biol. Evol.* 36(5), 966–973.

Dunn, C.W., Hejnol, A., Matus, D.Q., Pang, K., Browne, W.E., Smith, S.A., Seaver, E., Rouse, G.W., Obst, M., Edgecombe, G.D., 2008. Broad phylogenomic sampling improves resolution of the animal tree of life. *Nature* 452(7188), 745.

Duong, T.D., Erickson, C.A., 2004. MMP-2 plays an essential role in producing epithelial-mesenchymal transformations in the avian embryo. *Dev. Dyn.* 229(1), 42–53.

Ereskovsky, A.V., 2010. *The Comparative Embryology of Sponges.* Springer Science & Business Media.

Ereskovsky, A.V., Konyukov, P.Y., Tokina, D.B., 2010. Morphogenesis accompanying larval metamorphosis in *Plakina trilopha* (Porifera, Homoscleromorpha). *Zoomorphology* 129(1), 21–31.

Ereskovsky, A.V., Renard, E., Borchiellini, C., 2013. Cellular and molecular processes leading to embryo formation in sponges: Evidences for high conservation of processes throughout animal evolution. *Dev. Genes Evol.* 223(1–2), 5–22.

Erickson, C.A., 1987. Behavior of neural crest cells on embryonic basal laminae. *Dev. Biol.* 120(1), 38–49.

Fairchild, C.L., Conway, J.P., Schiffmacher, A.T., Taneyhill, L.A., Gammill, L.S., 2014. FoxD3 regulates cranial neural crest EMT via downregulation of tetraspanin18 independent of its functions during neural crest formation. *Mech. Dev.* 132, 1–12.

Fairchild, C.L., Gammill, L.S., 2013. Tetraspanin18 is a FoxD3-responsive antagonist of cranial neural crest epithelial-to-mesenchymal transition that maintains cadherin-6B protein. *J. Cell Sci.* 126(6), 1464–1476.

Fendrich, V., Waldmann, J., Feldmann, G., Schlosser, K., Konig, A., Ramaswamy, A., Bartsch, D.K., Karakas, E., 2009. Unique expression pattern of the EMT markers Snail, Twist and E-cadherin in benign and malignant parathyroid neoplasia. *Eur. J. Endocrinol.* 160(4), 695–703.

Ferronha, T., Rabadán, M.A., Gil-Guiñon, E., Le Dréau, G., de Torres, C., Martí, E., 2013. LMO4 is an essential cofactor in the Snail2-mediated epithelial-to-mesenchymal transition of neuroblastoma and neural crest cells. *J. Neurosci.* 33(7), 2773–2783.

Fritzenwanker, J.H., Saina, M., Technau, U., 2004. Analysis of forkhead and snail expression reveals epithelial-mesenchymal transitions during embryonic and larval development of *Nematostella vectensis*. *Dev. Biol.* 275(2), 389–402.

Gallin, W.J., 1998. Evolution of the "classical" cadherin family of cell adhesion molecules in vertebrates. *Mol. Biol. Evol.* 15(9), 1099–1107.

Gans, C., Northcutt, R.G., 1983. Neural crest and the origin of vertebrates: A new head. *Science* 220(4594), 268–274.

Garmon, T., Wittling, M., Nie, S., 2018. MMP14 regulates cranial neural crest epithelial-to-mesenchymal transition and migration. *Dev. Dyn.* 247(9), 1083–1092.

Gee, H., 1996. *Before the Backbone: Views on the Origin of the Vertebrates*. Springer, New York.

Genuth, M.A., Allen, C.D., Mikawa, T., Weiner, O.D., 2018. Chick cranial neural crest cells use progressive polarity refinement, not contact inhibition of locomotion, to guide their migration. *Dev. Biol.* 444 Supplement 1, S252–S261.

Gillis, J.A., Fritzenwanker, J.H., Lowe, C.J., 2012. A stem-deuterostome origin of the vertebrate pharyngeal transcriptional network. *Proc. R. Soc. Lond. B Biol. Sci.* 279(1727), 237–246.

Gouignard, N., Andrieu, C., Theveneau, E., 2018. Neural crest delamination and migration: Looking forward to the next 150 years. *Genesis* 56(6–7), e23107.

Green, S.A., Norris, R.P., Terasaki, M., Lowe, C.J., 2013. FGF signaling induces mesoderm in the hemichordate *Saccoglossus kowalevskii*. *Development* 140(5), 1024–1033.

Green, S.A., Simoes-Costa, M., Bronner, M.E., 2015. Evolution of vertebrates as viewed from the crest. *Nature* 520(7548), 474–482.

Green, S.A., Uy, B.R., Bronner, M.E., 2017. Ancient evolutionary origin of vertebrate enteric neurons from trunk-derived neural crest. *Nature* 544(7648), 88–91.

Groves, A.K., LaBonne, C., 2014. Setting appropriate boundaries: Fate, patterning and competence at the neural plate border. *Dev. Biol.* 389(1), 2–12.

Guaita, S., Puig, I., Franci, C., Garrido, M., Dominguez, D., Batlle, E., Sancho, E., Dedhar, S., De Herreros, A.G., Baulida, J., 2002. Snail induction of epithelial to mesenchymal transition in tumor cells is accompanied by MUC1 repression and ZEB1 expression. *J. Biol. Chem.* 277(42), 39209.

Halanych, K.M., Bacheller, J.D., Aguinaldo, A., Liva, S.M., Hillis, D.M., Lake, J.A., 1995. Evidence from 18S ribosomal DNA that the lophophorates are protostome animals. *Science* 267(5204), 1641–1641.

Hall, B.K., 2008. *The Neural Crest and Neural Crest Cells in Vertebrate Development and Evolution*. Springer, New York.

Hay, E.D., 1995. An overview of epithelio-mesenchymal transformation. *Acta Anat.* 154(1), 8–20.

Hay, E.D., 2005. The mesenchymal cell, its role in the embryo, and the remarkable signaling mechanisms that create it. *Dev. Dyn.* 233(3), 706–720.

Heeg-truesdell, E., Labonne, C., 2004. A slug, a fox, a pair of sox: Transcriptional responses to neural crest inducing signals. *Birth Defects Res. C. Embryo Today Rev.* 72(2), 124–139.

Heimberg, A.M., Cowper-Sal-Lari, R., Sémon, M., Donoghue, P.C.J., Peterson, K.J., 2010. MicroRNAs reveal the interrelationships of hagfish, lampreys, and gnathostomes and the nature of the ancestral vertebrate. *Proc. Natl. Acad. Sci. U.S.A.* 107(45), 19379–19383.

Hejnol, A., Obst, M., Stamatakis, A., Ott, M., Rouse, G.W., Edgecombe, G.D., Martinez, P., Baguñà, J., Bailly, X., Jondelius, U., 2009. Assessing the root of bilaterian animals with scalable phylogenomic methods. *Proc. R. Soc. Lond. B Biol. Sci.* 276(1677), 4261–4270.

Hemavathy, K., Ashraf, S.I., Ip, Y.T., 2000. Snail/slug family of repressors: Slowly going into the fast lane of development and cancer. *Gene* 257(1), 1–12.

Hemavathy, K., Meng, X., Ip, Y.T., 1997. Differential regulation of gastrulation and neuro-ectodermal gene expression by Snail in the *Drosophila embryo. Development* 124(19), 3683–3691.

Hill, A.S., Nishino, A., Nakajo, K., Zhang, G.X., Fineman, J.R., Selzer, M.E., Okamura, Y., Cooper, E.C., 2008. Ion channel clustering at the axon initial segment and node of Ranvier evolved sequentially in early chordates. *PLOS Genet.* 4(12), 15.

Hill, E., Broadbent, I.D., Chothia, C., Pettitt, J., 2001. Cadherin superfamily proteins in Caenorhabditis elegans and Drosophila melanogaster. *J. Mol. Biol.* 305(5), 1011–1024.

Holland, L., Laudet, V., Schubert, M., 2004. The chordate amphioxus: An emerging model organism for developmental biology. *CMLS, Cell. Mol. Life Sci.* 61(18), 2290–2308.

Holland, N.D., Panganiban, G., Henyey, E.L., Holland, L.Z., 1996. Sequence and develop-mental expression of AmphiDll, an amphioxus distal-less gene transcribed in the ecto-derm, epidermis and nervous system: Insights into evolution of craniate forebrain and neural crest. *Development* 122(9), 2911–2920.

Hopwood, N., Pluck, A., Gurdon, J., 1989. A Xenopus mRNA related to Drosophila twist is expressed in response to induction in the mesoderm and the neural crest. *Cell* 59(5), 893–903.

Huang, C., Kratzer, M.-C., Wedlich, D., Kashef, J., 2016. E-cadherin is required for cranial neural crest migration in *Xenopus laevis. Dev. Biol.* 411(2), 159–171.

Huang, X., Saint-Jeannet, J.-P., 2004. Induction of the neural crest and the opportunities of life on the edge. *Dev. Biol.* 275(1), 1–11.

Hulpiau, P., Van Roy, F., 2009. Molecular evolution of the cadherin superfamily. *Int. J. Biochem. Cell Biol.* 41(2), 349–369.

Hutchins, E.J., Bronner, M.E., 2018. Draxin acts as a molecular rheostat of canonical Wnt signaling to control cranial neural crest EMT. *J. Cell Biol.* 217(10), 3683–3697.

Hutchins, E.J., Bronner, M.E., 2019. Draxin alters laminin organization during base-ment membrane remodeling to control cranial neural crest EMT. *Dev. Biol.* 446(2), 151–158.

Hynes, R.O., 2012. The evolution of metazoan extracellular matrix. *J. Cell Biol.* 196(6), 671–679.

Jandzik, D., Garnett, A.T., Square, T.A., Cattell, M.V., Yu, J.K., Medeiros, D.M., 2015. Evolution of the new vertebrate head by co-option of an ancient chordate skeletal tis-sue. *Nature* 518(7540), 534–537.

Jeffery, W.R., 2006. Ascidian neural crest-like cells: Phylogenetic distribution, relationship to larval complexity, and pigment cell fate. *J. Exp. Zool. B Mol. Dev. Evol.* 306(5), 470–480.

Jeffery, W.R., Chiba, T., Krajka, F.R., Deyts, C., Satoh, N., Joly, J.-S., 2008. Trunk lateral cells are neural crest- like cells in the ascidian *Ciona intestinalis*: Insights into the ancestry and evolution of the neural crest. *Dev. Biol.* 324(1), 152–160.

Jeffery, W.R., Strickler, A.G., Yamamoto, Y., 2004. Migratory neural crest- like cells form body pigmentation in a urochordate embryo. *Nature* 431(7009), 696–699.

Jourdeuil, K., Taneyhill, L.A., 2018. Spatiotemporal expression pattern of connexin 43 during early chick embryogenesis. *Gene Expr. Patterns* 27, 67–75.

Kalcheim, C., 2018. Neural crest emigration: From start to stop. *Genesis* 56(6–7), e23090.

Kalluri, R., Weinberg, R.A., 2009. The basics of epithelial-mesenchymal transition. *J. Clin. Invest.* 119(6), 1420–1428.

Kaltenbach, S.L., Yu Jr, K., Holland, N.D., 2009. The origin and migration of the earliest-developing sensory neurons in the peripheral nervous system of amphioxus. *Evol. Dev.* 11(2), 142–151.

Katsamba, P.S., Carroll, K., Ahlsen, G., Bahna, F., Vendome, J., Posy, S., Rajebhosale, M., Price, S., Jessell, T.M., Ben-Shaul, A., 2009. Linking molecular affinity and cellular specificity in cadherin-mediated adhesion. *Proc. Natl. Acad. Sci. U.S.A.* 106(28), 11594–11599.

Kee, Y., Hwang, B.J., Sternberg, P.W., Bronner-Fraser, M., 2007. Evolutionary conservation of cell migration genes: From nematode neurons to vertebrate neural crest. *Genes Dev.* 21(4), 391.

Kemler, R., 1992. Classical cadherins. *Semin. Cell Biol.* 3(3), 149–155.

Kerosuo, L., Bronner-Fraser, M., 2012. What is bad in cancer is good in the embryo: Importance of EMT in neural crest development. *Semin. Cell Dev. Biol.* 23(3), 320–332.

Kim, J., Lo, L., Dormand, E., Anderson, D.J., 2003. SOX10 maintains multipotency and inhibits neuronal differentiation of neural crest stem cells. *Neuron* 38(1), 17–31.

Knecht, A.K., Bonner-Fraser, M., 2002. Induction of the neural crest: A multigene process. *Nat. Rev. Genet.* 3(6), 453–461.

Kraus, Y., Technau, U., 2006. Gastrulation in the sea anemone *Nematostella vectensis* occurs by invagination and immigration: An ultrastructural study. *Dev. Genes Evol.* 216(3), 119–132.

Kubota, Y., Ito, K., 2000. Chemotactic migration of mesencephalic neural crest cells in the mouse. *Dev. Dyn.* 217(2), 170–179.

Kulesa, P., Ellies, D.L., Trainor, P.A., 2004. Comparative analysis of neural crest cell death, migration, and function during vertebrate embryogenesis. *Dev. Dyn.* 229(1), 14–29.

Kulesa, P.M., Bailey, C.M., Kasemeier-Kulesa, J.C., McLennan, R., 2010. Cranial neural crest migration: New rules for an old road. *Dev. Biol.* 344(2), 543–554.

Lacalli, T.C., 2010. The emergence of the chordate body plan: Some puzzles and problems. *Acta Zool.* 91(1), 4–10.

Lai, S.L., Chang, C.N., Wang, P.J., Lee, S.J., 2005. Rho mediates cytokinesis and epiboly via ROCK in zebrafish. *Mol. Reprod. Dev.* 71(2), 186–196.

Lakiza, O., Miller, S., Bunce, A., Lee, E.M.J., McCauley, D.W., 2011. SoxE gene duplication and development of the lamprey branchial skeleton: Insights into development and evolution of the neural crest. *Dev. Biol.* 359(1), 149–161.

Lander, R., Nordin, K., LaBonne, C., 2011. The F-box protein Ppa is a common regulator of core EMT factors Twist, Snail, Slug, and Sip1. *J. Cell Biol.* 194(1), 17–25.

Langeland, J.A., Tomsa, J.M., Jackman, W.R., Kimmel, C.B., 1998. An amphioxus snail gene: Expression in paraxial mesoderm and neural plate suggests a conserved role in patterning the chordate embryo. *Dev. Genes Evol.* 208(10), 569–577.

Langer, E.M., Feng, Y., Zhaoyuan, H., Rauscher, F.J., Kroll, K.L., Longmore, G.D., 2008. Ajuba LIM proteins are snail/slug corepressors required for neural crest development in Xenopus. *Dev. Cell* 14(3), 424–436.

Laubichler, M.D., Maienschein, J., 2007. From embryology to evo-devo : A history of developmental evolution. MIT Press, Cambridge, MA.

Le Douarin, N., Kalcheim, C., 1999. *The Neural Crest*, 2nd ed. Cambridge University Press, New York.

Lee, E.M., Yuan, T., Ballim, R.D., Nguyen, K., Kelsh, R.N., Medeiros, D.M., McCauley, D.W., 2016. Functional constraints on SoxE proteins in neural crest development: The importance of differential expression for evolution of protein activity. *Dev. Biol.* 418(1), 166–178.

Lehmann, W., Mossmann, D., Kleemann, J., Mock, K., Meisinger, C., Brummer, T., Herr, R., Brabletz, S., Stemmler, M.P., Brabletz, T., 2016. ZEB1 turns into a transcriptional activator by interacting with YAP1 in aggressive cancer types. *Nat. Commun.* 7, 10498.

Leptin, M., 1999. Gastrulation in Drosophila: The logic and the cellular mechanisms. *EMBO J.* 18(12), 3187–3192.

Leptin, M., 2005. Gastrulation movements: The logic and the nuts and bolts. *Dev. Cell* 8(3), 305–320.

Lespinet, O., Nederbragt, A.J., Cassan, M., Dictus, W., van Loon, A.E., Adoutte, A., 2002. Characterisation of two snail genes in the gastropod mollusc *Patella vulgata*. Implications for understanding the ancestral function of the snail-related genes in Bilateria. *Dev. Genes Evol.* 212(4), 186–195.

Linker, C., Bronner-Fraser, M., Mayor, R., 2000. Relationship between gene expression domains of Xsnail, Xslug, and Xtwist and cell movement in the prospective neural crest of Xenopus. *Dev. Biol.* 224(2), 215–225.

Liu, J.A., Wu, M.-H., Yan, C.H., Chau, B.K., So, H., Ng, A., Chan, A., Cheah, K.S., Briscoe, J., Cheung, M., 2013. Phosphorylation of Sox9 is required for neural crest delamination and is regulated downstream of BMP and canonical Wnt signaling. *Proc. Natl. Acad. Sci. U.S.A.* 110(8), 2882–2887.

Liu, J.P., Jessell, T.M., 1998. A role for rhoB in the delamination of neural crest cells from the dorsal neural tube. *Development* 125(24), 5055.

Locascio, A., Manzanares, M., Blanco, M.J., Nieto, M.A., 2002. Modularity and reshuffling of Snail and Slug expression during vertebrate evolution. *Proc. Natl. Acad. Sci. U.S.A.* 99(26), 16841–16846.

Lowe, C.J., Wu, M., Salic, A., Evans, L., Lander, E., Stange-Thomann, N., Gruber, C.E., Gerhart, J., Kirschner, M., 2003. Anteroposterior patterning in hemichordates and the origins of the chordate nervous system. *Cell* 113(7), 853–865.

Lu, T.M., Luo, Y.J., Yu, J.K., 2012. BMP and Delta/Notch signaling control the development of amphioxus epidermal sensory neurons: Insights into the evolution of the peripheral sensory system. *Development* 139(11), 2020–2030.

Magie, C.R., Daly, M., Martindale, M.Q., 2007. Gastrulation in the cnidarian *Nematostella vectensis* occurs via invagination not ingression. *Dev. Biol.* 305(2), 483–497.

Magie, C.R., Martindale, M.Q., 2008. Cell-cell adhesion in the Cnidaria: Insights into the evolution of tissue morphogenesis. *Biol. Bull.* 214(3), 218–232.

Mani, S.A., Guo, W., Liao, M.-J., Eaton, E.N., Ayyanan, A., Zhou, A.Y., Brooks, M., Reinhard, F., Zhang, C.C., Shipitsin, M., 2008. The epithelial-mesenchymal transition generates cells with properties of stem cells. *Cell* 133(4), 704–715.

Mansfield, J.H., Haller, E., Holland, N.D., Brent, A.E., 2015. Development of somites and their derivatives in amphioxus, and implications for the evolution of vertebrate somites. *EvoDevo* 6, 21.

Manzanares, M., Locascio, A., Nieto, M.A., 2001. The increasing complexity of the Snail gene superfamily in metazoan evolution. *Trends Genet.* 17(4), 178–181.

Martik, M.L., Bronner, M.E., 2017. Regulatory logic underlying diversification of the neural crest. *Trends Genet.* 33(10), 715–727.

Martín-Durán, J.M., Amaya, E., Romero, R., 2010. Germ layer specification and axial patterning in the embryonic development of the freshwater planarian *Schmidtea polychroa*. *Dev. Biol.* 340(1), 145–158.

Martindale, M.Q., Henry, J.Q., 1999. Intracellular fate mapping in a basal metazoan, the ctenophore Mnemiopsis leidyi, reveals the origins of mesoderm and the existence of indeterminate cell lineages. *Dev. Biol.* 214(2), 243–257.

Martins-Green, M., Erickson, C.A., 1987. Basal lamina is not a barrier to neural crest cell emigration: Documentation by TEM and by immunofluorescent and immunogold labelling. *Development* 101(3), 517–533.

McCauley, D.W., Bronner-Fraser, M., 2003. Neural crest contributions to the lamprey head. *Development* 130(11), 2317–2327.

McCauley, D.W., Bronner-Fraser, M., 2004. Conservation and divergence of BMP2/4 genes in the lamprey: Expression and phylogenetic analysis suggest a single ancestral vertebrate gene. *Evol. Dev.* 6(6), 411–422.

McCauley, D.W., Bronner-Fraser, M., 2006. Importance of SoxE in neural crest development and the evolution of the pharynx. *Nature* 441(7094), 750–752.

McCauley, D.W., Docker, M.F., Whyard, S., Li, W., 2015. Lampreys as diverse model organisms in the genomics era. *BioScience* 65(11), 1046–1056.

McCusker, C., Cousin, H., Neuner, R., Alfandari, D., 2009. Extracellular cleavage of cadherin-11 by ADAM metalloproteases is essential for Xenopus cranial neural crest cell migration. *Mol. Biol. Cell* 20(1), 78–89.

McKeown, S.J., Lee, V.M., Bronner-Fraser, M., Newgreen, D.F., Farlie, P.G., 2005. Sox10 overexpression induces neural crest-like cells from all dorsoventral levels of the neural tube but inhibits differentiation. *Dev. Dyn.* 233(2), 430–444.

Merchant, B., Edelstein-Keshet, L., Feng, J.J., 2018. A Rho-GTPase based model explains spontaneous collective migration of neural crest cell clusters. *Dev. Biol.* 444 Supplement 1, S262–S273.

Metzstein, M.M., Horvitz, H.R., 1999. The C-elegans cell death. specification gene ces-1 encodes a snail family zinc finger protein. *Mol. Cell* 4(3), 309–319.

Meulemans, D.M., Bronner-Fraser, M., 2005. Central role of gene cooption in neural crest evolution. *J. Exp. Zool. B Mol. Dev. Evol.* 304B(4), 298–303.

Milet, C., Monsoro-Burq, A.H., 2012. Neural crest induction at the neural plate border in vertebrates. *Dev. Biol.* 366(1), 22–33.

Miller, D.J., Ball, E.E., 2008. Animal evolution: Trichoplax, trees, and taxonomic turmoil. *Curr. Biol.* 18(21), R1003–R1005.

Miller, J.R., McClay, D.R., 1997. Characterization of the role of cadherin in regulating cell adhesion during sea urchin development. *Dev. Biol.* 192(2), 323–339.

Modrell, M.S., Hockman, D., Uy, B., Buckley, D., Sauka-Spengler, T., Bronner, M.E., Baker, C.V., 2014. A fate-map for cranial sensory ganglia in the sea lamprey. *Dev. Biol.* 385(2), 405–416.

Monsoro-Burq, A.-H., Wang, E., Harland, R., 2005. Msx1 and Pax3 cooperate to mediate FGF8 and WNT signals during Xenopus neural crest induction. *Dev. Cell* 8(2), 167–178.

Montell, D.J., 1999. The genetics of cell migration in *Drosophila melanogaster* and *Caenorhabditis elegans* development. *Development* 126(14), 3035–3046.

Morales, A.V., Barbas, J.A., Nieto, M.A., 2005. How to become neural crest: From segregation to delamination. *Semin. Cell Dev. Biol.* 16(6), 655–662.

Moreno-Bueno, G., Cubillo, E., Sarrió, D., Peinado, H., Rodriguez-Pinilla, S.M., Villa, S., Bolos, V., Jorda, M., Fabra, A., Portillo, F., Palacios, J., Cano, A., 2006. Genetic profiling of epithelial cells expressing E-cadherin repressors reveals a distinct role for snail, slug, and E47 factors in epithelial-mesenchymal transition. *Cancer Res.* 66(19), 9543–9556.

Moroz, L.L., 2015. Convergent evolution of neural systems in ctenophores. *J. Exp. Biol.* 218(4), 598–611.

Muñoz, W.A., Trainor, P.A., 2015. Neural crest cell evolution: How and when did a neural crest cell become a neural crest cell. In: Trainor, P.A. (Ed.), *Neural Crest and Placodes (Current Topics in Developmental Biology).* Elsevier Academic Press Inc., San Diego, pp. 3–26.

Nakagawa, S., Takeichi, M., 1998. Neural crest emigration from the neural tube depends on regulated cadherin expression. *Development* 125(15), 2963–2971.

Nakanishi, N., Sogabe, S., Degnan, B.M., 2014. Evolutionary origin of gastrulation: Insights from sponge development. *BMC Biol.* 12, 26.

Neuner, R., Cousin, H., McCusker, C., Coyne, M., Alfandari, D., 2009. Xenopus ADAM19 is involved in neural, neural crest and muscle development. *Mech. Dev.* 126(3–4), 240–255.

Nielsen, C., 2012. *Animal Evolution: Interrelationships of the Living Phyla.* Oxford University Press, Oxford.

Nieto, M.A., 2002. The snail superfamily of zinc-finger transcription factors. *Nat. Rev. Mol. Cell Biol.* 3(3), 155–166.

Nieto, M.A., Sargent, M.G., Wilkinson, D.G., Cooke, J., 1994. Control of cell behavior during vertebrate development by Slug, a zinc-finger gene. *Science* 264(5160), 835–839.

Nikitina, N., Sauka-Spengler, T., Bronner-Fraser, M., 2008. Dissecting early regulatory relationships in the lamprey neural crest gene network. *Proc. Natl. Acad. Sci. U.S.A.* 105(51), 20083–20088.

Nobes, C.D., Hall, A., 1995. Rho, rac, and cdc42 GTPases regulate the assembly of multimolecular focal complexes associated with actin stress fibers, lamellipodia, and filopodia. *Cell* 81(1), 53–62.

Northcutt, R.G., Gans, C., 1983. The genesis of neural crest and epidermal placodes: A reinterpretation of vertebrate origins. *Q. Rev. Biol.* 58(1), 1–28.

Ochoa, S.D., Labonne, C., 2011. LMO4 modulates slug/snail function in neural crest development. *Dev. Biol.* 356(1), 260–260.

Ochoa, S.D., Salvador, S., Labonne, C., 2012. The LIM adaptor protein LMO4 is an essential regulator of neural crest development. *Dev. Biol.* 361(2), 313–325.

Oda, H., Akiyama-Oda, Y., Zhang, S., 2004. Two classic cadherin-related molecules with no cadherin extracellular repeats in the cephalochordate amphioxus: Distinct adhesive specificities and possible involvement in the development of multicell-layered structures. *J. Cell Sci.* 117(13), 2757.

Oda, H., Takeichi, M., 2011. Structural and functional diversity of cadherin at the adherens junction. *J. Cell Biol.* 193(7), 1137–1146.

Oda, H., Tsukita, S., Takeichi, M., 1998. Dynamic behavior of the cadherin-based cell–cell adhesion system during Drosophila gastrulation. *Dev. Biol.* 203(2), 435–450.

Oda, H., Wada, H., Tagawa, K., Akiyama-Oda, Y., Satoh, N., Humphreys, T., Zhang, S., Tsukita, S., 2002. A novel amphioxus cadherin that localizes to epithelial adherens junctions has an unusual domain organization with implications for chordate phylogeny. *Evol. Dev.* 4(6), 426–434.

Ohno, S., 1970. *Evolution by Gene Duplication.* Springer-Verlag, Berlin.

Parker, H.J., De Kumar, B., Green, S.A., Prummel, K.D., Hess, C., Kaufman, C.K., Mosimann, C., Wiedemann, L.M., Bronner, M.E., Krumlauf, R., 2019. A Hox-TALE regulatory circuit for neural crest patterning is conserved across vertebrates. *Nat. Commun.* 10(1), 1189.

Parker, H.J., Sauka-Spengler, T., Bronner, M., Elgar, G., 2014. A reporter assay in lamprey embryos reveals both functional conservation and elaboration of vertebrate enhancers. *PLOS ONE* 9(1), 7.

Peinado, H., Ballestar, E., Esteller, M., Cano, A., 2004a. Snail mediates E-cadherin repression by the recruitment of the Sin3A/histone deacetylase 1 (HDAC1)/HDAC2 complex. *Mol. Cell. Biol.* 24(1), 306–319.

Peinado, H., Portillo, F., Cano, A., 2004b. Transcriptional regulation of cadherins during development and carcinogenesis. *Int. J. Dev. Biol.* 48(5–6), 365–375.

Perez-Alcala, S., Nieto, M.A., Barbas, J.A., 2004. LSox5 regulates RhoB expression in the neural tube and promotes generation of the neural crest. *Development* 131(18), 4455–4465.

Perez-Moreno, M.A., Locascio, A., Rodrigo, I., Dhondt, G., Portillo, F., Nieto, M.A., Cano, A., 2001. A new role for E12/E47 in the repression of E- cadherin expression and epithelial-mesenchymal transitions. *J. Biol. Chem.* 276(29), 27424.

Philippe, H., Brinkmann, H., Copley, R.R., Moroz, L.L., Nakano, H., Poustka, A.J., Wallberg, A., Peterson, K.J., Telford, M.J., 2011. Acoelomorph flatworms are deuterostomes related to *Xenoturbella*. *Nature* 470(7333), 255.

Pla, P., Monsoro-Burq, A.H., 2018. The neural border: Induction, specification and maturation of the territory that generates neural crest cells. *Dev. Biol.* 444 Supplement 1, S36–S46.

Pla, P., Moore, R., Morali, O.G., Grille, S., Martinozzi, S., Delmas, V., Larue, L., 2001. Cadherins in neural crest cell development and transformation. *J. Cell. Physiol.* 189(2), 121–132.

Putnam, N.H., Butts, T., Ferrier, D.E.K., Furlong, R.F., Hellsten, U., Kawashima, T., Robinson-Rechavi, M., Shoguchi, E., Terry, A., Yu, J.-K., Benito-Gutiérrez, E.L., Dubchak, I., Garcia-Fernàndez, J., Gibson-Brown, J.J., Grigoriev, I.V., Horton, A.C., de Jong, P.J., Jurka, J., Kapitonov, V.V., Kohara, Y., Kuroki, Y., Lindquist, E., Lucas, S., Osoegawa, K., Pennacchio, L.A., Salamov, A.A., Satou, Y., Sauka-Spengler, T., Schmutz, J., Shin-I. T., Toyoda, A., Bronner-Fraser, M., Fujiyama, A., Holland, L.Z., Holland, P.W.H., Satoh, N., Rokhsar, D.S., 2008. The amphioxus genome and the evolution of the chordate karyotype. *Nature* 453(7198), 1064.

Rahimi, R.A., Allmond, J.J., Wagner, H., McCauley, D.W., Langeland, J.A., 2009. Lamprey snail highlights conserved and novel patterning roles in vertebrate embryos. *Dev. Genes Evol.* 219(1), 31–36.

Raible, D.W., Ragland, J.W., 2005. Reiterated Wnt and BMP signals in neural crest development. *Semin. Cell Dev. Biol.* 16(6), 673–682.

Reece-Hoyes, J.S., Deplancke, B., Barrasa, M.I., Hatzold, J., Smit, R.B., Arda, H.E., Pope, P.A., Gaudet, J., Conradt, B., Walhout, A.J.M., 2009. The *C. elegans* Snail homolog CES- 1 can activate gene expression in vivo and share targets with bHLH transcription factors. *Nucleic Acids Res.* 37(11), 3689.

Rogers, C.D., Nie, S., 2018. Specifying neural crest cells: From chromatin to morphogens and factors in between. *Wiley Interdiscip. Rev. Dev. Biol.* 7, e322.

Rogers, C.D., Saxena, A., Bronner, M., 2013. Sip1 mediates an E-cadherin-to-N-cadherin switch during cranial neural crest EMT. *J. Cell Biol.* 203(5), 835–847.

Ryan, J.F., Chiodin, M., 2015. Where is my mind? How sponges and placozoans may have lost neural cell types. *Philos. Trans. R. Soc. B* 370, 20150059.

Ryan, J.F., Pang, K., Schnitzler, C.E., Nguyen, A.-D., Moreland, R.T., Simmons, D.K., Koch, B.J., Francis, W.R., Havlak, P., Smith, S.A., 2013. The genome of the ctenophore *Mnemiopsis leidyi* and its implications for cell type evolution. *Science* 342(6164), 1242592.

Sadok, A., Marshall, C.J., 2014. Rho GTPases: Masters of cell migration. *Small GTPases* 5, e983878.

Sakai, D., Suzuki, T., Osumi, N., Wakamatsu, Y., 2006. Cooperative action of Sox9, Snail2 and PKA signaling in early neural crest development. *Development* 133(7), 1323–1333.

Salinas-Saavedra, M., Rock, A.Q., Martindale, M.Q., 2018. Germ layer-specific regulation of cell polarity and adhesion gives insight into the evolution of mesoderm. *eLife* 7, e36740.

Sánchez-Tilló, E., Lázaro, A., Torrent, R., Cuatrecasas, M., Vaquero, E., Castells, A., Engel, P., Postigo, A., 2010. ZEB1 represses E-cadherin and induces an EMT by recruiting the SWI/SNF chromatin-remodeling protein BRG1. *Oncogene* 29(24), 3490–3500.

Sato, T., Sasai, N., Sasai, Y., 2005. Neural crest determination by co-activation of Pax3 and Zic1 genes in *Xenopus* ectoderm. *Development* 132(10), 2355–2363.

Sauka-Spengler, T., Bronner-Fraser, M., 2008a. Evolution of the neural crest viewed from a gene regulatory perspective. *Genesis* 46(11), 673–682.

Sauka-Spengler, T., Bronner-Fraser, M., 2008b. Insights from a sea lamprey into the evolution of neural crest gene regulatory network. *Biol. Bull.* 214(3), 303–314.

Sauka-Spengler, T., Meulemans, D.M., Jones, M., Bronner-Fraser, M., 2007. Ancient evolutionary origin of the neural crest gene regulatory network. *Dev. Cell* 13(3), 405–420.

Saunders, L.R., McClay, D.R., 2014. Sub-circuits of a gene regulatory network control a developmental epithelial-mesenchymal transition. *Development* 141(7), 1503–1513.

Savagner, P., 2001. Leaving the neighborhood: Molecular mechanisms involved during epithelial-mesenchymal transition. *BioEssays* 23(10), 912–923.

Savagner, P., 2010. The epithelial-mesenchymal transition (EMT) phenomenon. *Ann. Oncol.* 21 Supplement 7, 89–92.

Scarpa, E., Szabo, A., Bibonne, A., Theveneau, E., Parsons, M., Mayor, R., 2015. Cadherin switch during EMT in neural crest cells leads to contact inhibition of locomotion via repolarization of forces. *Dev. Cell* 34(4), 421–434.

Schierwater, B., 2005. My favorite animal, *Trichoplax adhaerens*. *BioEssays* 27(12), 1294–1302.

Sela-Donenfeld, D., Kalcheim, C., 1999. Regulation of the onset of neural crest migration by coordinated activity of BMP4 and Noggin in the dorsal neural tube. *Development* 126(21), 4749–4762.

Shellard, A., Szabó, A., Trepat, X., Mayor, R., 2018. Supracellular contraction at the rear of neural crest cell groups drives collective chemotaxis. *Science* 362(6412), 339–343.

Shook, D., Keller, R., 2003. Mechanisms, mechanics and function of epithelial–mesenchymal transitions in early development. *Mech. Dev.* 120(11), 1351–1383.

Shoval, I., Ludwig, A., Kalcheim, C., 2007. Antagonistic roles of full-length N-cadherin and its soluble BMP cleavage product in neural crest delamination. *Development* 134(3), 491–501.

Shubin, N., Tabin, C., Carroll, S., 2009. Deep homology and the origins of evolutionary novelty. *Nature* 457(7231), 818.

Smallhorn, M., Murray, M.J., Saint, R., 2004. The epithelial-mesenchymal transition of the *Drosophila mesoderm* requires the Rho GTP exchange factor Pebble. *Development* 131(11), 2641–2651.

Smith, C.L., Varoqueaux, F., Kittelmann, M., Azzam, R.N., Cooper, B., Winters, C.A., Eitel, M., Fasshauer, D., Reese, T.S., 2014. Novel cell types, neurosecretory cells, and body plan of the early-diverging metazoan *Trichoplax adhaerens*. *Curr. Biol.* 24(14), 1565–1572.

Smith, J.J., Kuraku, S., Holt, C., Sauka-Spengler, T., Jiang, N., Campbell, M.S., Yandell, M.D., Manousaki, T., Meyer, A., Bloom, O.E., Morgan, J.R., Buxbaum, J.D., Sachidanandam, R., Sims, C., Garruss, A.S., Cook, M., Krumlauf, R., Wiedemann, L.M., Sower, S.A., Decatur, W.A., Hall, J.A., Amemiya, C.T., Saha, N.R., Buckley, K.M., Rast, J.P., Das, S., Hirano, M., McCurley, N., Guo, P., Rohner, N., Tabin, C.J., Piccinelli, P., Elgar, G., Ruffier, M., Aken, B.L., Searle, S.M.J., Muffato, M., Pignatelli, M., Herrero, J., Jones, M., Brown, C.T., Chung-Davidson, Y.W., Nanlohy, K.G., Libants, S.V., Yeh, C.Y., McCauley, D.W., Langeland, J.A., Pancer, Z., Fritzsch, B., de Jong, P.J., Zhu,

B.L., Fulton, L.L., Theising, B., Flicek, P., Bronner, M.E., Warren, W.C., Clifton, S.W., Wilson, R.K., Li, W.M., 2013. Sequencing of the sea lamprey (*Petromyzon marinus*) genome provides insights into vertebrate evolution. *Nat. Genet.* 45(4), 415–421.

Smith, J.J., Timoshevskaya, N., Ye, C., Holt, C., Keinath, M.C., Parker, H.J., Cook, M.E., Hess, J.E., Narum, S.R., Lamanna, F., 2018. The sea lamprey germline genome provides insights into programmed genome rearrangement and vertebrate evolution. *Nat. Genet.* 50(2), 270–277.

Square, T., Jandzik, D., Romášek, M., Cerny, R., Medeiros, D., 2016. The origin and diversification of the developmental mechanisms that pattern the vertebrate head skeleton. *Dev. Biol.* 427(2), 219–229.

Square, T., Romasek, M., Jandzik, D., Cattell, M.V., Klymkowsky, M., Medeiros, D.M., 2015. CRISPR/Cas9-mediated mutagenesis in the sea lamprey *Petromyzon marinus*: A powerful tool for understanding ancestral gene functions in vertebrates. *Development* 142(23), 4180–4187.

Srivastava, M., Begovic, E., Chapman, J., Putnam, N.H., Hellsten, U., Kawashima, T., Kuo, A., Mitros, T., Salamov, A., Carpenter, M.L., 2008. The Trichoplax genome and the nature of placozoans. *Nature* 454(7207), 955.

Srivastava, M., Simakov, O., Chapman, J., Fahey, B., Gauthier, M., Mitros, T., Richards, G., Conaco, C., Dacre, M., Hellsten, U., Larroux, C., Putnam, N., Stanke, M., Adamska, M., Darling, A., Degnan, S., Oakley, T., Plachetzki, D., Zhai, Y., Adamski, M., Calcino, A., Cummins, S., Goodstein, D., Harris, C., Jackson, D., Leys, S., Shu, S., Woodcroft, B., Vervoort, M., Kosik, K., Manning, G., Degnan, B., Rokhsar, D., 2010. The *Amphimedon queenslandica* genome and the evolution of animal complexity. *Nature* 466(7307), 720–726.

Stewart, R.A., Arduini, B.L., Berghmans, S., George, R.E., Kanki, J.P., Henion, P.D., Look, A.T., 2006. Zebrafish foxd3 is selectively required for neural crest specification, migration and survival. *Dev. Biol.* 292(1), 174–188.

Stolfi, A., Ryan, K., Meinertzhagen, I.A., Christiaen, L., 2015. Migratory neuronal progenitors arise from the neural plate borders in tunicates. *Nature* 527(7578), 371.

Szabó, A., Theveneau, E., Turan, M., Mayor, R., 2019. Neural crest streaming as an emergent property of tissue interactions during morphogenesis. *PLOS Comp. Biol.* 15(4), e1007002.

Tahtakran, S.A., Selleck, M.A.J., 2003. Ets-1 expression is associated with cranial neural crest migration and vasculogenesis in the chick embryo. *Gene Expr. Patterns* 3(4), 455–458.

Taneyhill, L.A., Bronner-Fraser, M., 2005. Dynamic alterations in gene expression after Wnt-mediated induction of avian neural crest. *Mol. Biol. Cell* 16(11), 5283–5293.

Taneyhill, L.A., Coles, E.G., Bronner-Fraser, M., 2007. Snail2 directly represses cadherin6B during epithelial-tomesenchymal transitions of the neural crest. *Development* 134(8), 1481–1490.

Tepass, U., Truong, K., Godt, D., Ikura, M., Peifer, M., 2000. Cadherins in embryonic and neural morphogenesis. *Nat. Rev. Mol. Cell Biol.* 1(2), 91.

Thellmann, M., Hatzold, J., Conradt, B., 2003. The Snail-like CES- 1 protein of *C. elegans* can block the expression of the BH3-only cell- death activator gene egl- 1 by antagonizing the function of bHLH proteins. *Development* 130(17), 4057.

Théveneau, E., Duband, J.-L., Altabef, M., 2007. Ets-1 confers cranial features on neural crest delamination. *PLOS ONE* 2(11), e1142.

Theveneau, E., Mayor, R., 2011. Collective cell migration of the cephalic neural crest: The art of integrating information. *Genesis* 49(4), 164–176.

Theveneau, E., Mayor, R., 2012a. Neural crest delamination and migration: From epithelium-to-mesenchyme transition to collective cell migration. *Dev. Biol.* 366(1), 34–54.

Theveneau, E., Mayor, R., 2012b. Neural crest migration: Interplay between chemorepellents, chemoattractants, contact inhibition, epithelial–mesenchymal transition, and collective cell migration. *Wiley Interdiscip. Rev. Dev. Biol.* 1(3), 435–445.

Thiery, J., Sleeman, J., 2006. Complex networks orchestrate epithelial-mesenchymal transitions. *Nat. Rev. Mol. Cell Biol.* 7(2), 131–142.

Thiery, J.P., 2003. Epithelial-mesenchymal transitions in development and pathologies. *Curr. Opin. Cell Biol.* 15(6), 740–746.

Trainor, P.A., 2013. *Neural Crest Cells: Evolution, Development and Disease.* Academic Press, Cambridge.

Tucker, R.P., 2004. Antisense knockdown of the beta 1 integrin subunit in the chicken embryo results in abnormal neural crest cell development. *Int. J. Biochem. Cell Biol.* 36(6), 1135–1139.

Tyler, S., 2003. Epithelium—The primary building block for metazoan complexity. *Integr. Comp. Biol.* 43(1), 55–63.

Vallin, J., Girault, J.-M., Thiery, J.P., Broders, F., 1998. Xenopus cadherin- 11 is expressed in different populations of migrating neural crest cells. *Mech. Dev.* 75(1–2), 171–174.

Vandewalle, C., Comijn, J., De Craene, B., Vermassen, P., Bruyneel, E., Andersen, H., Tulchinsky, E., Van Roy, F., Berx, G., 2005. SIP1/ZEB2 induces EMT by repressing genes of different epithelial cell-cell junctions. *Nucleic Acids Res.* 33(20), 6566–6578.

Vannier, C., Mock, K., Brabletz, T., Driever, W., 2013. Zeb1 regulates E- cadherin and Epcam (epithelial cell adhesion molecule) expression to control cell behavior in early zebrafish development. *J. Biol. Chem.* 288(26), 18643.

Vega, S., Morales, A.V., Ocana, O.H., Valdes, F., Fabregat, I., Nieto, M.A., 2004. Snail blocks the cell cycle and confers resistance to cell death. *Genes Dev.* 18(10), 1131–1143.

Wada, H., 2001. Origin and evolution of the neural crest: A hypothetical reconstruction of its evolutionary history. *Development, Growth, and Differentiation*, 43(5), 509–520.

Wada, H., Makabe, K., 2006. Genome duplications of early vertebrates as a possible chronicle of the evolutionary history of the neural crest. *Int. J. Biol. Sci.* 2(3), 133–141.

Wheelock, M.J., Shintani, Y., Maeda, M., Fukumoto, Y., Johnson, K.R., 2008. Cadherin switching. *J. Cell Sci.* 121(6), 727–735.

Willems, B., Tao, S., Yu, T., Huysseune, A., Witten, P.E., Winkler, C., 2015. The Wnt co-receptor Lrp5 is required for cranial neural crest cell migration in zebrafish. *PLOS ONE* 10(6), 21.

Wu, S.-Y., McClay, D.R., 2007. The Snail repressor is required for PMC ingression in the sea urchin embryo. *Development* 134(6), 1061.

Wu, S.-Y., Yang, Y.-P., McClay, D.R., 2008. Twist is an essential regulator of the skeletogenic gene regulatory network in the sea urchin embryo. *Dev. Biol.* 319(2), 406–415.

Wu, S.Y., Ferkowicz, M., McClay, D.R., 2007. Ingression of primary mesenchyme cells of the sea urchin embryo: A precisely timed epithelial mesenchymal transition. *Birth Defects Res. C Embryo Today* 81(4), 241–252.

Yang, J., Mani, S.A., Donaher, J.L., Ramaswamy, S., Itzykson, R.A., Come, C., Savagner, P., Gitelman, I., Richardson, A., Weinberg, R.A., 2004. Twist, a master regulator of morphogenesis, plays an essential role in tumor metastasis. *Cell* 117(7), 927–939.

Yang, M.-H., Wu, K.-J., 2008. TWIST activation by hypoxia inducible factor-1 (HIF-1): Implications in metastasis and development. *Cell Cycle* 7(14), 2090.

York, J.R., Lee, E.M., McCauley, D.W., 2019a. The lamprey as a model vertebrate in evolutionary developmental biology. In: Docker, M.F. (Ed.), *Lampreys: Biology, Conservation and Control.* Springer, Dordecht, pp. 481–526.

York, J.R., Zehnder, K., Yuan, T., Lakiza, O., McCauley, D.W., 2019b. Evolution of Snail-mediated regulation of neural crest and placodes from an ancient role in bilaterian neurogenesis. *Dev. Biol.* 453(2), 180–190.

York, J.R., Yuan, T., Lakiza, O., McCauley, D.W., 2018. An ancestral role for Semaphorin3F-Neuropilin signaling in patterning neural crest within the new vertebrate head. *Development* 145(14), dev164780.

York, J.R., Yuan, T., Zehnder, K., McCauley, D.W., 2017. Lamprey neural crest migration is Snail-dependent and occurs without a differential shift in cadherin expression. *Dev. Biol.* 428(1), 176–187.

Yu, J.-K., Meulemans, D., McKeown, S.J., Bronner-Fraser, M., 2008. Insights from the amphioxus genome on the origin of vertebrate neural crest. *Genome Res.* 18(7), 1127–1132.

Yuan, T., York, J.R., McCauley, D.W., 2018. Gliogenesis in lampreys shares gene regulatory interactions with oligodendrocyte development in jawed vertebrates. *Dev. Biol.* 441(1), 176–190.

Zheng, H., Kang, Y., 2014. Multilayer control of the EMT master regulators. *Oncogene* 33(14), 1755.

Zu, Y., Zhang, X., Ren, J., Dong, X., Zhu, Z., Jia, L., Zhang, Q., Li, W., 2016. Biallelic editing of a lamprey genome using the CRISPR/Cas9 system. *Sci. Rep.* 6, 23496.

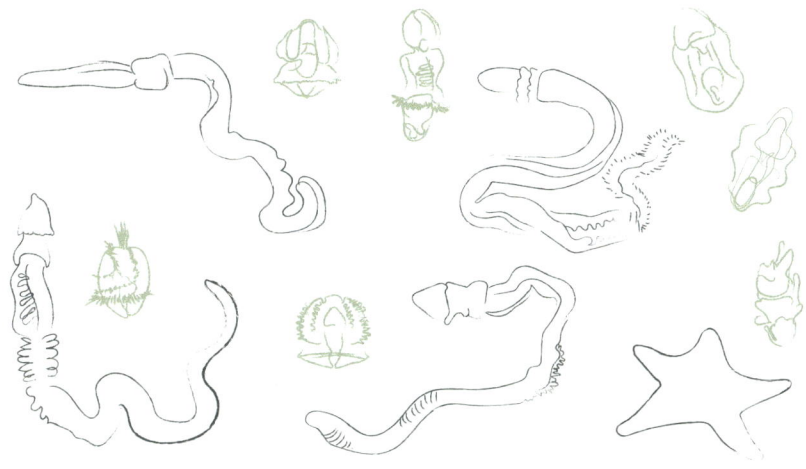

3 The Evolution of the Neural Border and Peripheral Nervous System—Insights from Invertebrate Deuterostome Animals

Jr-Kai Yu and Yi-Hsien Su

CONTENTS

3.1 INTRODUCTION

In vertebrate embryogenesis, the central nervous system (CNS) is derived from the dorsal ectoderm (the neural plate) and eventually includes both the brain and spinal cord. On the other hand, most of the sensory neurons and supporting glia of the vertebrate peripheral nervous system (PNS) are derived from two other embryonic tissues, the neural crest and the cranial placodes, both of which are specified at the neural plate boundary (Figure 3.1). In the rostral region, the panplacodal ectoderm directly abuts the neural plate, but in more caudal regions, the neural crest lies between the panplacodal ectoderm and the neural plate. The specification of the neural plate, the epidermal ectoderm, and the border regions between these embryonic territories are crucial processes in shaping the vertebrate nervous system. This arrangement of ectodermal territories is established during gastrula and early neurula stages, when signals (mainly Bmp, Wnt, and FGF) from both the ectoderm itself and adjacent tissues induce the dorsoventrally restricted expression of various transcription factors, which serve to specify different ectodermal fates in a combinatorial fashion (Groves and LaBonne, 2014; Maharana and Schlosser, 2018; Martik and Bronner, 2017; Meulemans and Bronner-Fraser, 2004) (and other articles in this volume).

To understand how vertebrate neural borders and the PNS have evolved, it is important to explore nervous system development in a wide range of animals. Interestingly, while there is remarkable uniformity in nerve-cell function throughout the animal kingdom, there is great diversity in the topological organization of nervous systems.

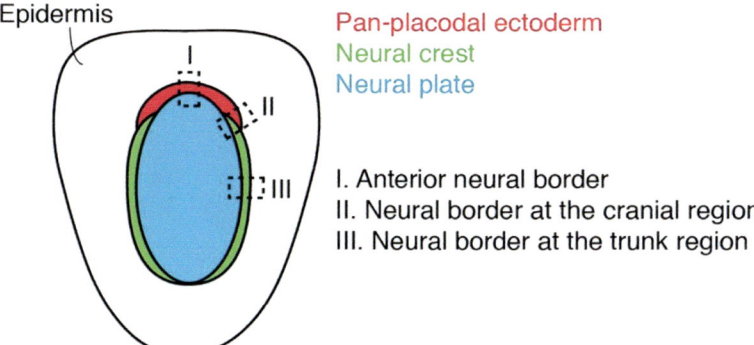

Epidermis

Pan-placodal ectoderm
Neural crest
Neural plate

I. Anterior neural border
II. Neural border at the cranial region
III. Neural border at the trunk region

FIGURE 3.1   Neural borders of vertebrate embryos. Before neurulation, the neural plate (blue) is specified on the dorsal side and segregated from the epidermis (white). Between the neural plate and the epidermis, the panplacodal ectoderm (red) forms in the rostral region, and neural crest (green) can be found in the trunk regions. In the anterior region, the neural border is established by only the panplacodal ectoderm (I); in the cranial region, both the panplacodal ectoderm and the neural crest establish the border (II), and in the trunk region, only the neural crest comprises the border (III).

For example, outside the vertebrate lineage, there is not necessarily a tight coupling between CNS and PNS development. Therefore, making morphological comparisons of nervous systems among distantly related taxa is often challenging. Based on the current view of the animal phylogeny (Figure 3.2), the group that is most closely related to vertebrates is urochordates (tunicates), and together with the cephalochordate (amphioxus), these three taxa constitute the phylum Chordata (Bourlat et al., 2006; Delsuc et al., 2006; Delsuc et al., 2008). Although invertebrate chordates (tunicates and amphioxus) lack definitive neural crest and placodes, the border regions between the neural plate and epidermal ectoderm are clearly present in some form. Understanding how these neural borders are set up in invertebrate chordates may provide important insights into the origins of the vertebrate neural crest and cranial placodes.

Outside the chordate lineage, the closest sister group is the Ambulacraria, which includes hemichordates and echinoderms. Together with chordates, these three taxa constitute the deuterostomes (Figure 3.2). The nervous systems of hemichordates and echinoderms are distinctively different at the larval and adult stages, and both arrangements have been considered in the context of understanding the origins of the chordate nervous system. However, due to the distinct morphological organizations of deuterostome nervous systems (Figure 3.2), direct comparisons have been extremely difficult. Thus, it had not been obvious why examining whether neural borders are present in non-chordate deuterostomes is informative. However, recent comparative studies using hemichordates and echinoderms revealed astounding similarities among deuterostome animals in terms of their ectodermal and neural patterning mechanisms, shedding new light on the evolution of nervous systems in the deuterostome lineage.

In this chapter, we review the developmental processes that give rise to the nervous systems in deuterostome animals, focusing on how the boundaries of neural

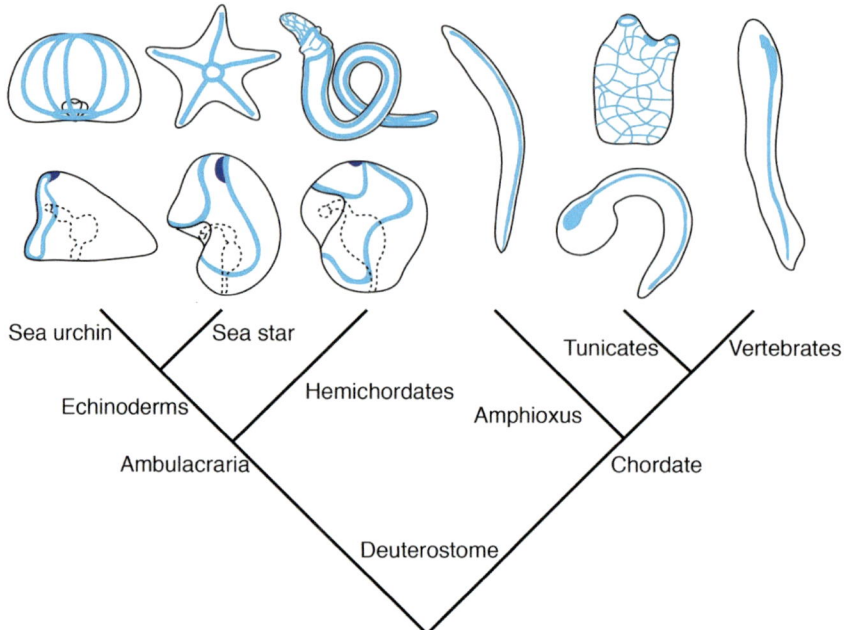

FIGURE 3.2   Phylogeny and nervous systems of the deuterostomes. The deuterostome superphylum includes two branches, the Ambulacraria (Echinoderms + Hemichordates) and the chordates. The nervous systems of the adult echinoderms (e.g., sea urchin and sea star) are pentaradially symmetric with five radial nerves. The indirect-developing echinoderm larvae (e.g., pluteus larva of sea urchin and bipinnaria larva of sea star) possess neurons that extend axon fibers along the ciliary bands (light blue) and a ganglionic structure called the *apical organ*, which contains serotonergic neurons (dark blue). The nervous system of the tornaria larva of indirect-developing acorn worms has a similar organization to those of echinoderm larvae. Adult acorn worms have dorsal and ventral nerve cords (thick blue lines), in addition to a diffuse nerve net (thin blue lines) in the ectoderm. The tripartite guts of the ambulacrarian larvae are contoured with dashed lines. All chordates, including amphioxus, tunicates (e.g., ascidian tadpole larva), and vertebrates (e.g., lamprey) have a dorsal, hollow neural tube. The nervous system of the adult ascidian is rearranged after metamorphosis, with a cerebral ganglion and a rich nerve net throughout the ectoderm.

and nonneural ectoderm are established and whether a distinct neural border region is recognizable. We will also review the development of the PNS (mainly on sensory neurons and ganglia) in these animals and discuss possible relationships with the vertebrate neural crest and/or placode-derived neurons.

3.2   NEURAL BORDERS AND THE PNS IN CHORDATES

3.2.1   Establishment of the Neural Borders in Vertebrates

In vertebrates, the neural plate is specified in the dorsal ectoderm after gastrulation. The border regions between the neural plate and the surrounding epidermal ectoderm are then gradually formed, as recognized by the expression of different sets of transcription factor genes in the anterior, the cranial, and the trunk regions (Figure 3.1). At the anterior border, no neural crest is formed; instead, a pan-placodal territory is specified. The pan-placodal ectoderm at the cranial region is sandwiched between the epidermal ectoderm and the neural crest cells. The different cranial ectodermal placodes are formed within the pan-placodal ectoderm after induction by signals released from neighboring tissues. In the trunk region, the neural crest cells mark the lateral boundary of the neural plate. This distinction between the anterior (cranial) and posterior (trunk) CNS border regions may reflect their respective progenitor properties, as suggested by recent discoveries that some posterior CNS tissue may come from a bipotent neuromesodermal progenitor that contribute to both the spinal cord and paraxial mesoderm (Henrique et al., 2015). In addition, analysis on enhancer usage by examining the chromatin accessibility profiles in differentiating neural progenitors reveled that cells commit to different axial identities prior to neural induction (Metzis et al., 2018), further highlighting the intrinsic differences of CNS progenitor cells along the anterior–posterior (AP) axis.

Detailed mechanisms that further specify the cranial and trunk CNS border regions into ectodermal placodes and neural crests during vertebrate development are reviewed elsewhere (e.g., Groves and LaBonne, 2014; Martik and Bronner, 2017; Pla and Monsoro-Burq, 2018), and we will not repeat the information here. We only point out that recent studies on the lamprey have shown that the ground state of the neural crest developmental gene regulatory network is conserved between jawless and jawed vertebrates, although some variations exist (Nikitina et al., 2008; Sauka-Spengler et al., 2007). In addition, detailed fate mapping and marker gene expression studies in lamprey embryos also demonstrated that the developmental mechanisms underlying the formation of neurogenic placodes and cranial sensory ganglia are highly conserved between jawless vertebrates and gnathostomes (Modrell et al., 2014). Together, these results suggest that the common ancestor of all vertebrates had already possessed cells with characteristic regulatory states that specify CNS border regions and PNS formation.

3.2.2 ESTABLISHMENT OF THE NEURAL BORDERS IN TUNICATES

Tunicates are a group of marine filter feeders, and during the larval stage, they have a typical tadpole-like chordate body plan that is characterized by a dorsal hollow neural tube, a notochord, and lateral muscle cells (Holland, 2016; Lemaire, 2011). While one class of tunicates (Appendicularia) retain their larval form in the planktonic adult stage, the other classes of tunicates undergo drastic metamorphosis, shedding their chordate body plan to become pelagic (Thaliacea) or sessile (Ascidiancea) adults. During tunicate metamorphosis, most parts of the larval CNS (except the tail nerve cord) are maintained and reorganized into an adult CNS—the cerebral ganglion (Horie et al., 2011). The cerebral ganglion is located between the incurrent and excurrent siphons, and together with a rich nerve net throughout the ectoderm, constitutes the adult nervous system (Manni et al., 2005; Sasakura et al., 2012). To date, studies on the developmental mechanisms of tunicate larval CNS have been mostly carried out using solitary ascidian species, such as *Ciona intestinalis* and *Halocynthia roretzi* (Hudson, 2016; Lemaire, 2011; Satoh, 2014); as such, we describe results obtained from those representative species herein.

3.2.2.1 Neural Patterning in Tunicate Embryos

Ascidian early embryogenesis is highly stereotypic, characterized by an invariant cleavage pattern with early fixation of cell lineages mediated largely by maternal determinants and short-range cell signaling interactions (Hudson, 2016; Lemaire, 2011; Satoh, 2014). The ascidian larval nervous system is derived from a-line (anterior animal), b-line (posterior animal), and A-line (anterior vegetal) blastomeres (Figure 3.3A). At the six-row neural plate stage, these cells develop into a grid-like pattern comprising six rows of cells along the AP axis and are arranged as three to five bilateral pairs of columns in the medial–lateral orientation (Figure 3.3B). During the formation of this cellular grid, the FGF/MEK/ERK signaling pathway controls neural induction and CNS specification along the AP axis, while Nodal and Delta-Notch signals pattern the neural plate across the lateral–medial direction, which represents the future dorsal–ventral (DV) axis of the neural tube (Hudson, 2016). Unlike vertebrates, there is no Spemann organizer–like structure during early ascidian embryogenesis, and it appears that neither BMP signaling nor Shh signaling play major roles in ascidian neural induction and DV (lateral–medial) patterning of the CNS (Hudson et al., 2011; Stolfi et al., 2011).

The mechanisms of ascidian neural induction and CNS cell fate specification have been examined extensively and are recently reviewed elsewhere (Hudson, 2016; here we focus on the formation of the neural plate borders, especially the anterior and lateral border regions (Figure 3.3C). Initially, at the 32-cell stage, FGF9/16/20 signals work with maternally expressed factors Ets1/2 and Gata-a to induce *Otx* expression in a-line neural precursors (Bertrand et al., 2003); subsequently, FGF signals from A-line cells repress the expression of *FoxC* and activate the expression of *ZicL* (*Zic-related.b*) in the adjacent a-line CNS precursors (row III and IV), while in the precursor cells of more anterior rows V and VI (also from the a-line), *FoxC* expression is retained, committing the cells to become the non-CNS anterior neural plate precursors (Imai et al., 2006; Wagner and Levine, 2012); the descendants of these precursors contribute to palps and anterior apical trunk epidermal sensory neurons (aATENs) (Hudson, 2016). Interestingly, it has been shown that cells in this anterior neural plate border co-express *Six1/2* and *Eya* and produce ciliated aATEN neurons, which can secrete GnRH (Abitua et al., 2015). Notably, inhibition of BMP signaling is required for the expression of *Six1/2* and the production of GnRH neurons. Because this non-CNS anterior neural plate border region appears to share a conserved developmental process with the anterior-most neurogenic placode (olfactory placode) in vertebrates, it has been suggested to be the proto-placodal ectoderm in ascidians (Abitua et al., 2015; Horie et al., 2018).

Row III and IV neural plate precursor cells are also derived from a-line cells and give rise to the anterior part of the larval brain (sensory vesicle); meanwhile, the more posterior row I and II cells are derived from the A-line and contribute to the lateral and ventral parts of the larval brain, as well as the trunk ganglion and tail nerve cord (Figure 3.3B–D). Although FGF signals are involved in cell fate determination for A-line cells, repression of FGF signaling (via Ephrin-Eph signaling) is required for A-line precursors to adopt a neural fate (Picco et al., 2007). Despite this difference, the lateral border regions of the a- and A-line neural

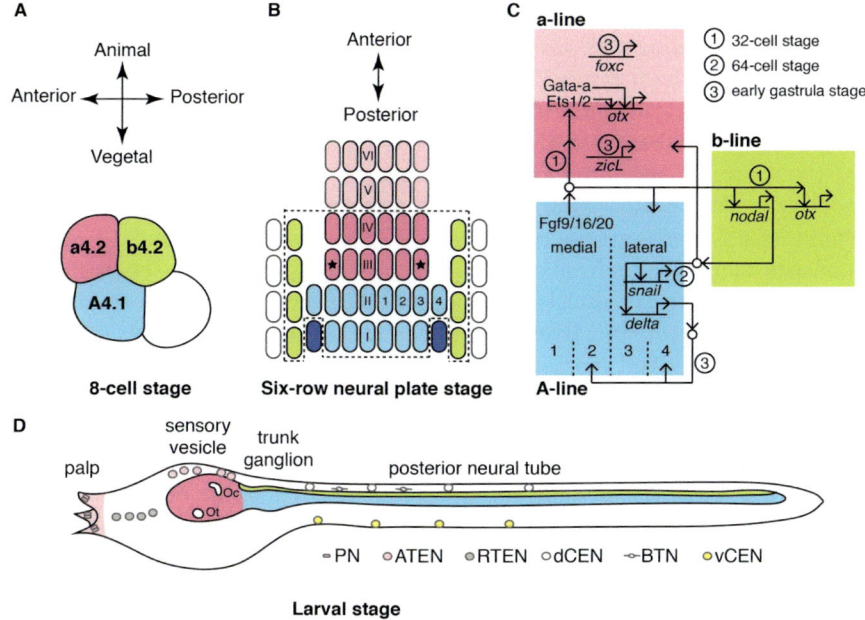

FIGURE 3.3   The nervous system of the ascidian tadpole larva. (A) One side of the bilaterally symmetric embryo at the eight-cell stage, showing cell lineages that form the nervous system. The axes of the embryo are indicated at the top of the illustration. The color code is as follows: pink, a-line lineage; blue, A-line lineage; green, b-line lineage. (B) The neural plate (contoured by a dashed line) and the surrounding cells at the six-row neural plate stage are shown with the anterior direction toward the top. The color code is the same as in (A), except for the light pink cells, which are also derived from the a-line lineage. The asterisks indicate precursors of the pigment cells located at the neural plate border. The dark blue cells are muscle cells from the A-line lineage. The six rows are named rows I to VI from posterior to anterior. The cells from medial to lateral are labelled 1–4. The light pink and white cells outside the neural plate (contoured with gray lines) are neural border cells that later develop into epidermal sensory neurons of the PNS. (C) CNS specification and neural plate patterning. The colors of boxes correspond to the cell lineages in (A) and (B). The stages at which neural induction and medial–lateral patterning of the neural plate occur are indicated. (D) The organizations of the ascidian CNS and PNS at the tadpole larval stage. The CNS (dorsal neural tube) can be divided into distinct regions including the sensory vesicle, the trunk ganglion, and the posterior neural tube. The color code corresponds to cell lineages in (A) and (B). The neural plate border cells of the a-line (asterisks in [B]) develop into the otolith (Ot) and the melanocyte of ocellus (Oc). The anterior neural border (light pink cells in [B]) develops into the palp, an adhesion organ that contains sensory palp neurons (PN), and apical trunk epidermal neurons (ATEN). Other neurons of the PNS that are derived from the neural borders include the rostral trunk epidermal neurons (RTEN), dorsal caudal epidermal neurons (dCEN), and bipolar tail neurons (BTN). Ventral caudal epidermal neurons (vCEN) arise from the ventral epidermis, far from the neural plate.

plate (columns 3 and 4) are specified by similar mechanisms, involving Nodal and Delta/Notch signaling pathways (Esposito et al., 2017; Hudson et al., 2007; Hudson and Yasuo, 2005; Imai et al., 2006). Nodal is required in the lateral domain of the neural plate to activate *Snail* (in both columns 3 and 4) and *Msxb* (in a-line column 3) expression, and these lateral genes repress expression of medial genes (*Gsx* and *Meis*) to segregate column 3 and 4 from column 1 and 2 precursors. Meanwhile, Nodal-dependent Delta/Notch signaling functions to specify column 4 fate over column 3 (in A-line), and the expression of Delta ligands in column 3 further distinguishes column 2 fate from column 1. Interestingly, it has been shown that a pigment cell lineage (a9.49, asterisks in Figure 3.3B) is derived from the lateral border of the a-line CNS, and this lineage exhibits several properties reminiscent to those of neural crest–derived pigment cells (Abitua et al., 2012). The *Ciona* a9.49 cell lineage expresses *ZicL, Msx-b, Pax3/7, Snail, Id*, and *MITF*. Eventually, a pair of the a9.49 progeny intercalate tandemly at the midline of the CNS, with the anterior cell differentiating into an otolith (a gravity receptor cell containing a melanin granule) and the posterior cell forming the melanocyte of the ocellus (a multicellular photosensory structure) (Nishida and Satoh, 1989). Experimental evidence shows that *Wnt7*, which is expressed in the dorsal midline of the neural tube just posterior to the ocellus, is responsible for this cell fate determination through its activation of *FoxD* expression in the presumptive ocellus precursor (Abitua et al., 2012); activated *FoxD* then inhibits *MITF* expression and specifies the ocellus fate. Based on the embryonic origin of this cell lineage in the ascidian neural plate border, the expression of conserved neural plate border genes, and its ability to give rise to pigmented cell types through conserved genetic mechanisms involving Wnt signaling and MITF/FoxD transcription factors, it has been proposed that the a9.49 cell lineage of ascidian embryos may represent a rudimentary cell type corresponding to the vertebrate neural crest (Abitua et al., 2012). However, unlike the vertebrate neural crest cells, this lineage lacks the ability to undergo long-range migration. Intriguingly, ectopic expression of *Twist* (of which the vertebrate homolog is an important transcription factor for mediating neural crest cell specification and epithelial-to-mesenchymal transition [Martik and Bronner, 2017]) in the *Ciona* a9.49 lineage triggers a mesenchymal phenotype, wherein the manipulated cells exhibit long-range migration and seem to gain the ability to produce mesoderm derivatives (Abitua et al., 2012). These studies suggest an evolutionary scenario in which the emergence of vertebrate neural crest cells might have been achieved through the cooption of one or more regulatory factors in certain evolutionary precursor cell types located at the neural plate border in the chordate ancestor.

3.2.2.2   Segregation of the B-Line Neural Lineages and Other Derivatives that Give Rise to Epidermal Sensory Neurons

The ascidian b-line CNS cells are located in the lateral region of the neural plate and mainly contribute to the dorsal part of the neural tube (Figure 3.3B and D). The b-line neural precursor cells are also induced by FGF signals at the 32-cell stage, and subsequently Nodal signaling is required to activate *Msxb, Snail, Pax3/7, Delta-like*, and *Chordin*, restricting cell fates to be the lateral neural borders (Figure 3.3C) (Hudson and Yasuo, 2005; Imai et al., 2006). In addition, the b-line neural precursor cells (the b6.5 cell pair) generate lateral non-CNS border cells. These non-CNS border cells also express the conserved neural border genes (including *Msxb, Dll-C, Snail,* and *Pax3/7*) and contribute to the dorsal midline epidermis (Imai et al., 2017; Roure and Darras, 2016; Wada et al., 1997), which produce the dorsal series of caudal epidermal sensory neurons (CENs) by a process involving Delta/Notch-mediated lateral inhibition (Pasini et al., 2006); the anterior bipolar tail neurons (BTNs) are also derived from this region (Stolfi et al., 2015). Interestingly, recent cell tracing experiments showed that the BTN precursors delaminate from the lateral borders of the neural plate and migrate anteriorly along the paraxial mesoderm on the lateral surface of the neural tube. Moreover, these BTNs express ion channels and display synaptic connections that are similar to those of vertebrate neural crest–derived dorsal root ganglia neurons (Stolfi et al., 2015). Thus, the ascidian BTNs may represent an intermediate cell population that possesses some neural crest–like characteristics.

In ascidians, patterning across the medial–lateral axis of the neural plate is achieved by distinct signaling molecules compared with that in vertebrates. Notably, ascidian Nodal signaling plays a primary role in setting up the border region of the neural plate (Hudson, 2016). The relative order of expression along this axis appears to be conserved for many transcription factor genes, such as lateral *Snail, Pax3/7* and *Msx*, intermediate *Gsx*, and ventral *FoxAa*. In contrast to vertebrates, ascidian BMP signaling does not play a major role in early neural induction and medial–lateral patterning of the neural plate, although BMP signaling does operate later in the lateral borders of ascidian CNS to specify particular cell fates, including the pigmented otolith (Darras and Nishida, 2001) and certain trunk epidermal sensory neurons (Ohtsuka et al., 2014). Importantly, the ascidian a-line and b-line neural border regions appear to be compartmentalized into distinct neural progenitor domains that generate different derivative cell types via distinct regulatory states. This arrangement is highly comparable to the vertebrate panplacodal ectoderm and neural crest (Abitua et al., 2015; Abitua et al., 2012; Horie et al., 2018; Stolfi et al., 2015). Interestingly, a recent study showed that misexpression of *FoxC*, a key determinant of the a-line nonCNS anterior neural plate border, in b-line neural border cells caused a change in the regulatory state during development, transforming the *Ciona* BTNs into palp sensory cells (Horie et al., 2018). This study demonstrated that the anterior and posterior neural plate border regions in ascidians possess common developmental capabilities and are able to generate sensory cell types that are highly comparable to vertebrate placode– and neural crest–derived cell types. It raises an evolutionary scenario that in the common ancestor of tunicates and vertebrates, the CNS border region may have been the source of several distinct but related PNS cell types, and this border region was probably already compartmentalized as distinct domains via gene regulatory interactions similar to those that exist in vertebrate neural crest and cranial placodes (Horie et al., 2018).

In addition to those PNS neurons originating from the CNS border regions, ascidians possess another population of epidermal sensory neurons, the ventral CENs (vCEN, Figure 3.3D), which arise far from the neural plate in the ventral midline epidermis during embryogenesis (Pasini et al., 2006). In contrast to the aforementioned specification process of dorsal CENs, which require FGF and Nodal signaling cascades, BMP signaling (mediated by ADMP and BMP2/4 ligands) is required in the ventral ectoderm to specify the vCEN progenitor field (Pasini et al., 2006; Waki et al., 2015). Subsequently, the conserved transcription factor, *Msxb*, is activated in the ventral midline to regulate downstream genes (including *Klf1/2/4, Achaete-scute a-like2, Cagf9/tox, Pou4* and *Delta.b*) for the differentiation of vCENs (Pasini et al., 2006; Roure and Darras, 2016; Tang et al., 2013; Waki et al., 2015). It is noteworthy that the patterning mechanism of vCENs is similar to mechanisms that have been described for the formation of epidermal sensory neurons in cephalochordates (Lu et al., 2012) and the peripheral sensory cells in certain protostomes (Denes et al., 2007; Rusten et al., 2002). Therefore, vCENs may represent an ancestral PNS population that predates the origin of chordates. We will further discuss this idea in the following sections.

3.2.3 ESTABLISHMENT OF THE NEURAL BORDERS IN AMPHIOXUS

3.2.3.1 The Amphioxus Nervous System

The CNS of amphioxus is highly similar to those found in vertebrates, except that the amphioxus CNS is much simpler and has fewer visible anatomical landmarks (Wicht and Lacalli, 2005). Furthermore, the molecular mechanisms that pattern the overall AP and DV axes of the CNS are largely conserved between amphioxus and vertebrates (Albuixech-Crespo et al., 2017; Kozmikova and Yu, 2017; Onai et al., 2010; Schubert et al., 2006; Yu et al., 2008; Zieger et al., 2018a), although the functions of certain secondary organizers of the vertebrate brain may have been modified or appear to be absent in the amphioxus lineage (Albuixech-Crespo et al., 2017; Holland et al., 2013; Pani et al., 2012). Using mid-neurula-stage amphioxus embryo as a model, detailed mapping for the expression patterns of many genes with known functions in vertebrate CNS patterning revealed a complex molecular compartmentalization of the amphioxus developing CNS (Albuixech-Crespo et al., 2017). The results of this molecular comparison suggest that the ancestral chordate CNS may derive from three major primodia along the AP axis: the rostral Hypothalamo-Prethalamic, the middle Di-Mesencephalic, and the caudal Rhombencephalo-Spinal primodium; and along the DV axis distinct floor, basal and alar plates were already exist (Albuixech-Crespo et al., 2017). Outside the CNS, the amphioxus PNS comprises sensory neurons and plexuses that are derived peripherally during embryonic and larval stages, which raises questions regarding to their evolutionary relationships to the vertebrate neural crest and placode-derived PNS cell types. Recent molecular studies in amphioxus have provided many important insights into these questions.

3.2.3.2 Neural Induction and Neural Border Patterning in Amphioxus Embryos

In contrast to tunicates, amphioxus has a dorsal organizer similar to that of vertebrates (Kozmikova and Yu, 2017; Le Petillon et al., 2017; Yu et al., 2007). The dorsal blastopore lip of the amphioxus gastrula expresses homologs of key signaling molecules (such as *Chordin, ADMP, Nodal, Lefty, Vg1* and *FGF8/17/18*) and transcription factors (including *Goosecoid* and *Lim1/5*) that are commonly found in Spemann's organizer of vertebrates (Bertrand et al., 2011; Kozmikova and Yu,

2017; Langeland et al., 2006; Neidert et al., 2000; Onai et al., 2010; Yu et al., 2007). The amphioxus neural plate is formed from the dorsal ectoderm during gastrulation, and the induction signals from the organizer mediate this process (Le Petillon et al., 2017; Yu et al., 2007). The BMP antagonist Chordin is one of the main inhibitors of BMP pathway functioning at the dorsal–medial ectoderm to form the amphioxus neural plate. Functional experiments demonstrated that elevation of BMP signaling before the blastula stage completely abolished amphioxus CNS induction and causes an overall ventralization/posteriolization of the body plan, suggesting that inhibition of BMP signaling in the dorsal ectoderm is necessary for amphioxus CNS induction (Kozmikova et al., 2013; Yu et al., 2007). On the other hand, elevating Nodal signaling before the blastula stage promotes anterior/dorsal development and upregulates expression of anterior CNS markers in amphioxus embryos (Onai et al., 2010). Interestingly, in addition to recent experiments demonstrating that Nodal signaling plays a major role in amphioxus neural induction, there is evidence that Nodal signaling also functions in vertebrate neural induction, which might have been overlooked in previous studies (Le Petillon et al., 2017). In contrast to vertebrates and ascidians, inhibition of FGF signaling during amphioxus embryogenesis does not block CNS formation, suggesting that FGF signaling may not play a major role in early neural induction of amphioxus (Bertrand et al., 2011).

During neural plate formation, amphioxus *SoxB1a* (previously named *Sox1/2/3*) begins to be expressed sometime around the early-to-mid gastrula stage and marks the entire future neural plate (Holland et al., 2000; Yu et al., 2008); at the same time, *Dlx*, which is originally expressed throughout the entire ectodermal domain from the early gastrula stage, is downregulated in the future neural plate, with strong *Dlx* expression marking the border region of the nonneural ectoderm (Holland et al., 1996; Yu et al., 2008). Outside the future neural plate, the entire amphioxus epidermal ectoderm expresses *Tfap2* (Meulemans and Bronner-Fraser, 2002); the ascidian homolog of this gene is also expressed in the nonneural ectoderm and specifies epidermal fate (Imai et al., 2017). Hence, that the ancestral function of *Tfap2* in chordates might have been to specify the formation of non-neural epidermis.

The developing amphioxus CNS has clear border regions that show distinct regulatory states. At the late gastrula stage, the amphioxus neural plate becomes flattened, and a series of transcription factor genes are expressed across the border region between the neural and non-neural ectoderm (Figure 3.4A, B). As previously mentioned, *Tfap2* and *Dlx* are expressed in the entire non-neural ectoderm, and strong *Dlx* expression marks the boundary of the non-neural ectoderm abutting the neural plate (Holland et al., 1996; Yu et al., 2008). *Msx* is also expressed in non-neural ectoderm (except the most anterior and posterior ectodermal domains), as well as in the lateral borders of the developing neural plate (Sharman et al., 1999; Yu et al., 2008). *Zic* and *Pax3/7* are expressed in the lateral borders of the amphioxus neural plate (Gostling and Shimeld, 2003; Holland et al., 1999; Yu et al., 2008); this part of the neural plate will eventually contribute to the dorsal and lateral cells of the amphioxus neural tube and continues to express *Msx*, *Zic*, and *Pax3/7* (Gostling and Shimeld, 2003; Holland et al., 1999; Sharman et al., 1999). The *Snail* homolog in amphioxus is expressed transiently in the lateral borders of the neural plate during later gastrula and early neurula stages; however, its expression is not confined to this border region and is extended to the entire CNS during subsequent developmental stages (Langeland et al., 1998; Yu et al., 2008). Thus, the graded expression patterns of these transcription factor genes (especially according to the medial boundaries of their expression) clearly divide the amphioxus dorsal ectoderm into distinct *epidermal* (expressing *Tfap2* and *Dlx*) and *neural* (expressing *Msx*, *Zic*, *Pax3/7* and *Snail*) boundary domains (Figure 3.4A, B).

It has been demonstrated that, similar to vertebrate embryos, BMP signaling is involved in setting up the amphioxus CNS boundary and the proper mediolateral expression domains of transcription factor genes across the dorsal ectoderm. Elevating BMP signaling level during the blastula to gastrula stages results in dose-dependent reduction of the neural plate and expansion of the epidermal ectoderm (Yu et al., 2008), and inhibiting BMP signaling causes the opposite effect (Kozmikova et al., 2013). It is interesting to note that when BMP signaling is elevated, *Dlx* and *Msx* become upregulated in the dorsal ectoderm, and their expression domains converge

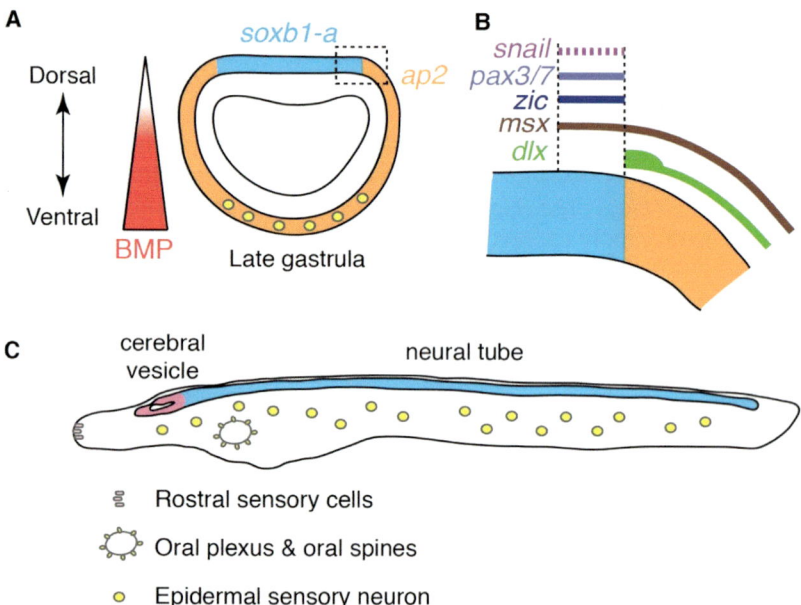

FIGURE 3.4   The amphioxus nervous system. (A) Cross-section of an amphioxus embryo at the late gastrula stage. BMP activity is high on the ventral and low on the dorsal side. The ectodermal expression of *Soxb1-a* marks the neural plate, while *ap2* is expressed in the epidermal ectoderm. The yellow circles in the ventral ectoderm represent the precursors of the epidermal sensory cells. The dashed square demarcates the neural border region, which is magnified in (B). (B) Expression patterns of neural border genes are color coded. The black dashed lines encompass the neural border region. The *snail* gene is transiently expressed in the neural border region (shown as purple dashed line). (C) The organizations of the amphioxus CNS and PNS at the larval stage. The CNS includes the cerebral vesicle (pink) and the posterior neural tube. The epidermal sensory neurons, which originate from the ventral side, together with the neuron surrounding the mouth (oral plexus) and the rostral sensory neurons, constitute the PNS.

to the dorsal midline in the treated amphioxus embryos; however under the same conditions, *Zic* and *Pax3/7* are downregulated or cease to be expressed in the dorsal ectoderm (Yu et al., 2008), suggesting that these two groups of transcription factor genes are likely regulated by different BMP signaling effectors. Meanwhile, amphioxus *FGF9/16/20* and *FGF8/17/18* are expressed in the neural plate and in the underlying axial mesendoderm, respectively (Bertrand et al., 2011; Yu et al., 2008); Wnt7b is expressed in the developing CNS with preferential expression in the lateral parts of the neural plate during the early neurula stage (Schubert et al., 2000a), and *Wnt8* is expressed in the dorsal paraxial mesoderm just below the lateral neural plate border (Schubert et al., 2000b; Yu et al., 2008). These expression patterns suggest that FGF and Wnt signaling pathways may also play roles in specifying the lateral boundaries of the amphioxus neural plate; however, there is not yet any experimental data available to support this speculation, and further functional studies are required to test this hypothesis.

3.2.3.3   Amphioxus Homologues of the Neural Crest-Specifier Genes Are Not Expressed at the Neural Border Regions during Embryogenesis

Unlike the lateral neural plate border in tunicates, so far there is no evidence that migratory neural crest–like cells exist in amphioxus. In addition, amphioxus homologs of most neural crest-specifier genes (Martik and Bronner, 2017; Meulemans and Bronner-Fraser, 2004) are not expressed in the lateral border of the neural plate, nor are they expressed in the dorsal nerve cord after neurulation. Interestingly, many of these genes, including *Twist*, *Id*, *FoxD*, and *SoxE*, are expressed mostly in mesodermal structures (somites and notochord) (Meulemans et al., 2003; Yasui et al., 1998; Yu et al., 2002; Yu et al., 2008). These observations led to the hypothesis that cooption of this group of the genes into the neural plate border cells may have conferred new cellular properties (such as pluripotency, delamination, and migration), enabling the evolution of neural crest cells (Meulemans and Bronner-Fraser, 2005; Yu, 2010). In support of this idea, recent comparative genomic analyses and cross-species experiments showed that after two rounds of whole-genome duplication in the early vertebrate lineage (Holland et al., 2008; Putnam et al., 2008), certain paralogs of the *FoxD*, *Tfap2*, and *SoxE* genes gained additional *cis*-regulatory elements and novel regulatory interactions that would facilitate the cooption of these genes into the neural plate border (Jandzik et al., 2015; Van Otterloo et al., 2012; Yu et al., 2008).

Besides *cis*-regulatory evolution, changes in protein-coding sequence within particular paralogs of the neural crest-specifier genes may have also contributed to their functional evolution. For instance, it has been shown that FoxD3, a vertebrate paralog of the FoxD subfamily, evolved a unique N-terminal domain compared to other FoxD members, and this domain was found to play an essential role in regulating the migration phenotype of vertebrate neural crest cells in cross-species experiments (Ono et al., 2014). Similarly, vertebrate SoxE paralogs also have divergent functions with regard to their abilities to regulate downstream target genes for specifying certain neural crest derivatives (Lee et al., 2016; Tai et al., 2016). Thus, the acquisition of new functions by neural crest-specifier genes in the vertebrate lineage (or even in the common ancestor of tunicates and vertebrates) probably involved changes in both *cis*-regulatory elements and protein-coding sequences. The significant increase in gene numbers caused by the two rounds of whole-genome duplication may have greatly facilitated the cooption of these genes into the lateral neural plate borders in the vertebrates, and in turn, this cooption may have led to the diversification of various neural crest-derivatives.

3.2.3.4   Amphioxus CNS-Derived Pigment Cells

Similar to tunicates and vertebrates, amphioxus also possess CNS-derived pigment cells, including the pigmented cup cell of the dorsal ocelli (organs of Hesse) and the pigment cells of the frontal eye (Pergner and Kozmik, 2017; Vopalensky et al., 2012; Wicht and Lacalli, 2005). However, unlike vertebrate neural crest–derived pigment cells, these amphioxus pigment cells are sedentary inside the neural tube (Wicht and Lacalli, 2005). Together with the lamellar body and Joseph cells, these pigment cells comprise the four distinctive types of photoreceptive organs found in the amphioxus larval and adult CNS (Lacalli, 2004; Pergner and Kozmik, 2017). The amphioxus frontal eye has long been considered to be homologous to the vertebrate paired eyes, based on its anatomical characteristics (Lacalli, 1996; Lacalli, 2004; Lacalli et al., 1994; Wicht and Lacalli, 2005), and recent molecular studies further support this idea (Pergner and Kozmik, 2017; Vopalensky et al., 2012). On the other hand, the evolutionary affiliation of amphioxus dorsal ocelli is still an open question. The first dorsal ocelli forms in the ventral position of the CNS at the level between somite 4 and 5 during the neurula stage, and the expression of genes involved in a conserved cascade for black-pigment melanin synthesis (including the transcription factor gene *Mitf* and melanin synthesizing enzyme genes *Tyrosinase*, *Tyrp-a*, and *Tyrp-b*) can be detected in its anlage prior to the production of actual pigment (Yu et al., 2008). Additional dorsal ocelli are formed in more posterior regions along the entire neural tube during larval development; eventually the number of ocelli can reach hundreds or even more than 1,000 in a mature amphioxus (Lacalli, 2004; Pergner and Kozmik, 2017). It has been proposed that the pigmented dorsal ocelli may represent the evolutionary precursor of vertebrate neural crest cells (Ivashkin and Adameyko, 2013). However, dorsal ocelli are situated near the ventral–lateral region of the nerve cord, and there is no evidence that these cells arise from the embryonic neural plate border region (Lacalli, 2004; Pergner and Kozmik, 2017; Wicht and Lacalli, 2005). This difference in embryonic origin makes it less likely that amphioxus ocelli are related to vertebrae neural crest cells. Despite the uncertainty about whether the relationship between amphioxus dorsal ocelli and neural crest–derived pigment cells is homologous, the widespread distribution of pigmented dorsal ocelli along the amphioxus neural tube seems to suggest that its CNS has the necessary differentiation genetic battery for pigment cell development. Cooption and/or restriction of this genetic battery to the CNS border cells in the vertebrate lineage might therefore have been a key step in the origin of neural crest–derived pigment cells.

3.2.3.5   Peripheral Sensory Neurons in Amphioxus

To date, there is no clear evidence to suggest that the epidermal border regions of the amphioxus dorsal ectoderm might be equivalent to the vertebrate preplacodal ectoderm. During amphioxus neurulation, the lateral border regions of the epidermal ectoderm detach from the neural plate and migrate medially, eventually fusing together at the dorsal midline to cover the forming CNS; however, there is no local thickening or further differentiation into neuronal cell types that can be observed in this dorsal epidermal ectoderm. While border markers, such as *Tfap2*, *Dlx*, and *Msx*, are expressed in the dorsal epidermal ectoderm, genes encoding homologs of the vertebrate preplacodal ectoderm markers (including *Six1/2*, *Six4/5* and their cofactor *Eya*) are not specifically expressed in this region (Kozmik et al., 2007); instead, some of these genes are expressed transiently in scattered epidermal sensory neurons (ESNs) during early larval stages (Figure 3.4C). In addition, some specific types of sensory neurons likely develop *in situ* at several positions in the early larva, such as those near the rostral tip and around the mouth (Figure 3.4C). These sensory neurons appear to have distinct neurotransmitter signatures (Candiani et al., 2012; Lacalli, 2004; Zieger et al., 2018a), but the exact stimuli to which these cells respond have not been identified. Unlike vertebrate placode-derived neurons, which are usually formed as clusters, these amphioxus ESNs are mostly solitary neurons (Lacalli, 2004). If these amphioxus ESNs are related to vertebrate placode-derived neurons, the ancestral vertebrates must have evolved means to enlarge the size and cell numbers of these peripheral sensory units.

The expression of putative placode markers in amphioxus ESNs raises some speculation about what (if any) evolutionary relationship these neurons might have with vertebrate placode-derived neurons (Lacalli, 2004; Mazet et al., 2004). However, unlike placode-derived neurons, most amphioxus ESNs originate from the ventral epidermal ectoderm (Benito-Gutierrez et al., 2005a; Lu et al., 2012; Satoh et al., 2001), not the lateral dorsal ectoderm surrounding the neural plate. During the gastrula stage, a high level of BMP signaling activates *Tlx* expression in a restricted domain of the ventral ectoderm in the amphioxus embryo, and subsequently, Notch signaling ligand *Delta* and the basic helix-loop-helix proneural gene *Ash* are coexpressed in a salt-and pepper pattern within this domain to specify individual ESNs from the epidermal ectoderm via lateral inhibition process (Lu et al., 2012; Rasmussen et al., 2007). Starting from the early neurula stage, these ESNs express the postmitotic neuron marker *Hu/Elav*, and then delaminate from the ventral ectoderm to enter the subepithelial space and migrate dorsally underneath the epidermal ectoderm, eventually re-erupting into the epithelium at the flanks of the early larva (Benito-Gutierrez et al., 2005b; Kaltenbach et al., 2009). During ESN development, several subpopulations can be recognized by their expression of distinct gene combinations, including *Six1/2*, *Eya*, *Islet*, *Coe*, *Soxb1c* and several *Hox* genes (Kozmik et al., 2007; Mazet et al., 2004; Meulemans and Bronner-Fraser, 2007; Schubert et al., 2004; Zieger et al., 2018b). Moreover, retinoic acid signaling plays an important role in patterning the distribution and abundance of these ESNs along the AP axis (Schubert et al., 2004; Zieger et al., 2018a; Zieger et al., 2018b). After the larval stage, different types of ESNs produce specific neurotransmitters (Candiani et al., 2012; Zieger et al., 2018a; Zieger et al., 2018b) and thus may perform distinct sensory functions. According to the embryonic location of their progenitors (ventral ectoderm) and the conservation of the BMP pathway in specifying this progenitor field, amphioxus ESNs share several key features with the ascidian vCENs described in the previous section. Additionally, both the amphioxus and ascidian sensory neurons employed conserved basic helix-loop-helix proneural genes and Delta/Notch-mediated lateral inhibition to specify their fate. Thus it has been suggested that amphioxus ESNs and ascidian vCENs likely share a common ancestry, and they may represent an ancient series of PNS sensory neurons that existed prior to the evolution of neural crest cells and placodes (Lu et al., 2012; Waki et al., 2015). From these comparisons, it is tempting to speculate that vertebrate neural crest- or placode-derived neurons might have evolved via cooption of original PNS gene circuitry to the neural plate border in the common ancestor of ascidians and vertebrates (Olfactores); subsequently, ascidians would have retained both lineages, with vertebrates losing the ventral lineage.

3.3   NERVOUS SYSTEMS OF NON-CHORDATE DEUTEROSTOMES

3.3.1   Nervous Systems of Echinoderms

3.3.1.1   Organization of the Echinoderm Nervous System

Adult echinoderms are pentaradially symmetric and so is their nervous system. The adult echinoderm nervous system is composed of five radial nerves extending from a central nerve ring that surrounds the esophagus (Figure 3.2). Branches of the radial nerves form an interconnected network that coordinates the movements of their appendages, such as tube feet and, in the case of sea urchins, spines. Most studies on echinoderm neurogenesis have focused on sea urchin and sea star larval nervous systems (Figure 3.5A), which appear to be somewhat unrelated to the adult nervous systems. Most larval neurons are located near or within the ciliary band, which is a series of ciliated ectodermal cells surrounding the oral field (Burke et al., 2006a; Burke et al., 2014). At the anterior end of the ciliary band, clusters of neurons, including two bilaterally symmetric clusters of serotonergic neurons, constitute a ganglion called the *apical organ* that is present in all echinoderm classes (Byrne et al., 2007). Additionally, neurons in the pharyngeal region of the sea urchin larva are derived directly from the endoderm (Murabe et al., 2008; Wei et al., 2011). It has been suggested that the apical organ represents the CNS, while the ciliary band neurons constitute the PNS (Burke et al., 2014). However, unlike the chordate nervous system, the distinction between CNS and PNS is less pronounced in echinoderm larvae. Although the neuronal connectivity between the apical organ and the ciliary band has not been investigated in detail, it is known that the axons of the sea urchin neurons located within the apical organ and within or near the ciliary band form extensively connected axon tracts that travel beneath the basal surface of the ciliary band cells, encircle the oral field, and extend toward the apical organ. This larval nervous system coordinates the ciliary beat for larval motility as well as muscle contraction around the pharynx.

3.3.1.2   Ectodermal Patterning and Neurogenesis in Echinoderm Embryos

Comparisons between the echinoderm and chordate nervous system have been difficult to make due to extreme morphological disparities. Nevertheless, there are many similar molecular mechanisms that underlie ectodermal patterning and neurogenesis in chordate and echinoderm embryos (Hinman and Burke, 2018). In sea urchin and chordate embryos, neurogenesis is intimately related to the axial patterning of the ectoderm. The AP axis of both sea urchin and sea star embryos is established by posterior Wnt signals. Wnt antagonists expressed in the apical regions and a series of feedback loops control the size and boundary of the apical neurogenic ectoderm (Angerer et al., 2011; Jarvela et al., 2016; Range, 2014). Maintenance of the size of the sea urchin apical plate also relies on the modulation of BMP activity through Fez, a zinc finger protein (Yaguchi et al., 2011). Along the DV axis, two TGFβ signals on opposite sides (Nodal in the ventral region and BMP in dorsal) delineate the positions of the ciliary bands, which are marked by the expression of the *onecut* gene, in

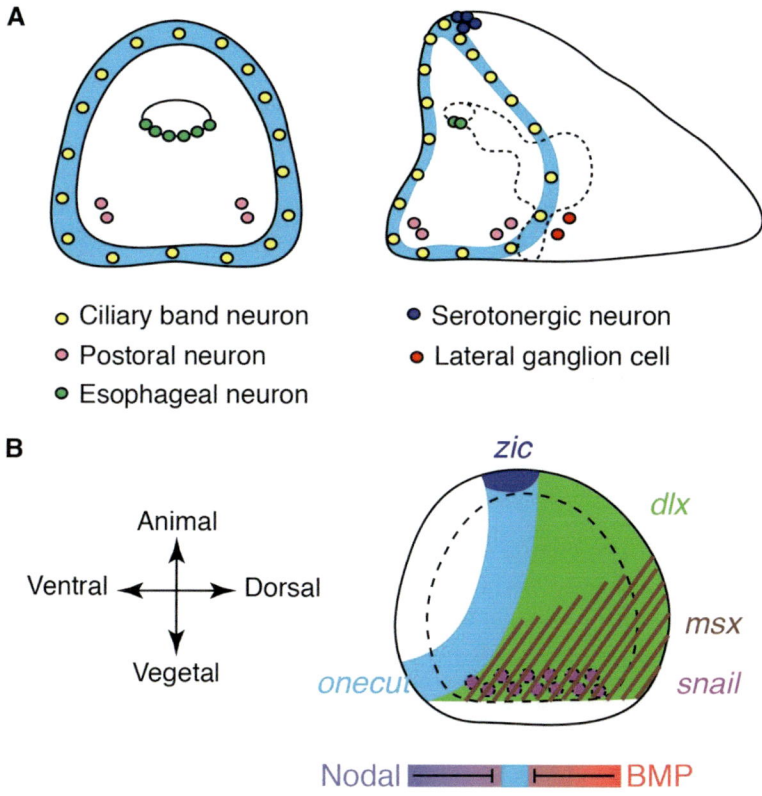

FIGURE 3.5 **Neurogenesis and the larval nervous system of sea urchins.** (A) The distribution of neurons in sea urchin larvae. The larva on the left is depicted from the ventral side, while the larva on the right is shown from a lateral view with ventral to the left. Different neuronal types are color coded as indicated. The tripartite gut is contoured with dashed lines. (B) Ectodermal patterning at the mesenchyme blastula stage. At this stage, the primary mesenchyme cells (purple circles), which expressed *snail*, are ingressed into the blastocoel. *Onecut* expression marks the position of the neurogenic ciliary band. *Dlx* and *msx* are both expressed in the dorsal ectoderm, adjacent to the *onecut* expression domain. The *msx* expression domain is more towards the vegetal side, while *dlx* expression covers the whole dorsal ectoderm. The *zic* gene is expressed in the animal neurogenic ectoderm. Positioning of the ciliary band between the ventral and dorsal ectoderm depends on two signals: Nodal on the ventral side and BMP on the dorsal side.

between the ventral and dorsal ectoderm (Figure 3.5B) (Molina et al., 2013; Yaguchi et al., 2010; Yankura et al., 2013).

Genes involved in sea urchin neurogenesis, such as *six3*, *soxC*, and *brn1/2/4*, are expressed in all neural progenitors. *Six3* is required for neurogenesis throughout the embryo (Wei et al., 2009) and acts upstream of *soxC* and *brn1/2/4*, which respectively mark the mitotic neural progenitors and postmitotic neural precursors (Garner et al., 2016; Wei et al., 2016). In addition, several transcription factor genes have been reported to be involved in the specification or differentiation of specific neuronal subtypes (Figure 3.5A). For example, the proneural bHLH factor *achaete-scute* is necessary and sufficient for the development of the serotonergic neurons in the apical plate (Slota and McClay, 2018), and a zinc finger homeobox transcription factor, zfhx1/z81, is also required for the differentiation of these serotonergic neurons (Yaguchi et al., 2012). *Neurogenin*, on the other hand, is necessary but not sufficient for the formation of the cholinergic neurons in the ciliary band (Slota and McClay, 2018). *Orthopedia* is not a proneural gene in the sea urchin embryo, although it is required for the differentiation of the postoral neurons that are born bilaterally in the ventral ectoderm near the ciliary band (Slota and McClay, 2018). Essentially, every step of neurogenesis in sea urchin embryos relies on Delta-Notch signaling (Garner et al., 2016; Slota and McClay, 2018; Wei et al., 2016). In the sea star embryo, a similar set of regulatory factors is involved in neurogenesis. Delta-Notch signaling enables the scattered expression pattern of *soxC* throughout the ectoderm (Yankura et al., 2013), and anterior subsets of the *soxC*-positive multipotent progenitors express *lhx2.9*; later the *lhx2.9*-positive cells in the apical pole domain differentiate into the serotonergic neurons (Jarvela et al., 2016). *SoxC* is also required for the development of the ciliary band neurons in the sea star embryo (Yankura et al., 2013).

3.3.1.3   Is There a "Neural Border" in Echinoderm Embryos?

In order to consider whether neural borders were present before the emergence of the chordate lineage, it is important to first understand how the neurogenic and epidermal ectoderm are segregated and how the orthologs of the chordate neural border genes are deployed in echinoderm embryos. The anterior neuroectoderm of the sea urchin embryo is initially marked by the expression of *six3* and *foxq2* (Wei et al., 2009; Yaguchi et al., 2008). Three different Wnt signaling pathways, Wnt/ß-catenin, Wnt/JNK, and Wnt/PKC, are involved in positioning the anterior neural ectoderm (Range et al., 2013). The anterior neurogenic territory is then resolved into two concentric domains by interactions between *foxq2*, *six3*, and Wnt modulators *sfrp1/5* and *dkk3* (Range and Wei, 2016). The sea urchin regulatory circuitries that are necessary to define the borders of the anterior neuroectoderm are similar to those of chordate embryos, suggesting that these mechanisms existed in the last common ancestor of deuterostomes (Range, 2014; Range and Wei, 2016). The second sea urchin neurogenic territory is the ciliary band, near or within which ciliary band neurons differentiate throughout the larval stage. This territory is marked by the expression of the *onecut* gene before gastrulation (Otim et al., 2004; Poustka et al., 2004), and the boundaries between the ciliary band and neighboring ectodermal cells are controlled by four regulatory circuits comprising about ten transcriptional repressors (Barsi et al., 2015). One of these repressors is *msx*, an ortholog of the chordate neural border marker gene. Sea urchin *msx* is normally expressed in the vegetal dorsal ectoderm, adjacent to the ciliary band territory (Figure 3.5B) (Chen et al., 2011). When *msx* is knocked down, the expression of the animal lateral ciliary band marker, *tlx*, expands to the dorsal ectoderm (Barsi et al., 2015), suggesting that *msx* represses *tlx* expression in the dorsal ectoderm to restrict *tlx* expression to the ciliary band. Interestingly, the amphioxus *tlx* gene is expressed specifically in a ventral epidermal domain corresponding to the putative progenitor field for epidermal sensory cells (Lu et al., 2012). The expression domain of amphioxus *tlx* is expanded when BMP activity is elevated (Lu et al., 2012). In sea urchin embryos, BMP signaling positively regulates *msx* (Saudemont et al., 2010; Su et al., 2009), which in turn represses *tlx* expression. Therefore, while the sea urchin *tlx* gene is indirectly repressed by BMP signaling via Msx, the amphioxus *tlx* gene is activated by BMP signaling. The amphioxus *msx* gene is expressed throughout the ventral ectoderm (Yu et al., 2008), and it is therefore possible that BMP signaling activates the *tlx* gene via *msx*. Further studies deciphering the *cis*-regulatory control of these genes will be necessary to identify possible molecular changes that led to opposite transcriptional controls of the *tlx* genes by BMP signaling and/or the *msx* genes.

Among the orthologs of other chordate genes that delineate the neural borders, the sea urchin *dlx* is expressed in a broad region of the dorsal ectoderm (Figure 3.5B) (Chen et al., 2011), although its function in shaping the ciliary band has not been examined. The sea urchin and sea star *zic* genes are expressed in the anterior neural ectoderm, where the apical organ forms (Materna et al., 2006; Yankura et al., 2010). Interestingly, the sea urchin *snail* gene seems to have acquired a novel function during evolution, as it is expressed in the primary mesenchyme cells, an echinoid-specific cell type that participates in skeletogenesis (Wu and McClay, 2007). Moreover, a *bona fide* ortholog of *pax3/7* is not identified in the sea urchin genome (Howard-Ashby et al., 2006). Based on the expression patterns of these genes, echinoderm embryos do not have a distinct neural border region. Indeed, the ciliary band closely abuts the ventral and dorsal ectoderm, and no neural border region with a distinct regulatory state has been described. Nevertheless, at least two types of neurons, the postoral and the lateral ganglion neurons (Figure 3.5A), develop in regions within the respective ventral and dorsal ectoderm adjacent to the ciliary band (Burke et al., 2014). Notably, the neurogenic region in which lateral ganglion neurons are formed is situated between the neurogenic ciliary band and the epithelial dorsal ectoderm, and it may be considered to be a potential neural border, comparable to that of chordates. It is still unclear how this neurogenic region is established, and more studies are needed to identify genes involved in specifying this region.

3.3.2   NERVOUS SYSTEMS OF HEMICHORDATES

3.3.2.1   Organization of the Hemichordate Nervous System

Hemichordates are divided into two monophyletic groups, the solitary enteropneusts (acorn worms) and the colonial tube-dwelling pterobranchs (Cannon et al., 2014; Rottinger and Lowe, 2012). Due to the limited availability of pterobranch embryos, most developmental and molecular studies have been conducted on enteropneusts. Adult enteropneusts display a tripartite body organization with an anterior proboscis, a middle collar region, and a posterior trunk region. The nervous system of these animals is comprised of two nerve cords, one dorsal and one ventral, along with an elaborated epidermal nerve net that is concentrated anteriorly (Figure 3.2). The two nerve cords are composed of dense agglomerations of neurons and are considered to be the CNS, while elements of the PNS are found in the epidermis (Nomaksteinsky et al., 2009). The nerve cords are superficial in the trunk, but in the collar region, the dorsal nerve cord is internalized to form a hollow neural tube called the *collar cord*. It has been shown that the collar cord develops from the dorsal neural plate in a process very similar to neurulation in chordates (Kaul and Stach, 2010; Luttrell et al., 2012). In chordates, hedgehog signals from the notochord induce the floor plate, the medial region of the neural plate. Intriguingly, it has been shown that during metamorphosis, *hedgehog* is expressed in the dorsal endoderm of the buccal tube that is located ventral to the forming collar cord (Miyamoto and Wada, 2013). However, although the expression of neural patterning genes of the collar cord is highly conserved among different enteropneust species, the deployment of these genes is dissimilar to the chordate neural tube (Kaul-Strehlow et al., 2017). Therefore, despite the apparent similarities in ontogeny and the potential utilization of hedgehog as an induction signal, the distinct mechanisms for neural patterning do not support the idea that a homologous relationship exists between the enteropneust collar cord and the chordate neural tube.

The most common enteropneusts used for developmental studies are *Saccoglossus kowalevskii* and *Ptychodera flava* (Rottinger and Lowe, 2012; Simakov et al., 2015). *S. kowalevskii* is a direct developer, exhibiting an adult body plan within a few days after fertilization, while *P. flava* develops indirectly into a tornaria larva that remains in the water column for several months before metamorphosing into a benthic juvenile. Neurogenesis in *S. kowalevskii* begins early during embryogenesis (Cunningham and Casey, 2014). Maternal transcripts of neural progenitor markers are distributed broadly, and the expression of panneural markers begins in the late blastula to early gastrula stages, with concentrated expression of neural markers along the ventral and dorsal midlines observed at the end of gastrulation. Serotonergic neurons are distributed in the proboscis of *S. kowalevskii* embryos but not in the apical region, where these neurons can be found in the echinoderm larvae and tornaria larva of the indirect-developing enteropneusts (Cunningham and Casey, 2014; Kaul-Strehlow et al., 2015). Organization of the nervous system in the tornaria larva is very similar to that in echinoderm larvae (Figure 3.2). Serotonergic neurons are concentrated in the apical organ of the tornaria larva (Nakajima et al., 2004; Nielsen and Hay-Schmidt, 2007; Tagawa et al., 2001), and non-serotonergic neurons are distributed along the ciliary bands and around the esophagus, as revealed by staining with a monoclonal antibody against sea urchin synaptotagmin B (a membrane Ca^{2+} sensor found in pre-synaptic axon terminals) (Burke et al., 2006b; Nakajima et al., 2004). During metamorphosis, the larval nervous system of *Balanoglossus simodensis*, another indirect-developing enteropneust, is not incorporated into the adult nervous system (Miyamoto and Saito, 2010). Despite the distinct organizations of the adult nervous systems, the similarities between the echinoderm and hemichordate larval nervous systems strengthen the idea that indirect development is ancestral in Ambulacraria (Nielsen, 1999). Nevertheless, most current evolutionary theories focus on comparisons between the hemichordate adult nervous system and the chordate CNS. Although it remains controversial whether the dorsal or ventral nerve cord is homologous to the dorsal nerve cord of chordates or, alternatively, if the deuterostome ancestor had a nerve net (Holland, 2015; Holland et al., 2015; Lowe et al., 2015), examining how the hemichordate ectoderm is patterned into neural and non-neural ectoderm may provide insight into the origin of the chordate neural border.

3.3.2.2   Ectodermal Patterning in Enteropneusts

Studies in *S. kowalevskii* detailing the expression of genes orthologous to those involved in patterning the chordate CNS have revealed impressive conservation between this species and the chordates. These genes are expressed in the ectoderm in an AP arrangement similar to that found in the chordate CNS (Lowe et al., 2003). However, the expression patterns of most of these genes are circular epidermal domains without clear DV polarity. Similar to other bilaterians, the AP axis is patterned by canonical Wnt signaling, which represses anterior fates and activates mid-axial ectodermal fates (Darras et al., 2018). Intriguingly, genetic programs that are homologous to three vertebrate neuroectodermal signaling centers—the anterior neural ridge (ANR), zona limitans intrathalamica (ZLI), and isthmic organizer (IsO)—are present in *S. kowalevskii* (Pani et al., 2012). The signaling genes marking the three signaling centers, *fgf8/17/18* in ANR, *hh* in ZLI, and *wnt1* in IsO, are expressed in the ectoderm without DV polarity. DV patterning in *S. kowalevskii* is similar to echinoderms, chordates, and a variety of other bilaterians in that it is mediated by BMP signaling. The BMP activity is high on the dorsal side and elevation or depletion of BMP causes dorsalization or ventralization of the embryo, respectively (Lowe et al., 2006). Notably, BMP signaling does not repress neural gene expression in *S. kowalevskii*, which is dissimilar to the chordate CNS and sea urchin larval nervous system development (Lowe et al., 2006). These results leave uncertainty as to whether the hemichordate adult diffuse nerve net, dorsal nerve cord, or ventral nerve cord could be homologous to the chordate neural tube.

Ectodermal patterning of indirect-developing acorn worms has been explored to a lesser extent. It was shown that the tornaria larva of *Schizocardium californicum* is transcriptionally similar to the anterior region of the direct developing *S. kowalevskii* (Gonzalez et al., 2017). Unlike the direct developer, several ectodermal genes display differential expression along the DV axis of the indirect developers. In *P. flava*, another indirect developer, the DV patterning mechanism is similar to that of sea urchin embryos (Rottinger and Martindale, 2011), and the high BMP on the dorsal side controls ectodermal gene expressions along the DV axis (Chang et al., 2016; Rottinger et al., 2015). Additionally, similar to the sea urchin and sea star larvae (Otim et al., 2004; Poustka et al., 2004; Yankura et al., 2013), the *P. flava* DV ectodermal patterning mechanism positions *onecut* expression in the ciliary bands, in between the dorsal and ventral ectoderm (Su et al., 2019). Moreover, neurogenesis of the *P. flava* larval nervous system is repressed by BMP signaling (Su et al., 2019), a regulatory feature that is comparable to the chordate CNS and the sea urchin larval nervous system, supporting the theory that indirect development with a larval nervous system is ancestral in deuterostomes.

3.3.2.3 Is There a "Neural Border" in Acorn Worms?

Given that the homology between the hemichordate and chordate nervous systems is still ambiguous, it is difficult, if not impossible, to discuss whether a neural border is present in acorn worms. Nevertheless, several genes that are orthologous to those that mark the chordate neural border have been examined in both direct and indirect-developing enteropneusts (Figure 3.6). Comparing the expression patterns of these orthologous genes among hemichordates, echinoderms, and chordates may provide some hints about how the evolution of the chordate neural plate border occurred. At the gastrula stage of *S. kowalevskii*, *dlx* is expressed in two patches of the dorsal ectoderm, and the space between the patches gives rise to the *telotroch*, the posterior ciliary band (Figure 3.6A) (Lowe et al., 2006). *Dlx* transcripts are also distributed in spots in the entire proboscis and in a single line in the ventral ectoderm (Figure 3.6B) (Lowe et al., 2003). The dorsal ectodermal expression of *dlx* is activated by BMP signaling, while the anterior spotted expression is BMP-independent (Lowe et al., 2006). Expression of *msx* partially overlaps with that of *dlx*, although *msx* transcripts are found in a circular pattern at the gastrula stage (Lowe et al., 2006) (Figure. 3.6A).

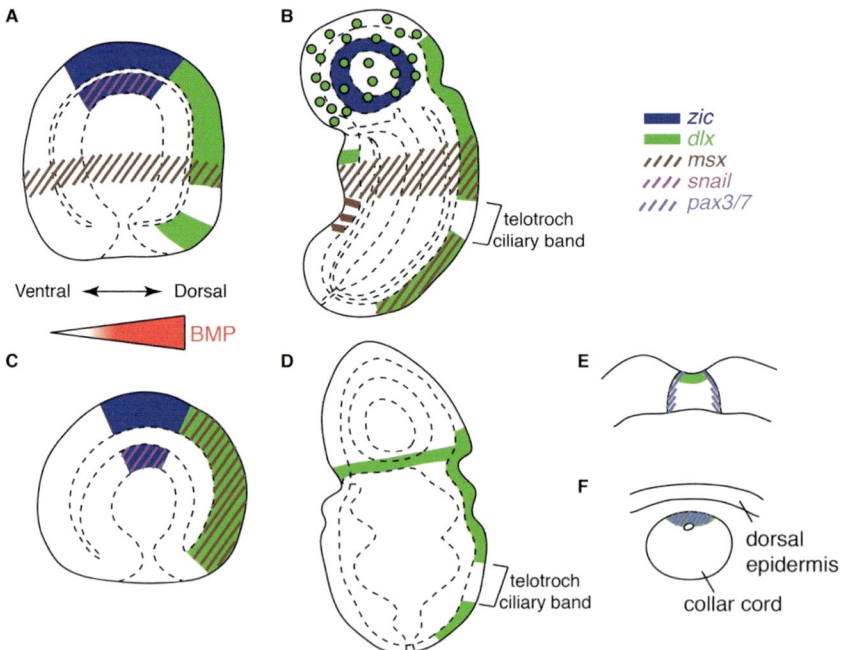

FIGURE 3.6 **Ectodermal patterning in enteropneusts.** Expression chordate neural border gene homologs, including *zic*, *dlx*, *msx*, *snail*, and *pax3/7*, are shown in the gastrula (A) and late neurula (B) stages of *S. kowalevskii*. Expression patterns of these genes are shown in indirect-developing enteropneusts at the gastrula stage (C) and in metamorphosing larva (D). Cross-sections show the neural plate and collar cord before (E) and after (F) neurulation, respectively. Summaries of indirect-developer data were compiled from several species, including *P. flava*, *Balanoglossus simodensis*, *B. misakiensis*, and *Schizocardium californicum*. Expression patterns are color coded as indicated.

At the late gastrula stage, *msx* expression displays a clear dorsal–ventral polarity, with a dorsal ectodermal patch overlapping with the *dlx* expression domain posterior to the telotroch (Fig. 6B). Similar to the sea urchin embryo, *zic* is expressed in the apical ectoderm (Green et al., 2013). Additionally, *zic* transcripts are detected in the presumptive mesoderm, where *snail* is also expressed (Green et al., 2013). The *pax3/7* expression pattern has not been described in *S. kowalevskii*.

Expression of these chordate border genes have also been studied to some extent in several species of the indirect-developing enteropneusts. Generally, *zic* and *snail* expression patterns are similar to that of *S. kowalevskii* (Figure 3.6C) (Fan et al., 2018; Gonzalez et al., 2017), but unlike in *S. kowalevskii*, *msx* shows clear DV polarity in the indirect-developers with expression in the dorsal ectoderm at the gastrula stage (Rottinger et al., 2015). *Dlx* is also expressed in the dorsal ectoderm at the gastrula stage (Gonzalez et al., 2017; Harada et al., 2001; Rottinger and Martindale, 2011). During metamorphosis, *dlx* transcripts are detected in the dorsal midline with an interruption at the telotroch, in addition to being detected in a ring at the posterior end of the proboscis (Figure 3.6D) (Kaul-Strehlow et al., 2017). During the formation of the collar cord, the expression of *dlx* in the dorsal midline marks the region of the collar cord after neurulation (Figure 3.6E and F) (Miyamoto and Wada, 2013). *Pax3/7* is expressed bilaterally in the neural plate, and after neurulation it is coexpressed with *dlx* (Miyamoto and Wada, 2013).

Comparing gene expression patterns of direct and indirect-developing enteropneusts with echinoderms is informative for reconstructing ancestral patterns. For example, in both echinoderms and hemichordates, BMP signaling is high on the dorsal side and is required for *dlx* expression in the dorsal ectoderm; consistently, elevation of BMP signaling expands the dorsal expression of *dlx* (Ben-Tabou de-Leon et al., 2013; Lowe et al., 2006; Saudemont et al., 2010). These similarities suggest that the dorsal ectodermal expression domain of *dlx* is ancient, while the spotted expression in the proboscis of *S. kowalevskii* is possibly a derived trait. In chordates, BMP signaling is high on the ventral side, and elevation of BMP signaling in the amphioxus embryo expands *dlx* expression throughout the embryo (Yu et al., 2008). While this patterning mechanism is similar between chordates and ambulacrarian embryos, the DV axis is inverted in the chordate lineage. Given this DV inversion, the dorsally located collar cord may not be homologous to the chordate neural tube, despite the similar ontogenic process. The expression of *pax3/7* in the border between the neural plate and epidermis before formation of the collar cord (Figure 3.6E) (Miyamoto and Wada, 2013) may result from recruiting a regulatory circuit that is required for cell shape changes, leading to internalization of the neural plate. It remains unclear how the neurogenic and epidermal ectoderm are segregated in direct-developing hemichordates. In indirect-developing hemichordates and sea urchins, perturbation of BMP levels is sufficient to modulate the positioning of the larval nervous systems. Transient overactivation of BMP signaling during gastrulation can block mouth formation and centralize the nervous system to the ventral ectoderm, and the resulting neurogenic ventral ectoderm displays a medial-to-lateral organization similar to that of the chordate neural plate (Su et al., 2019). These morphological changes are similar to a proposed intermediate stage leading to the emergence of chordates with a dorsally located CNS in an inverse DV body axis relative to their ancestors (Nielsen, 1999). Moreover, the centralized larval nervous system provides a platform for investigating that to what extent a neural border is present in these non-chordate deuterostomes. Further studies in deciphering the regulatory state demarcating the ventral neural ectoderm from the dorsal non-neural ectoderm in these BMP-perturbed embryos would reveal whether any rudimentary neural border was implemented before the emergence of chordates.

3.4 CONCLUSIONS AND PERSPECTIVES

In this chapter, we have summarized what is known about how the CNS is specified, how the neural border regions are patterned by the interactions between various signaling pathways and transcription factors, and what developmental fates can be attributed to cells from these neural border regions in invertebrate chordates (tunicates and amphioxus). We have also reviewed current knowledge of how the nervous system develops in Ambulacrarian animals (echinoderms and hemichordates), which represent the closest sister groups to chordates. This information is critical for understanding the evolutionary origin of the chordate CNS, and it is also necessary for the identification of possible homologous relationships among diverse sensory cell types of the PNS in different deuterostome animals. Due to the great morphological disparity among Ambulacrarian nervous systems, it remains challenging to reconstruct the ancestral situation of the deuterostome nervous system. Nevertheless, comparative studies among different deuterostome lineages seem to suggest an evolutionary trend whereby the ectodermal domain (from which peripheral sensory neurons are generated) appears to have gradually shifted after the emergence of the chordate CNS to occupy its current position in the anterior and lateral neural plate border. Several neural crest specifier genes were likely coopted at the neural plate border in the vertebrate lineage, and this cooption may have enabled the emergence of migrating neural crest cells. Some CNS-derived neurons in tunicates may have already exhibit some parts of this coopted genetic cascade, and thus, these cells likely represent the evolutionary precursors of cranial placode and neural crest cells. As suggested by several review papers (Meulemans and Bronner-Fraser, 2005; Schlosser, 2017; Schlosser et al., 2014; Yu, 2010), this evolutionary transition could have been achieved by utilizing the conserved AP- and DV-axis patterning mechanisms as a *trans*-regulatory platform to integrate the coopted transcription factors, thus conferring new cellular properties to the neural plate border regions.

Further tracing the evolution of nervous systems to cover a greater phylogenetic distance among bilaterian animals becomes even more problematic. For example, most protostome animals contain a single medially located nerve cord on the ventral side, while others have multiple paired nerve cords located at different DV levels along the AP axis (Hejnol and Lowe, 2015). Furthermore, Xenacoelomorphs, which represent an early branching group of bilateria, have diffuse basiepidermal nerve nets that are similar to those in cnidarians (Hejnol and Pang, 2016). Given that the organization and morphology of nervous systems is extremely diverse in the bilaterian animals, EvoDevo studies in the past decades have turned to molecular data in an attempt to identify deep homology. Previous studies in fly and vertebrates, and later in other animals, have shown that BMP signaling patterns the DV axis, and the neuroectoderm is specified on the side with the lowest BMP signaling activity (De Robertis, 2008). The staggered expression of a suite of conserved homeobox genes, including *nk2.1/nk2.2, nkx6, pax6, pax3/7* and *msx*, in the medial and lateral regions of neuroectoderm of several bilaterian species has been regarded as evidence of deep conservation of bilaterian nerve cords (Arendt, 2018; Holland et al., 2013; Mizutani and Bier, 2008). Moreover, similar to vertebrates, at least one of the conserved neural border genes (*msx*) in the nematode *Caenorhabditis elegans* specifies lateral neuro-blasts and regulates the development of mechanosensory neurons (Li et al., 2017). These results strengthen the related ideas that the bilaterian central nervous systems have a common evolutionary origin and that the mechanisms used to set the neu-ral borders are conserved. However, a recent study examining expression of these genes in several species of Xenacoelomorpha, Rotifera, Nemertea, Brachiopoda, and Annelida discovered that the mediolateral staggered gene expression was either not observed or it was not related to the neuroanatomy of these species, suggesting convergent evolution of bilaterian nerve cords (Martin-Duran et al., 2018). Clearly, broader sampling and deeper understanding of the mechanisms used for neural pat-terning and neurogenesis among diverse animal species will be required to resolve these conflicting ideas.

ACKNOWLEDGMENTS

We would like to thank Brian F. Eames, Igor Adameyko, and Daniel Meulemans Medeiros for organizing this book volume; we are greatly in debt to them for the continuing encouragement and immense patience. Research in our laboratories are supported by funding from ICOB, Academia Sinica, and the Ministry of Science and Technology, Taiwan.

REFERENCES

Abitua, P. B., Gainous, T. B., Kaczmarczyk, A. N., Winchell, C. J., Hudson, C., Kamata, K., Nakagawa, M., Tsuda, M., Kusakabe, T. G., Levine, M., 2015. The pre-vertebrate ori-gins of neurogenic placodes. *Nature* 524(7566), 462–465.

Abitua, P. B., Wagner, E., Navarrete, I. A., Levine, M., 2012. Identification of a rudimentary neural crest in a non-vertebrate chordate. *Nature* 492(7427), 104–107.

Albuixech-Crespo, B., Lopez-Blanch, L., Burguera, D., Maeso, I., Sanchez-Arrones, L., Moreno-Bravo, J. A., Somorjai, I., Pascual-Anaya, J., Puelles, E., Bovolenta, P., Garcia-Fernandez, J., Puelles, L., Irimia, M., Ferran, J. L., 2017. Molecular regionalization of the developing amphioxus neural tube challenges major partitions of the vertebrate brain. *PLOS Biol* 15(4), e2001573.

Angerer, L. M., Yaguchi, S., Angerer, R. C., Burke, R. D., 2011. The evolution of nervous system patterning: Insights from sea urchin development. *Development* 138(17), 3613–3623.

Arendt, D., 2018. Animal evolution: Convergent nerve cords? *Curr Biol* 28(5), R225–R227.

Barsi, J. C., Li, E., Davidson, E. H., 2015. Geometric control of ciliated band regulatory states in the sea urchin embryo. *Development* 142(5), 953–961.

Ben-Tabou de-Leon, S., Su, Y. H., Lin, K. T., Li, E., Davidson, E. H., 2013. Gene regulatory control in the sea urchin aboral ectoderm: Spatial initiation, signaling inputs, and cell fate lockdown. *Dev Biol* 374(1), 245–254.

Benito-Gutierrez, E., Illas, M., Comella, J. X., Garcia-Fernandez, J., 2005a. Outlining the nascent nervous system of *Branchiostoma floridae* (amphioxus) by the pan-neural marker AmphiElav. *Brain Res Bull* 66(4–6), 518–521.

Benito-Gutierrez, E., Nake, C., Llovera, M., Comella, J. X., Garcia-Fernandez, J., 2005b. The single AmphiTrk receptor highlights increased complexity of neurotrophin signalling in vertebrates and suggests an early role in developing sensory neuroepidermal cells. *Development* 132(9), 2191–2202.

Bertrand, S., Camasses, A., Somorjai, I., Belgacem, M. R., Chabrol, O., Escande, M. L., Pontarotti, P., Escriva, H., 2011. Amphioxus FGF signaling predicts the acquisition of vertebrate morphological traits. *Proc Natl Acad Sci U S A* 108(22), 9160–9165.

Bertrand, V., Hudson, C., Caillol, D., Popovici, C., Lemaire, P., 2003. Neural tissue in ascidian embryos is induced by FGF9/16/20, acting via a combination of maternal GATA and Ets transcription factors. *Cell* 115(5), 615–627.

Bourlat, S. J., Juliusdottir, T., Lowe, C. J., Freeman, R., Aronowicz, J., Kirschner, M., Lander, E. S., Thorndyke, M., Nakano, H., Kohn, A. B., Heyland, A., Moroz, L. L., Copley, R. R., Telford, M. J., 2006. Deuterostome phylogeny reveals monophyletic chordates and the new phylum Xenoturbellida. *Nature* 444(7115), 85–88.

Burke, R. D., Angerer, L. M., Elphick, M. R., Humphrey, G. W., Yaguchi, S., Kiyama, T., Liang, S., Mu, X., Agca, C., Klein, W. H., Brandhorst, B. P., Rowe, M., Wilson, K., Churcher, A. M., Taylor, J. S., Chen, N., Murray, G., Wang, D., Mellott, D., Olinski, R., Hallbook, F., Thorndyke, M. C., 2006a. A genomic view of the sea urchin nervous system. *Dev Biol* 300(1), 434–460.

Burke, R. D., Moller, D. J., Krupke, O. A., Taylor, V. J., 2014. Sea urchin neural development and the metazoan paradigm of neurogenesis. *Genesis* 52(3), 208–221.

Burke, R. D., Osborne, L., Wang, D., Murabe, N., Yaguchi, S., Nakajima, Y., 2006b. Neuron-specific expression of a synaptotagmin gene in the sea urchin *Strongylocentrotus purpuratus*. *J Comp Neurol* 496(2), 244–251.

Byrne, M., Nakajima, Y., Chee, F. C., Burke, R. D., 2007. Apical organs in echinoderm larvae: Insights into larval evolution in the Ambulacraria. *Evol Dev* 9(5), 432–445.

Candiani, S., Moronti, L., Ramoino, P., Schubert, M., Pestarino, M., 2012. A neurochemical map of the developing amphioxus nervous system. *BMC Neurosci* 13.

Cannon, J. T., Kocot, K. M., Waits, D. S., Weese, D. A., Swalla, B. J., Santos, S. R., Halanych, K. M., 2014. Phylogenomic resolution of the hemichordate and echinoderm clade. *Curr Biol* 24(23), 2827–2832.

Chang, Y. C., Pai, C. Y., Chen, Y. C., Ting, H. C., Martinez, P., Telford, M. J., Yu, J. K., Su, Y. H., 2016. Regulatory circuit rewiring and functional divergence of the duplicate admp genes in dorsoventral axial patterning. *Dev Biol* 410(1), 108–118.

Chen, J. H., Luo, Y. J., Su, Y. H., 2011. The dynamic gene expression patterns of transcription factors constituting the sea urchin aboral ectoderm gene regulatory network. *Dev Dyn* 240(1), 250–260.

Cunningham, D., Casey, E. S., 2014. Spatiotemporal development of the embryonic nervous system of *Saccoglossus kowalevskii*. *Dev Biol* 386(1), 252–263.

Darras, S., Fritzenwanker, J. H., Uhlinger, K. R., Farrelly, E., Pani, A. M., Hurley, I. A., Norris, R. P., Osovitz, M., Terasaki, M., Wu, M., Aronowicz, J., Kirschner, M., Gerhart, J. C., Lowe, C. J., 2018. Anteroposterior axis patterning by early canonical Wnt signaling during hemichordate development. *PLOS Biol* 16(1).

Darras, S., Nishida, H., 2001. The BMP/chordin antagonism controls sensory pigment cell specification and differentiation in the ascidian embryo. *Dev Biol* 236(2), 271–288.

De Robertis, E. M., 2008. Evo-devo: Variations on ancestral themes. *Cell* 132(2), 185–195.

Delsuc, F., Brinkmann, H., Chourrout, D., Philippe, H., 2006. Tunicates and not cephalochordates are the closest living relatives of vertebrates. *Nature* 439(7079), 965–968.

Delsuc, F., Tsagkogeorga, G., Lartillot, N., Philippe, H., 2008. Additional molecular support for the new chordate phylogeny. *Genesis* 46(11), 592–604.

Denes, A. S., Jekely, G., Steinmetz, P. R., Raible, F., Snyman, H., Prud'homme, B., Ferrier, D. E., Balavoine, G., Arendt, D., 2007. Molecular architecture of annelid nerve cord supports common origin of nervous system centralization in bilateria. *Cell* 129(2), 277–288.

Esposito, R., Yasuo, H., Sirour, C., Palladino, A., Spagnuolo, A., Hudson, C., 2017. Patterning of brain precursors in ascidian embryos. *Development* 144(2), 258–264.

Fan, T. P., Ting, H. C., Yu, J. K., Su, Y. H., 2018. Reiterative use of FGF signaling in mesoderm development during embryogenesis and metamorphosis in the hemichordate *Ptychodera flava*. *BMC Evol Biol* 18(1), 120.

Garner, S., Zysk, I., Byrne, G., Kramer, M., Moller, D., Taylor, V., Burke, R. D., 2016. Neurogenesis in sea urchin embryos and the diversity of deuterostome neurogenic mechanisms. *Development* 143(2), 286–297.

Gonzalez, P., Uhlinger, K. R., Lowe, C. J., 2017. The adult body plan of indirect developing hemichordates develops by adding a hox-patterned trunk to an anterior larval territory. *Curr Biol* 27(1), 87–95.

Gostling, N. J., Shimeld, S. M., 2003. Protochordate Zic genes define primitive somite compartments and highlight molecular changes underlying neural crest evolution. *Evol Dev* 5(2), 136–144.

Green, S. A., Norris, R. P., Terasaki, M., Lowe, C. J., 2013. FGF signaling induces mesoderm in the hemichordate *Saccoglossus kowalevskii*. *Development* 140(5), 1024–1033.

Groves, A. K., LaBonne, C., 2014. Setting appropriate boundaries: Fate, patterning and competence at the neural plate border. *Dev Biol* 389(1), 2–12.

Harada, Y., Okai, N., Taguchi, S., Shoguchi, E., Tagawa, K., Humphreys, T., Satoh, N., 2001. Embryonic expression of a hemichordate distal-less gene. *Zool Sci* 18(1), 57–61.

Hejnol, A., Lowe, C. J., 2015. Embracing the comparative approach: How robust phylogenies and broader developmental sampling impacts the understanding of nervous system evolution. *Philos Trans R Soc Lond B Biol Sci* 370(1684).

Hejnol, A., Pang, K., 2016. Xenacoelomorpha's significance for understanding bilaterian evolution. *Curr Opin Genet Dev* 39, 48–54.

Henrique, D., Abranches, E., Verrier, L., Storey, K. G., 2015. Neuromesodermal progenitors and the making of the spinal cord. *Development* 142(17), 2864–2875.

Hinman, V. F., Burke, R. D., 2018. Embryonic neurogenesis in echinoderms. *Wiley Interdiscip Rev Dev Biol* 7(4).

Holland, L. Z., 2015. The origin and evolution of chordate nervous systems. *Philos Trans R Soc Lond B Biol Sci* 370(1684).

Holland, L. Z., 2016. Tunicates. *Curr Biol* 26(4), R146–R152.

Holland, L. Z., Albalat, R., Azumi, K., Benito-Gutierrez, E., Blow, M. J., Bronner-Fraser, M., Brunet, F., Butts, T., Candiani, S., Dishaw, L. J., Ferrier, D. E., Garcia-Fernandez, J., Gibson-Brown, J. J., Gissi, C., Godzik, A., Hallbook, F., Hirose, D., Hosomichi, K., Ikuta, T., Inoko, H., Kasahara, M., Kasamatsu, J., Kawashima, T., Kimura, A., Kobayashi, M., Kozmik, Z., Kubokawa, K., Laudet, V., Litman, G. W., McHardy, A. C., Meulemans, D., Nonaka, M., Olinski, R. P., Pancer, Z., Pennacchio, L. A., Pestarino, M., Rast, J. P., Rigoutsos, I., Robinson-Rechavi, M., Roch, G., Saiga, H., Sasakura, Y., Satake, M., Satou, Y., Schubert, M., Sherwood, N., Shiina, T., Takatori, N., Tello, J., Vopalensky, P., Wada, S., Xu, A., Ye, Y., Yoshida, K., Yoshizaki, F., Yu, J. K., Zhang, Q., Zmasek, C. M., de Jong, P. J., Osoegawa, K., Putnam, N. H., Rokhsar, D. S., Satoh, N., Holland, P. W., 2008. The amphioxus genome illuminates vertebrate origins and cephalochordate biology. *Genome Res* 18(7), 1100–1111.

Holland, L. Z., Carvalho, J. E., Escriva, H., Laudet, V., Schubert, M., Shimeld, S. M., Yu, J. K., 2013. Evolution of bilaterian central nervous systems: A single origin? *EvoDevo* 4(1), 27.

Holland, L. Z., Schubert, M., Holland, N. D., Neuman, T., 2000. Evolutionary conservation of the presumptive neural plate markers AmphiSox1/2/3 and AmphiNeurogenin in the invertebrate chordate amphioxus. *Dev Biol* 226(1), 18–33.

Holland, L. Z., Schubert, M., Kozmik, Z., Holland, N. D., 1999. AmphiPax3/7, an amphioxus paired box gene: Insights into chordate myogenesis, neurogenesis, and the possible evolutionary precursor of definitive vertebrate neural crest. *Evol Dev* 1(3), 153–165.

Holland, N. D., Holland, L. Z., Holland, P. W., 2015. Scenarios for the making of vertebrates. *Nature* 520(7548), 450–455.

Holland, N. D., Panganiban, G., Henyey, E. L., Holland, L. Z., 1996. Sequence and developmental expression of AmphiDll, an amphioxus distal-less gene transcribed in the ectoderm, epidermis and nervous system: Insights into evolution of craniate forebrain and neural crest. *Development* 122(9), 2911–2920.

Horie, R., Hazbun, A., Chen, K., Cao, C., Levine, M., Horie, T., 2018. Shared evolutionary origin of vertebrate neural crest and cranial placodes. *Nature* 560(7717), 228–232.

Horie, T., Shinki, R., Ogura, Y., Kusakabe, T. G., Satoh, N., Sasakura, Y., 2011. Ependymal cells of chordate larvae are stem-like cells that form the adult nervous system. *Nature* 469(7331), 525–528.

Howard-Ashby, M., Materna, S. C., Brown, C. T., Chen, L., Cameron, R. A., Davidson, E. H., 2006. Identification and characterization of homeobox transcription factor genes in *Strongylocentrotus purpuratus*, and their expression in embryonic development. *Dev Biol* 300(1), 74–89.

Hudson, C., 2016. The central nervous system of ascidian larvae. *Wiley Interdiscip Rev Dev Biol* 5(5), 538–561.

Hudson, C., Ba, M., Rouviere, C., Yasuo, H., 2011. Divergent mechanisms specify chordate motoneurons: Evidence from ascidians. *Development* 138(8), 1643–1652.

Hudson, C., Lotito, S., Yasuo, H., 2007. Sequential and combinatorial inputs from Nodal, Delta2/Notch and FGF/MEK/ERK signalling pathways establish a grid-like organisation of distinct cell identities in the ascidian neural plate. *Development* 134(19), 3527–3537.

Hudson, C., Yasuo, H., 2005. Patterning across the ascidian neural plate by lateral Nodal signalling sources. *Development* 132(6), 1199–1210.

Imai, K. S., Hikawa, H., Kobayashi, K., Satou, Y., 2017. Tfap2 and Sox1/2/3 cooperatively specify ectodermal fates in ascidian embryos. *Development* 144(1), 33–37.

Imai, K. S., Levine, M., Satoh, N., Satou, Y., 2006. Regulatory blueprint for a chordate embryo. *Science* 312(5777), 1183–1187.

Ivashkin, E., Adameyko, I., 2013. Progenitors of the protochordate ocellus as an evolutionary origin of the neural crest. *EvoDevo* 4(1).

Jandzik, D., Garnett, A. T., Square, T. A., Cattell, M. V., Yu, J. K., Medeiros, D. M., 2015. Evolution of the new vertebrate head by co-option of an ancient chordate skeletal tissue. *Nature* 518(7540), 534–537.

Jarvela, A. M. C., Yankura, K. A., Hinman, V. F., 2016. A gene regulatory network for apical organ neurogenesis and its spatial control in sea star embryos. *Development* 143(22), 4214–4223.

Kaltenbach, S. L., Yu, J. K., Holland, N. D., 2009. The origin and migration of the earliest-developing sensory neurons in the peripheral nervous system of amphioxus. *Evol Dev* 11(2), 142–151.

Kaul, S., Stach, T., 2010. Ontogeny of the collar cord: Neurulation in the hemichordate *Saccoglossus kowalevskii*. *J Morphol* 271(10), 1240–1259.

Kaul-Strehlow, S., Urata, M., Minokawa, T., Stach, T., Wanninger, A., 2015. Neurogenesis in directly and indirectly developing enteropneusts: Of nets and cords. *Organ Divers Evol* 15(2), 405–422.

Kaul-Strehlow, S., Urata, M., Praher, D., Wanninger, A., 2017. Neuronal patterning of the tubular collar cord is highly conserved among enteropneusts but dissimilar to the chordate neural tube. *Sci Rep* 7(1).

Kozmik, Z., Holland, N. D., Kreslova, J., Oliveri, D., Schubert, M., Jonasova, K., Holland, L. Z., Pestarino, M., Benes, V., Candiani, S., 2007. Pax-six-Eya-Dach network during amphioxus development: Conservation in vitro but context specificity in vivo. *Dev Biol* 306(1), 143–159.

Kozmikova, I., Candiani, S., Fabian, P., Gurska, D., Kozmik, Z., 2013. Essential role of Bmp signaling and its positive feedback loop in the early cell fate evolution of chordates. *Dev Biol* 382(2), 538–554.

Kozmikova, I., Yu, J. K., 2017. Dorsal-ventral patterning in amphioxus: Current understanding, unresolved issues, and future directions. *Int J Dev Biol* 61(10–11), 601–610.

Lacalli, T. C., 1996. Frontal eye circuitry, rostral sensory pathways and brain organization in amphioxus larvae: Evidence from 3D reconstructions. *Philos Trans R Soc Lond B* 351(1337), 243–263.

Lacalli, T. C., 2004. Sensory systems in amphioxus: A window on the ancestral chordate condition. *Brain Behav Evol* 64(3), 148–162.

Lacalli, T. C., Holland, N. D., West, J. E., 1994. Landmarks in the anterior central-nervous-system of amphioxus larvae. *Philos Trans R Soc Lond B* 344(1308), 165–185.

Langeland, J. A., Holland, L. Z., Chastain, R. A., Holland, N. D., 2006. An amphioxus LIM-homeobox gene, AmphiLim1/5, expressed early in the invaginating organizer region and later in differentiating cells of the kidney and central nervous system. *Int J Biol Sci* 2(3), 110–116.

Langeland, J. A., Tomsa, J. M., Jackman, W. R., Jr., Kimmel, C. B., 1998. An amphioxus snail gene: Expression in paraxial mesoderm and neural plate suggests a conserved role in patterning the chordate embryo. *Dev Genes Evol* 208(10), 569–577.

Le Petillon, Y., Luxardi, G., Scerbo, P., Cibois, M., Leon, A., Subirana, L., Irimia, M., Kodjabachian, L., Escriva, H., Bertrand, S., 2017. Nodal/activin pathway is a conserved neural induction signal in chordates. *Nat Ecol Evol* 1(8), 1192–1200.

Lee, E. M., Yuan, T., Ballim, R. D., Nguyen, K., Kelsh, R. N., Medeiros, D. M., McCauley, D. W., 2016. Functional constraints on SoxE proteins in neural crest development: The importance of differential expression for evolution of protein activity. *Dev Biol* 418(1), 166–178.

Lemaire, P., 2011. Evolutionary crossroads in developmental biology: The tunicates. *Development* 138(11), 2143–2152.

Li, Y., Zhao, D., Horie, T., Chen, G., Bao, H., Chen, S., Liu, W., Horie, R., Liang, T., Dong, B., Feng, Q., Tao, Q., Liu, X., 2017. Conserved gene regulatory module specifies lateral neural borders across bilaterians. *Proc Natl Acad Sci U S A* 114(31), E6352–E6360.

Lowe, C. J., Clarke, D. N., Medeiros, D. M., Rokhsar, D. S., Gerhart, J., 2015. The deuterostome context of chordate origins. *Nature* 520(7548), 456–465.

Lowe, C. J., Terasaki, M., Wu, M., Freeman, R. M., Jr., Runft, L., Kwan, K., Haigo, S., Aronowicz, J., Lander, E., Gruber, C., Smith, M., Kirschner, M., Gerhart, J., 2006. Dorsoventral patterning in hemichordates: Insights into early chordate evolution. *PLOS Biol* 4(9), e291.

Lowe, C. J., Wu, M., Salic, A., Evans, L., Lander, E., Stange-Thomann, N., Gruber, C. E., Gerhart, J., Kirschner, M., 2003. Anteroposterior patterning in hemichordates and the origins of the chordate nervous system. *Cell* 113(7), 853–865.

Lu, T. M., Luo, Y. J., Yu, J. K., 2012. BMP and Delta/Notch signaling control the development of amphioxus epidermal sensory neurons: Insights into the evolution of the peripheral sensory system. *Development* 139(11), 2020–2030.

Luttrell, S., Konikoff, C., Byrne, A., Bengtsson, B., Swalla, B. J., 2012. Ptychoderid hemichordate neurulation without a notochord. *Integr Comp Biol* 52(6), 829–834.

Maharana, S. K., Schlosser, G., 2018. A gene regulatory network underlying the formation of pre-placodal ectoderm in *Xenopus laevis*. *BMC Biol* 16(1), 79.

Manni, L., Agnoletto, A., Zaniolo, G., Burighel, P., 2005. Stomodeal and neurohypophysial placodes in *Ciona intestinalis*: Insights into the origin of the pituitary gland. *J Exp Zool B Mol Dev Evol* 304(4), 324–339.

Martik, M. L., Bronner, M. E., 2017. Regulatory logic underlying diversification of the neural crest. *Trends Genet* 33(10), 715–727.

Martin-Duran, J. M., Pang, K., Borve, A., Le, H. S., Furu, A., Cannon, J. T., Jondelius, U., Hejnol, A., 2018. Convergent evolution of bilaterian nerve cords. *Nature* 553(7686), 45–50.

Materna, S. C., Howard-Ashby, M., Gray, R. F., Davidson, E. H., 2006. The C2H2 zinc finger genes of *Strongylocentrotus purpuratus* and their expression in embryonic development. *Dev Biol* 300(1), 108–120.

Mazet, F., Masood, S., Luke, G. N., Holland, N. D., Shimeld, S. M., 2004. Expression of AmphiCoe, an amphioxus COE/EBF gene, in the developing central nervous system and epidermal sensory neurons. *Genesis* 38(2), 58–65.

Metzis, V., Steinhauser, S., Pakanavicius, E., Gouti, M., Stamataki, D., Ivanovitch, K., Watson, T., Rayon, T., Mousavy Gharavy, S. N., Lovell-Badge, R., Luscombe, N. M., Briscoe, J., 2018. Nervous system regionalization entails axial allocation before neural differentiation. *Cell* 175(4), 1105–1118 e1117.

Meulemans, D., Bronner-Fraser, M., 2002. Amphioxus and lamprey AP-2 genes: Implications for neural crest evolution and migration patterns. *Development* 129(21), 4953–4962.

Meulemans, D., Bronner-Fraser, M., 2004. Gene-regulatory interactions in neural crest evolution and development. *Dev Cell* 7(3), 291–299.

Meulemans, D., Bronner-Fraser, M., 2005. Central role of gene cooption in neural crest evolution. *J Exp Zool B Mol Dev Evol* 304(4), 298–303.

Meulemans, D., Bronner-Fraser, M., 2007. The amphioxus SoxB family: Implications for the evolution of vertebrate placodes. *Int J Biol Sci* 3(6), 356–364.

Meulemans, D., McCauley, D., Bronner-Fraser, M., 2003. Id expression in amphioxus and lamprey highlights the role of gene cooption during neural crest evolution. *Dev Biol* 264(2), 430–442.

Miyamoto, N., Saito, Y., 2010. Morphological characterization of the asexual reproduction in the acorn worm *Balanoglossus simodensis*. *Dev Growth Differ* 52(7), 615–627.

Miyamoto, N., Wada, H., 2013. Hemichordate neurulation and the origin of the neural tube. *Nat Commun* 4, 2713.

Mizutani, C. M., Bier, E., 2008. EvoD/Vo: The origins of BMP signalling in the neuroectoderm. *Nat Rev Genet* 9(9), 663–677.

Modrell, M. S., Hockman, D., Uy, B., Buckley, D., Sauka-Spengler, T., Bronner, M. E., Baker, C. V., 2014. A fate-map for cranial sensory ganglia in the sea lamprey. *Dev Biol* 385(2), 405–416.

Molina, M. D., de Croze, N., Haillot, E., Lepage, T., 2013. Nodal: Master and commander of the dorsal-ventral and left-right axes in the sea urchin embryo. *Curr Opin Genet Dev* 23(4), 445–453.

Murabe, N., Hatoyama, H., Hase, S., Komatsu, M., Burke, R. D., Kaneko, H., Nakajima, Y., 2008. Neural architecture of the brachiolaria larva of the starfish, *Asterina pectinifera*. *J Comp Neurol* 509(3), 271–282.

Nakajima, Y., Humphreys, T., Kaneko, H., Tagawa, K., 2004. Development and neural organization of the tornaria larva of the *Hawaiian hemichordate, Ptychodera flava*. *Zool Sci* 21(1), 69–78.

Neidert, A. H., Panopoulou, G., Langeland, J. A., 2000. Amphioxus goosecoid and the evolution of the head organizer and prechordal plate. *Evol Dev* 2(6), 303–310.

Nielsen, C., 1999. Origin of the chordate central nervous system – And the origin of chordates. *Dev Genes Evol* 209(3), 198–205.

Nielsen, C., Hay-Schmidt, A., 2007. Development of the enteropneust *Ptychodera flava*: Ciliary bands and nervous system. *J Morphol* 268(7), 551–570.

Nikitina, N., Sauka-Spengler, T., Bronner-Fraser, M., 2008. Dissecting early regulatory relationships in the lamprey neural crest gene network. *Proc Natl Acad Sci U S A* 105(51), 20083–20088.

Nishida, H., Satoh, N., 1989. Determination and regulation in the pigment cell lineage of the ascidian embryo. *Dev Biol* 132(2), 355–367.

Nomaksteinsky, M., Rottinger, E., Dufour, H. D., Chettouh, Z., Lowe, C. J., Martindale, M. Q., Brunet, J. F., 2009. Centralization of the deuterostome nervous system predates chordates. *Curr Biol* 19(15), 1264–1269.

Ohtsuka, Y., Matsumoto, J., Katsuyama, Y., Okamura, Y., 2014. Nodal signaling regulates specification of ascidian peripheral neurons through control of the BMP signal. *Development* 141(20), 3889–3899.

Onai, T., Yu, J. K., Blitz, I. L., Cho, K. W. Y., Holland, L. Z., 2010. Opposing Nodal/Vg1 and BMP signals mediate axial patterning in embryos of the basal chordate amphioxus. *Dev Biol* 344(1), 377–389.

Ono, H., Kozmik, Z., Yu, J. K., Wada, H., 2014. A novel N-terminal motif is responsible for the evolution of neural crest-specific gene-regulatory activity in vertebrate FoxD3. *Dev Biol* 385(2), 396–404.

Otim, O., Amore, G., Minokawa, T., McClay, D. R., Davidson, E. H., 2004. SpHnf6, a transcription factor that executes multiple functions in sea urchin embryogenesis. *Dev Biol* 273(2), 226–243.

Pani, A. M., Mullarkey, E. E., Aronowicz, J., Assimacopoulos, S., Grove, E. A., Lowe, C. J., 2012. Ancient deuterostome origins of vertebrate brain signalling centres. *Nature* 483(7389), 289–294.

Pasini, A., Amiel, A., Rothbacher, U., Roure, A., Lemaire, P., Darras, S., 2006. Formation of the ascidian epidermal sensory neurons: Insights into the origin of the chordate peripheral nervous system. *PLOS Biol* 4(7), e225.

Pergner, J., Kozmik, Z., 2017. Amphioxus photoreceptors – Insights into the evolution of vertebrate opsins, vision and circadian rhythmicity. *Int J Dev Biol* 61(10–11), 665–681.

Picco, V., Hudson, C., Yasuo, H., 2007. Ephrin-Eph signalling drives the asymmetric division of notochord/neural precursors in Ciona embryos. *Development* 134(8), 1491–1497.

Pla, P., Monsoro-Burq, A. H., 2018. The neural border: Induction, specification and maturation of the territory that generates neural crest cells. *Dev Biol* 444 Suppl 1, S36–S46.

Poustka, A. J., Kuhn, A., Radosavljevic, V., Wellenreuther, R., Lehrach, H., Panopoulou, G., 2004. On the origin of the chordate central nervous system: Expression of onecut in the sea urchin embryo. *Evol Dev* 6(4), 227–236.

Putnam, N. H., Butts, T., Ferrier, D. E., Furlong, R. F., Hellsten, U., Kawashima, T., Robinson-Rechavi, M., Shoguchi, E., Terry, A., Yu, J. K., Benito-Gutierrez, E. L., Dubchak, I., Garcia-Fernandez, J., Gibson-Brown, J. J., Grigoriev, I. V., Horton, A. C., de Jong, P. J., Jurka, J., Kapitonov, V. V., Kohara, Y., Kuroki, Y., Lindquist, E., Lucas, S., Osoegawa, K., Pennacchio, L. A., Salamov, A. A., Satou, Y., Sauka-Spengler, T., Schmutz, J., Shin, I. T., Toyoda, A., Bronner-Fraser, M., Fujiyama, A., Holland, L. Z., Holland, P. W., Satoh, N., Rokhsar, D. S., 2008. The amphioxus genome and the evolution of the chordate karyotype. *Nature* 453(7198), 1064–1071.

Range, R., 2014. Specification and positioning of the anterior neuroectoderm in deuterostome embryos. *Genesis* 52(3), 222–234.

Range, R. C., Angerer, R. C., Angerer, L. M., 2013. Integration of canonical and noncanonical Wnt signaling pathways patterns the neuroectoderm along the anterior-posterior axis of sea urchin embryos. *PLOS Biol* 11(1), e1001467.

Range, R. C., Wei, Z., 2016. An anterior signaling center patterns and sizes the anterior neuroectoderm of the sea urchin embryo. *Development* 143(9), 1523–1533.

Rasmussen, S. L., Holland, L. Z., Schubert, M., Beaster-Jones, L., Holland, N. D., 2007. Amphioxus AmphiDelta: Evolution of delta protein structure, segmentation, and neurogenesis. *Genesis* 45(3), 113–122.

Rottinger, E., DuBuc, T. Q., Amiel, A. R., Martindale, M. Q., 2015. Nodal signaling is required for mesodermal and ventral but not for dorsal fates in the indirect developing hemichordate, *Ptychodera flava*. *Biol Open* 4(7), 830–842.

Rottinger, E., Lowe, C. J., 2012. Evolutionary crossroads in developmental biology: Hemichordates. *Development* 139(14), 2463–2475.

Rottinger, E., Martindale, M. Q., 2011. Ventralization of an indirect developing hemichordate by NiCl suggests a conserved mechanism of dorso-ventral (D/V) patterning in Ambulacraria (hemichordates and echinoderms). *Dev Biol* 354(1), 173–190.

Roure, A., Darras, S., 2016. Msxb is a core component of the genetic circuitry specifying the dorsal and ventral neurogenic midlines in the ascidian embryo. *Dev Biol* 409(1), 277–287.

Rusten, T. E., Cantera, R., Kafatos, F. C., Barrio, R., 2002. The role of TGF beta signaling in the formation of the dorsal nervous system is conserved between Drosophila and chordates. *Development* 129(15), 3575–3584.

Sasakura, Y., Mita, K., Ogura, Y., Horie, T., 2012. Ascidians as excellent chordate models for studying the development of the nervous system during embryogenesis and metamorphosis. *Dev Growth Differ* 54(3), 420–437.

Satoh, G., Wang, Y., Zhang, P. J., Satoh, N., 2001. Early development of amphioxus nervous system with special reference to segmental cell organization and putative sensory cell precursors: A study based on the expression of pan-neuronal marker gene Hu/elav. *J Exp Zool B* 291(4), 354–364.

Satoh, N., 2014. *Developmental Genomics of Ascidians*. Wiley-Blackwell.

Saudemont, A., Haillot, E., Mekpoh, F., Bessodes, N., Quirin, M., Lapraz, F., Duboc, V., Rottinger, E., Range, R., Oisel, A., Besnardeau, L., Wincker, P., Lepage, T., 2010. Ancestral regulatory circuits governing ectoderm patterning downstream of Nodal and BMP2/4 revealed by gene regulatory network analysis in an echinoderm. *PLOS Genet* 6(12), e1001259.

Sauka-Spengler, T., Meulemans, D., Jones, M., Bronner-Fraser, M., 2007. Ancient evolutionary origin of the neural crest gene regulatory network. *Dev Cell* 13(3), 405–420.

Schlosser, G., 2017. From so simple a beginning - What amphioxus can teach us about placode evolution. *Int J Dev Biol* 61(10–11), 633–648.

Schlosser, G., Patthey, C., Shimeld, S. M., 2014. The evolutionary history of vertebrate cranial placodes II. Evolution of ectodermal patterning. *Dev Biol* 389(1), 98–119.

Schubert, M., Holland, L. Z., Holland, N. D., 2000a. Characterization of two amphioxus Wnt genes (AmphiWnt4 and AmphiWnt7b) with early expression in the developing central nervous system. *Dev Dyn* 217(2), 205–215.

Schubert, M., Holland, L. Z., Panopoulou, G. D., Lehrach, H., Holland, N. D., 2000b. Characterization of amphioxus AmphiWnt8: Insights into the evolution of patterning of the embryonic dorsoventral axis. *Evol Dev* 2(2), 85–92.

Schubert, M., Holland, N. D., Escriva, H., Holland, L. Z., Laudet, V., 2004. Retinoic acid influences anteroposterior positioning of epidermal sensory neurons and their gene expression in a developing chordate (amphioxus). *Proc Natl Acad Sci U S A* 101(28), 10320–10325.

Schubert, M., Holland, N. D., Laudet, V., Holland, L. Z., 2006. A retinoic acid-Hox hierarchy controls both anterior/posterior patterning and neuronal specification in the developing central nervous system of the cephalochordate amphioxus. *Dev Biol* 296(1), 190–202.

Sharman, A. C., Shimeld, S. M., Holland, P. W., 1999. An amphioxus Msx gene expressed predominantly in the dorsal neural tube. *Dev Genes Evol* 209(4), 260–263.

Simakov, O., Kawashima, T., Marletaz, F., Jenkins, J., Koyanagi, R., Mitros, T., Hisata, K., Bredeson, J., Shoguchi, E., Gyoja, F., Yue, J. X., Chen, Y. C., Freeman, R. M., Jr., Sasaki, A., Hikosaka-Katayama, T., Sato, A., Fujie, M., Baughman, K. W., Levine, J., Gonzalez, P., Cameron, C., Fritzenwanker, J. H., Pani, A. M., Goto, H., Kanda, M., Arakaki, N., Yamasaki, S., Qu, J., Cree, A., Ding, Y., Dinh, H. H., Dugan, S., Holder, M., Jhangiani, S. N., Kovar, C. L., Lee, S. L., Lewis, L. R., Morton, D., Nazareth, L. V., Okwuonu, G., Santibanez, J., Chen, R., Richards, S., Muzny, D. M., Gillis, A., Peshkin, L., Wu, M., Humphreys, T., Su, Y. H., Putnam, N. H., Schmutz, J., Fujiyama, A., Yu, J. K., Tagawa, K., Worley, K. C., Gibbs, R. A., Kirschner, M. W., Lowe, C. J., Satoh, N., Rokhsar, D. S., Gerhart, J., 2015. Hemichordate genomes and deuterostome origins. *Nature* 527(7579), 459–465.

Slota, L. A., McClay, D. R., 2018. Identification of neural transcription factors required for the differentiation of three neuronal subtypes in the sea urchin embryo. *Dev Biol* 435(2), 138–149.

Stolfi, A., Ryan, K., Meinertzhagen, I. A., Christiaen, L., 2015. Migratory neuronal progenitors arise from the neural plate borders in tunicates. *Nature* 527(7578), 371–374.

Stolfi, A., Wagner, E., Taliaferro, J. M., Chou, S., Levine, M., 2011. Neural tube patterning by ephrin, FGF and Notch signaling relays. *Development* 138(24), 5429–5439.

Su, Y. H., Chen, Y. C., Ting, H. C., Fan, T. P., Lin, C. Y., Wang, K. T., Yu, J. K., 2019. BMP controls dorsoventral and neural patterning in indirect-developing hemichordates providing insight into a possible origin of chordates. *Proc Natl Acad Sci U S A* 116(26), 12925–12932.

Su, Y. H., Li, E., Geiss, G. K., Longabaugh, W. J., Kramer, A., Davidson, E. H., 2009. A perturbation model of the gene regulatory network for oral and aboral ectoderm specification in the sea urchin embryo. *Dev Biol* 329(2), 410–421.

Tagawa, K., Satoh, N., Humphreys, T., 2001. Molecular studies of hemichordate development: A key to understanding the evolution of bilateral animals and chordates. *Evol Dev* 3(6), 443–454.

Tai, A., Cheung, M., Huang, Y. H., Jauch, R., Bronner, M. E., Cheah, K. S., 2016. SOXE neofunctionalization and elaboration of the neural crest during chordate evolution. *Sci Rep* 6, 34964.

Tang, W. J., Chen, J. S., Zeller, R. W., 2013. Transcriptional regulation of the peripheral nervous system in *Ciona intestinalis*. *Dev Biol* 378(2), 183–193.

Van Otterloo, E., Li, W., Garnett, A., Cattell, M., Medeiros, D. M., Cornell, R. A., 2012. Novel Tfap2-mediated control of soxE expression facilitated the evolutionary emergence of the neural crest. *Development* 139(4), 720–730.

Vopalensky, P., Pergner, J., Liegertova, M., Benito-Gutierrez, E., Arendt, D., Kozmik, Z., 2012. Molecular analysis of the amphioxus frontal eye unravels the evolutionary origin of the retina and pigment cells of the vertebrate eye. *Proc Natl Acad Sci U S A* 109(38), 15383–15388.

Wada, H., Holland, P. W., Sato, S., Yamamoto, H., Satoh, N., 1997. Neural tube is partially dorsalized by overexpression of HrPax-37: The ascidian homologue of Pax-3 and Pax-7. *Dev Biol* 187(2), 240–252.

Wagner, E., Levine, M., 2012. FGF signaling establishes the anterior border of the Ciona neural tube. *Development* 139(13), 2351–2359.

Waki, K., Imai, K. S., Satou, Y., 2015. Genetic pathways for differentiation of the peripheral nervous system in ascidians. *Nat Commun* 6, 8719.

Wei, Z., Angerer, L. M., Angerer, R. C., 2016. Neurogenic gene regulatory pathways in the sea urchin embryo. *Development* 143(2), 298–305.

Wei, Z., Angerer, R. C., Angerer, L. M., 2011. Direct development of neurons within foregut endoderm of sea urchin embryos. *Proc Natl Acad Sci U S A* 108(22), 9143–9147.

Wei, Z., Yaguchi, J., Yaguchi, S., Angerer, R. C., Angerer, L. M., 2009. The sea urchin animal pole domain is a Six3-dependent neurogenic patterning center. *Development* 136(7), 1179–1189.

Wicht, H., Lacalli, T. C., 2005. The nervous system of amphioxus: Structure, development, and evolutionary significance. *Can J Zool Rev Canadienne Zool* 83(1), 122–150.

Wu, S. Y., McClay, D. R., 2007. The Snail repressor is required for PMC ingression in the sea urchin embryo. *Development* 134(6), 1061–1070.

Yaguchi, J., Angerer, L. M., Inaba, K., Yaguchi, S., 2012. Zinc finger homeobox is required for the differentiation of serotonergic neurons in the sea urchin embryo. *Dev Biol* 363(1), 74–83.

Yaguchi, S., Yaguchi, J., Angerer, R. C., Angerer, L. M., 2008. A Wnt-FoxQ2-nodal pathway links primary and secondary axis specification in sea urchin embryos. *Dev Cell* 14(1), 97–107.

Yaguchi, S., Yaguchi, J., Angerer, R. C., Angerer, L. M., Burke, R. D., 2010. TGFbeta signaling positions the ciliary band and patterns neurons in the sea urchin embryo. *Dev Biol* 347(1), 71–81.

Yaguchi, S., Yaguchi, J., Wei, Z., Jin, Y. H., Angerer, L. M., Inaba, K., 2011. Fez function is required to maintain the size of the animal plate in the sea urchin embryo. *Development* 138(19), 4233–4243.

Yankura, K. A., Koechlein, C. S., Cryan, A. F., Cheatle, A., Hinman, V. F., 2013. Gene regulatory network for neurogenesis in a sea star embryo connects broad neural specification and localized patterning. *Proc Natl Acad Sci U S A* 110(21), 8591–8596.

Yankura, K. A., Martik, M. L., Jennings, C. K., Hinman, V. F., 2010. Uncoupling of complex regulatory patterning during evolution of larval development in echinoderms. *BMC Biol* 8, 143.

Yasui, K., Zhang, S. C., Uemura, M., Aizawa, S., Ueki, T., 1998. Expression of a twist-related gene, Bbtwist, during the development of a lancelet species and its relation to cephalochordate anterior structures. *Dev Biol* 195(1), 49–59.

Yu, J. K., 2010. The evolutionary origin of the vertebrate neural crest and its developmental gene regulatory network--Insights from amphioxus. *Zoology (Jena)* 113(1), 1–9.

Yu, J. K., Holland, N. D., Holland, L. Z., 2002. An amphioxus winged helix/forkhead gene, AmphiFoxD: Insights into vertebrate neural crest evolution. *Dev Dyn* 225(3), 289–297.

Yu, J. K., Meulemans, D., McKeown, S. J., Bronner-Fraser, M., 2008. Insights from the amphioxus genome on the origin of vertebrate neural crest. *Genome Res* 18(7), 1127–1132.

Yu, J. K., Satou, Y., Holland, N. D., Shin, I. T., Kohara, Y., Satoh, N., Bronner-Fraser, M., Holland, L. Z., 2007. Axial patterning in cephalochordates and the evolution of the organizer. *Nature* 445(7128), 613–617.

Zieger, E., Candiani, S., Garbarino, G., Croce, J. C., Schubert, M., 2018a. Roles of retinoic acid signaling in shaping the neuronal architecture of the developing amphioxus nervous system. *Mol Neurobiol* 55(6), 5210–5229.

Zieger, E., Garbarino, G., Robert, N. S. M., Yu, J. K., Croce, J. C., Candiani, S., Schubert, M., 2018b. Retinoic acid signaling and neurogenic niche regulation in the developing peripheral nervous system of the cephalochordate amphioxus. *Cell Mol Life Sci* 75(13), 2407–2429.

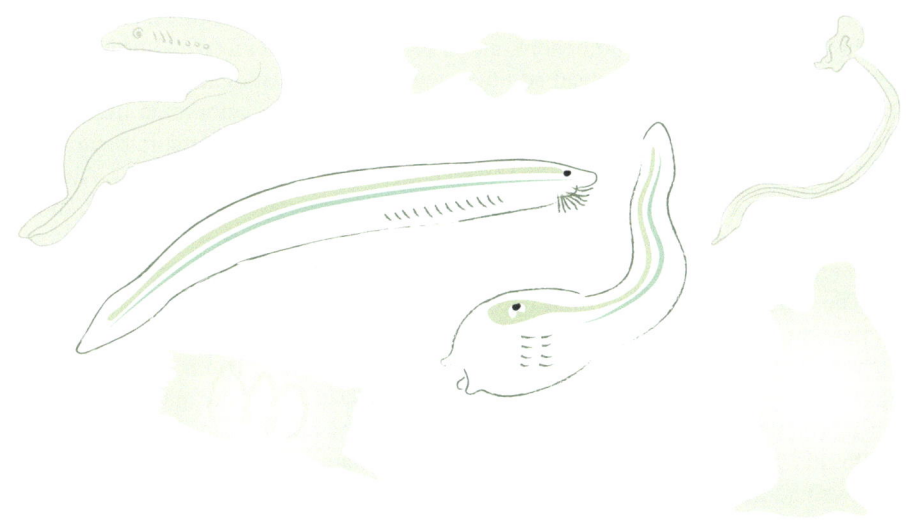

4 The Hunt for Neural Crest in Invertebrate Chordates

Philip B. Abitua

CONTENTS

4.1 INTRODUCTION: BEFORE THE VERTEBRATE

Chordates are a phylum that includes cephalochordates (e.g. amphioxus), tunicates (e.g. sea squirts), and vertebrates (e.g. amphibians, amniotes, and mammals)

(Figure 4.1). These three subphyla all share common defining features, such as a dorsal hollow nerve chord, a notochord, and a segmented axial muscle (Northcutt et al., 1983). Vertebrates have evolved additional structures, most of which can be traced back to neural crest cells and neurogenic placodes. These embryonic tissues were unlikely to arise *de novo*, and conceivably have origins in invertebrate chordates, such as cephalochordates or tunicates, often referred to as *protochordates*. Although definitive neural crest and placodes are not present in invertebrate chordates, molecular biology and the scrutiny of the gene regulatory network (GRN) has allowed researchers to ask if any of the rudiments of these tissues exist in lower chordates.

The phylogenetic relationship between invertebrate chordates and jawless vertebrates, such as lamprey, makes the former group particularly well-suited for investigating the origins of neural crest (Figure 4.1). Originally, it was widely assumed that cephalochordates were more closely related to vertebrates than tunicates, based on morphological features and a limited amount of sequencing data. However, in 2006 this view was revised once whole genomic data sets were compared, and now it is well accepted that tunicates are the sister group to the vertebrates (Delsuc et al., 2006) (Box 1).

Since neural crest cells are found in stem vertebrates, such as hagfish and lamprey, and not in any invertebrate chordates, it is assumed that neural crest is a monophyletic trait, unique to the vertebrate clade. Alternatively, neural crest cells may have once been present in invertebrate chordates, but were secondarily lost during evolution. However, a scenario in which there was a loss of neural crest cells is unlikely because there is no evidence for neural crest in invertebrate chordates, neither in the fossil record nor in any extant invertebrate species. Assuming the former scenario to be true, it is believed that neural crest cells evolved in the last common ancestor of tunicates and stem vertebrates. Seeing how the last common ancestor has long gone extinct, researchers use extant model animals as a proxy for what the GRN looked like before *bona fide* neural crest evolved. Using this approach, one can ask what changes were necessary to facilitate this monumental leap. In the next section, we will examine the research gathered on *Branchiostoma floridae*, the most represented model of the cephalochordates.

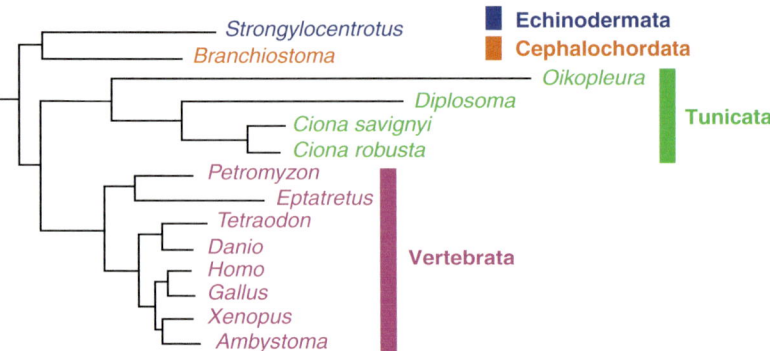

FIGURE 4.1   Phylogenic relationship of chordates. Tree topology indicates that tunicates are more closely related to vertebrates than cephalochordates.

4.2   CEPHALOCHORDATA

Amphioxus was first recognized as a model for primitive chordate development by the famous Russian embryologist Alexander Kovalevsky. In his two most important works on the subject, Kovalevsky meticulously describes the development of amphioxus, including a description of its gill slits, which astonished his contemporaries (Kovalevskij, 1867; Kowalevsky, 2009). The discovery of vertebrate-like features in an invertebrate laid the foundation for the study of modern evolutionary developmental biology. Here we will explore the efforts of researchers to find neural crest–like homologs in *Branchiostoma floridae*, with a focus on gene expression and its comparison to our knowledge of the vertebrate neural crest GRN.

The search for neural crest-like cells in amphioxus begins with defining a homologous neural plate border territory, similar to what we observe in a typical vertebrate. To draw comparisons, researchers focused on the neurula, the embryonic stage after gastrulation, prior to neurulation. Unlike other chordates, the morphological process of neural tube closure in amphioxus is radically different. In *B. floridae*, the epidermis flanking the neural plate collectively migrates medially to cover the future neural tube, dissimilar to the convergence of neural folds exhibited by most vertebrates (N.D. Holland et al., 1996). Despite this difference, the genes that demarcate the border between the neural and nonneural epidermis are largely conserved between cephalochordates and vertebrates. This region of interest is the most likely place to find neural crest-like cells in amphioxus.

When we consider the neural plate border module of the neural crest GRN, many of the transcription factors that pattern this territory are shared between amphioxus and other chordates. For example, *pax3/7* is expressed in the neural plate border of neurula stage *B. floridae* embryos (Meulemans and Bronner-Fraser, 2002; Yu et al., 2008). Likewise, this expression pattern is observed in lamprey and higher vertebrates (Sauka-Spengler et al., 2007). Although no neural crest cell homologs arise from the neural plate border in amphioxus, many of the genes that mark this territory are important for the establishment of the border itself and not the downstream activation of neural crest specifiers (de Crozé et al., 2011). These findings likely reflect an ancient role of neural plate border genes in defining neural versus non-neural epidermis, prior to the emergence of true neural crest cells. Additionally, a comprehensive examination of the neural plate border tier of the gene regulatory network reveals that the entire class of genes, including *dlx*, *msx*, *zic*, and *pax3/7*, are involved in patterning this boundary (Yu et al., 2008), further bolstering the previous point that these genes are not solely involved in development of neural crest cells.

Conversely, most neural crest specifiers are absent from the neural plate region of amphioxus; only the transient expression of *snail* is observed (Langeland et al., 1998; Yu et al., 2008). This is an interesting exception because *snail* is well known for promoting epithelial-to-mesenchymal transitions (EMT) (Nieto, 2002) and in this context, the *snail* expressing cells have not yet been observed to delaminate or migrate, perhaps due to the difficulty of performing lineage tracing in *B. floridae* embryos. The role of *snail* in the neural plate of amphioxus is possibly involved in the morphogenesis of the neural tube. This lack of neural crest specifier expression in the neural plate border of amphioxus suggests that important regulatory inputs for their expression may not have evolved in cephalochordates. In amphioxus, there seems to be a general disconnect between the neural plate border module and the neural crest specifier genes, and the gradual establishment of this hierarchical relationship was likely a pivotal step in neural crest evolution. In the next section, we will see that more of these connections were already present in tunicates before true neural crest arose.

4.3 TUNICATES

Tunicates, commonly referred to as *sea squirts* for their ability to forcefully eject water out of their siphons, have historically been somewhat of an enigma. When tunicates were first described, there was a considerable amount of confusion over their classification in the animal kingdom; much of this stems from their sessile filter-feeding lifestyle as adults, post metamorphosis. In the early 19th century, Georges Cuvier greatly advanced our understanding of tunicate anatomy, but he mistakenly characterized them as being related to bivalve mollusks (Cuvier, n.d.), a belief shared by many of his contemporaries. Tunicates were not recognized as chordates until the publication of Kovalevsky's embryological description in 1866 (Kovalevskij, 1866), a somewhat controversial view that was slow to be accepted. Despite the reluctance to accept tunicates as a close relative to vertebrates, they quickly became an important model for understanding the origin of neural crest in the modern era of molecular biology.

4.3.1 THE RESEMBLANCE OF TRUNK LATERAL CELLS TO NEURAL CREST

One of the first descriptions of neural crest–like cells in tunicates was published in 2004 by William Jeffery (Jeffery et al., 2004). In this manuscript, researchers studied the colonial mangrove tunicate, *Ecteinascidia turbinata*, a niche model that naturally produces the anti-tumor compound Trabectedin (Incalci et al., 2014). By injection of a vital dye, Jeffery *et al.* labeled a population of cells adjacent to the neural tube that migrate from this territory into the body wall where they differentiate into pigmented cells. Furthermore, these neural crest–like cells were shown to express the neural plate border homolog *zicl*, and stained positive for HNK-1(Jeffery et al., 2004), a carbohydrate expressed by migrating neural crest cells (García-Castro et al., 2002). From these findings, the authors concluded that neural crest–like cells were already present in tunicates, and that these migratory pigment cell types served to protect the sessile animal from harmful UV radiation. They proposed that these unipotent cells were further elaborated to give rise to multipotent, *bona fide* neural crest cells in early radiating vertebrates.

The presence of neural crest–like cells in tunicates was an astounding claim; unfortunately, there are issues with many of the supporting lines of evidence. For example, the dye injections used to trace the neural crest–like cells were done at the early tailbud stage, a time point at which injection accuracy is nearly impossible due to the small cell size late in development. In addition, directly beneath the neural plate border at this stage are anteriorly migrating pockets of mesenchyme called the *trunk lateral cells*. HNK-1 itself is not a definitive marker of neural crest, as it also labels other cell types, such as myoblasts (Nagase et al., 2000). Furthermore, the expression of *zicl* in these pigmented cells is much later than the neural plate border expression seen in vertebrates, possibly indicative of an unrelated function. With additional information, William Jeffery would revise his theory about neural crest evolution in tunicates in a series of subsequent papers.

In 2006, HNK-1 activity was examined in 12 different tunicates (Jeffery, 2006). From this extensive study, the authors found that HNK-1 marked a population of trunk mesenchyme cells in every species they investigated. In 2008, using the well annotated cell lineage of the tunicate model *Ciona*, Jeffery *et al.* determined that these HNK-1 positive cells originated from the trunk lateral cell lineage (A7.6) (Jeffery et al., 2008). At the tailbud stage, when the injections were performed in *Ecteinascidia turbinate*, these cells pass directly beneath the neural tube (Figure 4.2). Considering this body of literature, it seems likely that the neural crest–like cells originally described in 2004 were not of ectodermal origin, but rather derived from the trunk lateral cell lineage.

In line with this idea, Jeffery adapted his theory on neural crest evolution to suggest that the trunk lateral cells, which produce mesoderm, were homologous to the neural crest of vertebrates. This proposal was based on the shared expression of a handful of genes. Neural crest cells and the trunk lateral cell lineage of *Ciona* both express *twist, foxd, myc*, and *ap2*. However, the mesoderm of vertebrates, such as zebrafish, also express *twist, foxd*, and *myc* homologs (Kotkamp et al., 2014; Odenthal et al., 1996; Odenthal and Nüsslein-Volhard, 1998). Therefore, the most parsimonious explanation is that the genes that pattern the mesoderm of vertebrates and invertebrates do have evolutionarily conserved regulatory networks, and that any similarities with neural crest are likely because cranial neural crest can produce cell types that typically originate from mesoderm, such as cartilage and bone.

Jeffery *et al.* theorized that the trunk lateral cells of tunicates are homologs to neural crest in vertebrates, and that a primordial neural crest was present in the last common ancestor. Furthermore, he proposed that the primordial neural crest expressed some of the neural crest specifiers but none of the neural plate border genes. It was envisioned that these primordial neural crest cells originally differentiated next to

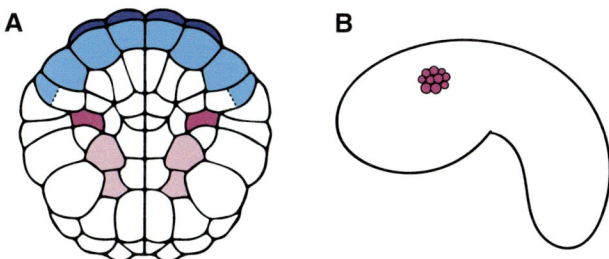

FIGURE 4.2 **Ontogeny of the trunk lateral cells (TLC) in *Ciona robusta*.** (A) Magenta indicates the A7.6 TLC lineage at the 112-cell stage (vegetal view). Light pink labels the other two lineages that contribute to the mesenchyme. Light blue and dark blue indicate the A-line neural lineage and posterior neural plate, respectively. (B) Magenta indicates the TLC derivatives during the tailbud stage. There are two pockets of TLC mesenchyme at the tailbud stage that flank the midline; only the left side is shown in the diagram for simplicity.

the neural plate and were incorporated into the folding neural tube in the vertebrate lineage, leading to the subsequent cooption of neural plate border specifiers. This scenario is unlikely because the mesodermally derived trunk lateral cell lineage of *Ciona* and neural crest cells come from completely different germ layers. It is, however, more likely that part of the mesoderm gene regulatory network was instead coopted into the neural plate border, and not vice versa. In the next section, we will examine research in *Ciona* that focuses on the neural plate border as a source for the origins of neural crest.

4.3.2   RUDIMENTARY NEURAL CREST IN THE CEPHALIC PIGMENTED CELL LINEAGE

The potential for neural crest to form cell types that are normally produced by the mesodermal germ layer, such as bone, muscle, and connective tissue, is a remarkable phenomenon. The importance of this peculiarity was recognized by Gans and Northcutt in their seminal manuscript *Neural Crest and the Origin of Vertebrates: A New Head*:

> *However, it is more difficult to understand why the neural crest should function like mesoderm in the anterior head. This is clearly the most striking change coincident with vertebrate origins; to make a crude analogy, the vertebrate head may be conceived as an addition to the existing body plan of protochordates* (Gans and Northcutt, 1983).

Neural plate border cells acquiring the ability to produce mesodermal-like cells, known as the *ectomesenchyme* (mesenchyme of ectodermal origin), was a turning point in neural crest evolution. Before vertebrates evolved, the ectodermal germ layer had the potential to form sensory neurons and melanocytes (Kaltenbach et al., 2009), but never mesoderm. This novelty would allow for the evolution of important anatomical innovations such as the jaw and most of the skull vault (Bildsoe et al., 2009) (Figure 4.3). To understand how this may have evolved, researchers again turned to the model *Ciona robusta* (formally known as *Ciona intestinalis*) to look for possible neural crest homologs.

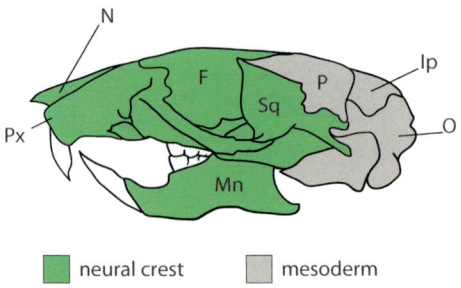

| neural crest | mesoderm |

FIGURE 4.3   Neural crest and mesodermal contributions to the mouse skull. Neural crest cell form premaxilla (Px), mandible (Mn), nasal (N), frontal (F), and squamosal (Sq) bones of the skull. The parietal (P), interparietal (Ip), and occipital (O) bones are derived from the mesoderm.

Like amphioxus, the neural plate border of tunicates expresses the core set of neural plate border specifiers (Imai et al., 2006; Russo et al., 2004; Tassy et al., 2010), but none of these cells have been observed migrating long distances from the neural tube nor forming mesodermal derivatives. However, the pigmented cell lineage (a9.49) that gives rise to the melanocytes of the otolith and ocellis—the respective gravity and light-sensing organs of the larval tunicate—does share some similarities with neural crest. For one, this bilaterally symmetric lineage does arise from the ectodermal neural plate border, an important distinction from the previous attempts at drawing homology. Secondly, not only do they express the neural plate border module, but they also express some neural crest specifier genes such as *ap2*, *id*, *snail*, and *foxd* (Abitua et al., 2012). Finally, although they are not multipotent, as is the case for true neural crest, they do form melanocytes, a sizable neural crest cell derivative. Considering all of these factors, the pigmented cell lineage of *Ciona* is a reasonable candidate for a neural crest homolog, which warranted a closer examination.

Wnt signaling is important for the induction of neural crest in vertebrates, and promotes the formation of melanocytes over neurons or glia (Dorsky et al., 1998). Abitua *et al.* sought to understand if Wnt signaling was also involved in specifying the pigmented cell lineage of *Ciona* (Abitua et al., 2012). It was known that this lineage expressed TCF/LEF, the coactivator of *β*-catenin, which acts downstream of Wnt signaling (Squarzoni et al., 2011). In addition to being competent to signal through Wnt, it was found that the pigmented cell lineage developed just anterior to a Wnt7 signal emanating from the neural tube. The overexpression of *wnt7* in the pigmented cell lineage, using a tissue specific *mitf* driver, converted both melanocyte cells into the more posterior ocelli type. Conversely, by overexpressing a dominant negative form of TCF/LEF, which blocks Wnt signaling, both cells adopted otolith type pigmentation (Abitua et al., 2012). This data suggested that Wnt signaling is important for determining the fate of the ocellus versus the otolith. The researchers then asked if any neural crest specifiers were expressed specifically in the ocellus in response to Wnt7.

Interestingly, in zebrafish, after the early role of *foxd3* in premigratory neural crest specification, the gene is redeployed in the pigment cell lineage to regulate the decision between specifying melanophores versus iridophores. Initially, the bipotent precursor of both cell types express *mitf*, the master regulator of melanogenesis. At 24 hours post fertilization, a portion of these precursor cells express *foxd3*, which represses *mitf*, resulting in iridescent iridophores. The other cells that do not express *foxd3* continue to express *mitf*, which leads to melanin production and the darker appearance of melanophores (Curran et al., 2010). In *Ciona*, *foxd* is specifically expressed in the future pigmented cell of the ocellus in response to Wnt7. When *foxd* is misexpressed in the pigment cell precursor of both the otolith and ocellus, *mitf* expression is abolished (Abitua et al., 2012). Therefore, the function of Foxd in repressing *mitf* appears to be conserved in both the pigmented cell lineage of tunicates and the neural crest–derived melanophores of zebrafish. It is important to note, however, that the role of *foxd3* in the specification of melanophores and iridophores is independent from the earlier function of *foxd3* as a general neural crest specifier. Nevertheless, the conservation of *foxd3* in distinguishing bipotent melanocyte fates supports the idea that the pigmented cell lineage of *Ciona* and neural crest in vertebrates are homologous.

Regardless of the similarities of the pigmented cell lineage to neural crest, the resulting melanocytes never exhibit long-range migration or produce mesodermal-like derivatives, like the ectomesenchyme. Utilizing the known vertebrate GRN (Meulemans and Bronner-Fraser, 2004), Abitua *et al.* asked what neural crest specifiers, absent from the pigmented cell lineage in tunicates, were likely to have been important for acquiring mesenchymal potential. Interestingly, the craniofacial mesenchyme of vertebrates that arises from the primary mesoderm and neural crest–derived ectomesenchyme (Figure 4.3), both express the transcription factor Twist (Bildsoe et al., 2009). The deep conservation of *twist* in the formation of bone from these tissues with distinct developmental ontogenies makes it a promising candidate gene, one that was potentially coopted into the incipient neural crest gene regulatory network, at the base of the vertebrate lineage.

To test this hypothesis, Abitua *et al.* specifically expressed *twist* in the pigmented cell lineage of *Ciona*, using the cis-regulatory sequence of *mitf*. Remarkably, the misexpression of *twist*, which is normally only found in the mesodermal-derived mesenchyme, resulted in cells migrating away from the neural tube (Figure 4.4). These manipulated cells developed protrusive activity and become proliferative, a phenotype not exhibited by the misexpression of other related basic helix-loop-helix transcription factors. Furthermore, the reprogrammed pigmented cell lineage shows expression of downstream mesenchymal genes, such as *erg*, which is observed in the mesoderm and neural crest (Vlaeminck-Guillem et al., 2000). At the juvenile stage, these transformed cells can be seen populating the cellulose-based tunic of the animal. Notably, the endogenous *twist* expressing mesoderm of *Ciona* differentiates

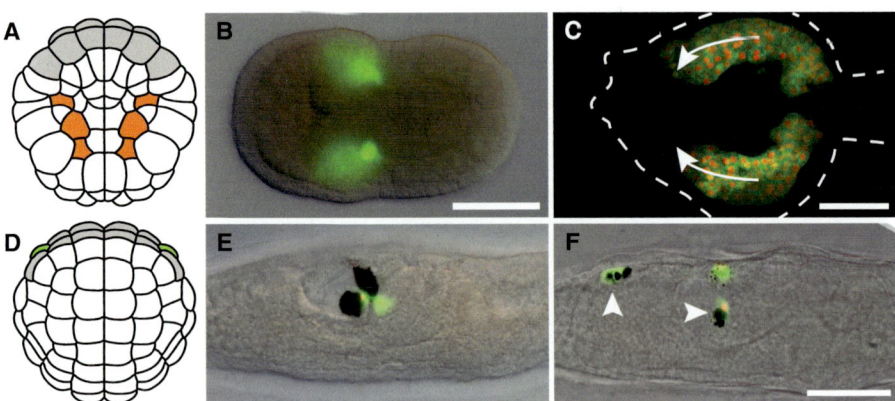

FIGURE 4.4   Twist reprograms the pigmented cell lineage in *Ciona robusta.* (A) *Twist* is endogenously expressed in the mesoderm lineages (orange) at the 112-cell stage (vegetal view). Neural plate progenitors are shown in gray. (B) *Twist* expression during the neurula stage. (C) During the tailbud stage, *twist* expressing cells proliferate and migrate anteriorly (arrows). (D) The pigmented cell lineage (green) comes from the neural plate progenitors (gray) at the 112-cell stage (animal view). (E) Derivatives of the pigmented cell lineage are labeled by the *mitf* enhancer shown in green. (F) Misexpression of *twist* by the *mitf* enhancer causes the pigmented cell lineage to migrate away from the neural tube (arrowheads). Scale bars represent 50 μm.

into a number of cell types, such as body wall muscle, hemocytes, and specialized tunic cells, and these populations are severely reduced when *twist* is knocked down (Tokuoka et al., 2005). The reprogramming of the pigmented cell lineage by the misexpression of a single transcription factor demonstrates that Twist is sufficient to produce ectomesenchyme-like cells.

From this study, the authors conclude that a mesenchymal transcription factor, such as Twist or a similar determinate, was coopted into rudimentary neural crest cells that developed at the neural plate border, and that this event was crucial for the emergence of ectomesenchyme. Of course, the exact scenario of how cells originally acquired the ability to migrate away from the neural tube and differentiate into mesoderm-like cells is impossible to reconstruct. Furthermore, the process of gaining migrating mesenchymal behavior and forming diverse ectomesenchymal derivatives may have been a more gradual stepwise progression. Alternatively, although unlikely, it is possible that the last common ancestor of tunicates and vertebrates had neural crest cells, but were subsequently lost in the tunicate lineage. Without direct evidence, such as transitional fossils, we cannot know for certain what changes led to the appearance of true neural crest.

Others have suggested that the pigmented cell lineage of *Ciona* may be homologous to the vertebrate retinal pigmented epithelium (RPE) (Green and Bronner, 2013). This comparison is based on the expression of *opsin1* and *rx* in the photoreceptors closely associated with the ocellus (D'Aniello et al., 2006; Kusakabe et al., 2001). There is little doubt that the ocellus of tunicates is a light-sensing organ; however, function should never be used to draw homology. In vertebrates, the RPE comes from the medial diencephalon (Lamb et al., 2007), not from the neural plate border like the pigmented cell lineage of tunicates. Additionally, the pigmented cell of the ocellus lacks RPE markers such as *RPE65*, an isomerase which is abundantly expressed in the lateral eyes, which are used for vision (Gollapalli et al., 2003). *RPE65* is instead expressed in the neural gland (primitive brain) of the adult tunicate (Oonuma et al., 2016). In theory, the RPE and neural crest–derived melanocytes could have evolved from an ancestral invertebrate chordate cell type, which may have segregated in the vertebrate lineage (Arendt, 2008).

The shared ontogeny and similar gene expression profile between the pigmented cell lineage and neural crest does support the idea that these cell populations may be evolutionarily related. Furthermore, the acquisition of mesodermal cell fates, which perplexed Gans and Northcutt, and the migration away from the neural tube may have occurred with the evolution of one or a few new regulatory interactions. From this study and the previous work on amphioxus, we can definitively conclude that neural crest evolved from an ancestral chordate gene regulatory network, and not *de novo* in vertebrates.

4.3.3 THE BIPOLAR TAIL NEURONS AS A FORERUNNER OF THE DORSAL ROOT GANGLION

In a complementary study, Stolfi *et al.* proposed that the bipolar tail neurons of *Ciona* are homologous to the neural crest–derived dorsal root ganglia (Stolfi et al., 2015). The authors show that this lineage originates from the posterior neural plate border, delaminates and migrates anteriorly along the paraxial mesoderm, and differentiates into afferent neurons that synapse with both epidermal sensory neurons and the motor ganglia. Furthermore, the bipolar tail neurons also express two genes of the neural crest neural differentiation circuit, *neurogenin* and *islet*. Once differentiated, the bipolar tail neurons of *Ciona* have a split branching morphology, reminiscent of the so-called pseudounipolar shape exhibited by the dorsal root ganglia.

The regulation of cell adhesion is critical for the delamination of neural crest cells. For cells to successfully undergo an EMT, type II cadherins that are less adhesive are upregulated, while type I cadherins, which are found in stable epithelia are downregulated. Stolfi *et al.* examined the expression of adhesion proteins in the bipolar tail neurons and surrounding neural tube. They found that *cadherin-7* (referred to as *cadherin.b* in the manuscript), a type II classic cadherin predominately expressed in the developing neural tube, was absent from the neurogenin positive bipolar tail neurons. They also observed that the lineage also lacked expression of *protocadherin β-15-like* (referred to as *protocadherin.c* in the manuscript), which is expressed in nonmigrating epidermal sensory neurons. Moreover, the misexpression of *protocadherin β-15-like* in bipolar tail neurons blocked delamination and subsequent migration. Although changes in adhesion are likely important for normal bipolar tail neuron development, it is difficult to interpret how these results relate to neural crest formation due to sequence divergence between tunicates and vertebrate cadherins (Noda and Satoh, 2008).

By using electron microscopy, Stolfi *et al.* confirmed that the cell body of the bipolar tail neurons lies outside of the neural tube. These cells are completely covered by epidermis and do not possess cilia, which suggest that they are not directly sensory. The authors show that, instead, the distal processes of the bipolar tail neurons synapse with the ciliated epidermal sensory neurons that decorate the larval tail and are believed to be mechanosensory (Tang et al., 2013). The proximal processes of the bipolar tail neurons form synaptic connections with the motor neurons, which function to control the muscle movements of the tail. This circuitry is reminiscent of the slowly adapting type I dorsal root ganglion neurons of mammals that relay mechanosensory information from Merkel cells (Maksimovic et al., 2014).

The bipolar tail neurons also share some similarity with Rohon-beard neurons, which also arise from the neural plate border and express *neurogenin* and *islet*. In fact, these neurons, found in fish and amphibians, are embryonic cell types that undergo programmed cell death and are functionally replaced by the dorsal root ganglion (Lamborghini, 1987). In zebrafish, there is evidence to support that specification of these two cell types both require the transcription factor *prdm1a*, which further supports their relatedness (Rossi et al., 2009). Fritzsch and Northcutt proposed that dorsal root ganglion evolved from Rohon-beard-like sensory neurons that acquired the ability to migrate away from the neural tube (Fritzsch and Northcutt, 1993). Stolfi *et al.* argue that the bipolar tail neurons of *Ciona* are more similar to dorsal root ganglion than Rohon-beard neurons because of their bipolar morphology, the expression of *ASIC* (acid-sensing ion channel), and their migration along the paraxial mesoderm (Stolfi et al., 2010).

The last common ancestor of tunicates and vertebrates likely possessed a neural plate border that produced sensory neurons with limited migratory capabilities, as supported by Stolfi *et al.* This ancestral sensory neuron population may have ultimately diversified into the Rohon-beard and dorsal root ganglion cells found in vertebrates. The description of the bipolar tail neurons in tunicates suggests that the cells that evolved into neural crest–derived sensory neurons may have already had the potential to migrate before vertebrates evolved. However, the migration of the bipolar tail neurons in tunicates is modest compared to the migration of the dorsal root ganglion in vertebrates, and likely differs at the mechanistic level. The hypothetical ancestral sensory neuron population may have been imbued with long-range migration capabilities seen in neural crest through the cooption of existing genes or by *de novo* gene evolution. In mice, sensory neuron progenitors express the chemokine *CXCR4* and respond to the chemoattractant SDF-1 (Belmadani, 2005). The *Ciona* genome contains many unclassified chemokine receptor-like genes, but the sequences are too diverged to support phylogenetic analysis. The lack of definitive chemokine receptor homologs suggests that evolution of this GPCR family evolved in the vertebrate lineage (Kamesh et al., 2008). This warrants future studies to compare the functional mechanisms by which the bipolar tail neurons and dorsal root ganglion migrate.

Together, the study of the pigmented cell lineage and bipolar tail neurons in *Ciona* suggest that the neural plate border of the last common ancestor of tunicates and vertebrates was primed for the evolution of neural crest. Although multipotent, migratory cells do not arise from this embryonic territory, most of the neural plate border module and some of the downstream neural crest genes are present in the proposed neural crest–like homologs (Figure 4.5). Notably, homologs of *sox9* and *sox10*

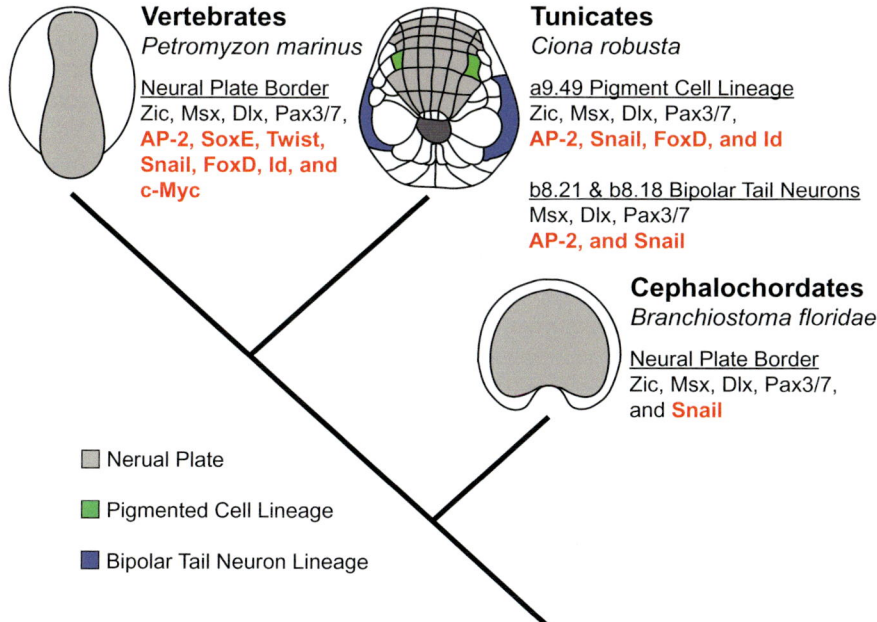

FIGURE 4.5 **Summary of neural crest homolog gene expression patterns in cephalochordates, tunicates, and vertebrates.** In each organism, gray represents the neural plate. In tunicates, the pigmented cell lineage is shown in green, and the bipolar tail neuron lineage is shown in blue. The neural crest specifier genes are highlighted in red (Abitua et al., 2012; Sauka-Spengler et al., 2007; Stolfi et al., 2010; Yu et al., 2008).

that belong to the SoxE family of transcription factors are absent from the neural plate border in *Ciona*. However, a single *soxE* gene exists in the *Ciona* genome but has no reported role in development. The last common ancestor likely had a single copy of *soxE* that was duplicated in the vertebrate lineage and subsequently diversified into *sox8*, *sox9*, and *sox10*. This process of gene duplication likely had a major impact on wiring of the nascent neural crest gene regulatory network and is the topic of discussion in the next section.

4.4 DID WHOLE GENOME DUPLICATIONS FUEL NEURAL CREST EVOLUTION?

Neural crest cells may have had their humble beginnings in invertebrate chordates, but they evolved considerably at the base of vertebrates. The neural plate border that once produced only a few derivatives, like melanocytes and sensory cells, suddenly produced the multitude of cell types we associate with neural crest. In addition, these cells acquired complex migration patterns and populated territories far from their developmental origin. This abrupt endowment of multipotency and long-range migration is hard to reconcile without the knowledge of the large-scale genomic modifications that took place directly prior to these changes. It is now widely accepted that two rounds of whole genome duplication occurred at the base of the vertebrate lineage (Dehal and Boore, 2005). It is believed that these events provided the raw genetic material that ultimately led to the formation of the vertebrate cranium and intricate nervous system, features largely derived from neural crest cells.

The idea of genome duplication as a major driving force of evolution was popularized by Susumu Ohno in his iconic book on the subject (Ohno, 1970). Here, Ohno explains that as long as a given function of an organism is controlled by a single gene locus, natural selection will not permit the accumulation of deleterious mutations in a critical region of that gene. He argues that once redundancy is introduced by gene duplication, previously *forbidden* mutations are allowed and new gene functions can arise. This concept is now referred to as *neofunctionalization*, a duplicated gene that gained a new function either by a change in its protein sequence or a change in where or when it is expressed. Alternatively, *subfunctionalization* occurs when mutations in a duplicated gene partitions the ancestral function into each new paralog, again either at the protein coding level or in cis-regulatory sequences (Figure 4.6). On paper, evolution by gene duplication is an intriguing hypothesis; however, providing evidence that these large-scale events took place was difficult until the genomic era.

The two rounds of whole genome duplication, the so-called 2R hypothesis, was intensely debated, largely because remnants of these events are hard to detect in the DNA of extant species. After genome duplication, only a small fraction of genes experience neofunctionalization or subfunctionalization before quickly (in less than a few million years) accumulating deleterious mutations that result in gene loss. Clear evidence of whole genome duplications is further obscured by the duplication of single genes and small genomic segments, which occur more frequently than previously appreciated, at a rate of 0.01 per gene per million years (Lynch and Conery, 2000). However, by mapping the position of only the genes duplicated at the base

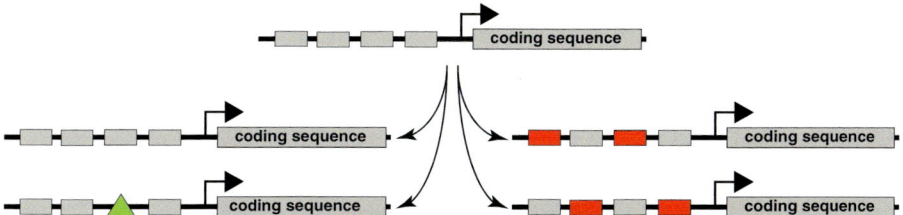

FIGURE 4.6   **Neofunctionalization or subfunctionalization after gene duplication.**
When a gene duplicates into two paralogs, occasionally one copy will gain a new function,
as represented on the left by the green triangle; this is therefore called *neofunctionalization*.
Alternatively, duplicated paralogs can acquire reciprocal mutations that result in a partitioning of the ancestral function, as represented on the right by the red rectangles, known as
subfunctionalization.

of the vertebrates to the human genome, a clear pattern of four paralogous gene sets
emerged, which confirmed the two rounds of whole genome duplication hypothesis
(Dehal and Boore, 2005).

Several genes of the neural crest gene regulatory network are paralogs that are
found as a single gene copy in invertebrate chordates. For example, *pax3* and *pax7*
are vertebrate paralogs that are represented as the single gene *pax3/7* prior to their
duplication at the base of the vertebrates. Both *pax3* and *pax7* are important for
neural crest formation, albeit sometimes in nonredundant ways depending on the
species. In frogs, *pax3* is abundantly expressed in the neural folds at the onset of
neural crest induction, while *pax7* is absent (Maczkowiak et al., 2010). Interestingly,
in chick, the inverse is true; *pax7* appears to have a crucial role in neural crest development, while *pax3* is dispensable (Basch et al., 2006). Regardless of the species-
specific differences, the ancestral gene *pax3/7* was already expressed in the neural
plate border in both amphioxus and tunicates (Wada et al., 1997; Yu et al., 2008). A
similar trend is seen with other neural crest border specifier paralogs, which argues
against genome duplications being important at this tier of the neural crest gene
regulatory network.

However, whole genome duplications may have had a significant impact on the
wiring of the neural crest gene regulatory network downstream of the neural plate
border module. *Sox9* and *sox10* are paralogs with divergent functions in neural crest
that arose from the invertebrate homolog *soxE*. In amphioxus, *soxE* is expressed in
the notochord, mesoderm, and medial neural plate, but not at the neural plate border
where neural crest originates in vertebrates. It is conceivable that after duplication,
a *soxE* paralog gained new cis-regulatory binding sites for neural plate border transcription factors, which allowed its cooption into the neural crest gene network. In
all vertebrates examined so far, the expression of *soxE* family transcription factors
in neural crest is controlled by AP2 (Van Otterloo et al., 2012), which makes it
the most likely candidate responsible for the evolution of this novel gene regulatory
interaction.

Consistent with this scenario, Jandzik *et al.* introduced the amphioxus *soxE* locus
into zebrafish and found that it was expressed in tissues similar to the endogenous
sites observed in amphioxus, and not in neural crest like *sox10*. Since *sox9* and *sox10*

expression is dependent on AP2, the authors wondered if the amphioxus *soxE* expression they observed in their interspecies assay was also reliant on AP2. The expression of the amphioxus *soxE* locus was similar in either wildtype or *ap2* knockdown zebrafish. Altogether, this provides evidence that the cis-regulatory sequences of *soxE* paralogs diverged after they were duplicated in the vertebrate lineage. The cis-regulatory sequences of *sox9* and *sox10* gained new AP2 binding sites that resulted in their assimilation into the neural crest gene regulatory network.

In addition to the two rounds of whole genome duplication at the base of the vertebrates, there was an additional fish-specific genome duplication (FSGD) event. Teleost are arguably the most successful clade of vertebrates, roughly comprising half of all 64,000 vertebrate species (Santini et al., 2009). Yet it is debated how important the FSGD was, due to the lag time between when the event occurred and when the majority of fish-specific diversification arose. It is estimated that the FSGD happened 316–226 million years ago, but 88% of extant teleost diversity emerged during more recent radiations of freshwater (Ostariophysi) and marine (Percomorpha) fish, independent of whole genome duplications (Santini et al., 2009). Despite the lack of correlation between FSGD and diversification, there is still precedent for FSGD being a driving force for teleost evolution. For example, pigmentation genes were preferentially maintained after the FSGD, which resulted in a approximant 30% increase in pigmentation genes in teleosts compared to tetrapods (Braasch et al., 2009a). In addition to whole-genome duplication events, other mechanisms of gene duplication such as DNA replication errors or retrotranspostions ultimately added to the raw genetic material that paved the way for the evolutionary success of the vertebrates (Cañestro et al., 2013).

All of the transcription factors that make up the known neural crest GRN were already present in the genomes of our invertebrate chordate relatives. This observation makes it unlikely that completely *de novo* gene synthesis was a major contributor to neural crest evolution, although its impact is not entirely negligible, as certain gene families such as endothelins have no invertebrate homologs (Braasch et al., 2009b). Nevertheless, Ohno's theory of a polyploidy ancestor is a compelling and somewhat provocative explanation of how neural crest evolved so precipitously. The gene duplication events at the base of the vertebrates may have been a sort of evolutionary experiment that led to the formation of new regulatory network connections, which ultimately resulted in neural crest. As new biotechnology evolves, so will our understanding of the status of the neural crest gene regulatory prior to the emergence of vertebrates. With this knowledge, we can better reconstruct the evolutionary history that culminated in the creation of neural crest.

BOX 4.1   LOOKS CAN BE DECEIVING:
A REVISED CHORDATE MORPHOLOGY

Up until the age of genomics, cephalochordates were considered the closest living relatives to the vertebrates. This traditionally accepted view was largely based on shared morphological features that are far from definitive.

For example, the presence of metameric segmentation, such as the repeating blocks of mesoderm (e.g. somites), were used to infer that cephalochordates and vertebrates were sister groups, because tunicates lack this feature (Schaeffer, 1987). However, metameric segmentation is now theorized to be an ancient morphological trait, one that some speculate arose in the ancestor of all bilaterian animals (Peel and Akam, 2003). The view grouping cephalochordates and vertebrates into a single clade was also supported by the phylogenetic analysis of a limited number of rRNA genes (Winchell et al., 2002). The alternative theory, which places tunicates and vertebrates together in a clade known as *Olfactores* (Jefferies, 1991), is backed by the phylogenetic analysis of classic cadherin genes (Oda et al., 2004). However, these two opposing views were not conclusive due to the limited genes considered in each study.

In 2005, a study that evaluated 59 proteins recovered significant support for the theory that tunicates were the closest relative to the vertebrates (Blair and Hedges, 2005). In the same study, the authors note that cephalochordates can be erroneously grouped with vertebrates due to long-branch attraction artifacts, in agreement with an independent group (Philippe et al., 2005). This error occurs in phylogeny analysis when rapidly evolving lineages are assumed to be closely related, irrespective of their actual evolutionary histories (Siddall and Whiting, 1999). The tunicate branch length was 50% longer when compared to its more slowly evolving vertebrate counterpart. However, these studies suffered from a very limited sampling of chordate diversity, but as more genomes became sequenced, better phylogenic inferences could be made.

In 2008, a comprehensive report was published that included sequence analysis from 13 chordate species, including four different tunicates and early-branching, jawless vertebrates. Importantly, this study included *Oikopleura dioica*, a representative model of the free-swimming larvaceans, which are significantly divergent from other tunicates (Denoeud et al., 2010). Multiple methods of phylogenetic analysis of 146 genes from this diverse group of chordates all converged on the same tree topology (Figure 4.1), which placed tunicates as the true sister group to the vertebrates.

This revised chordate phylogeny has had a significant impact on the interpretation of previous assumptions. For example, the metameric segments that were used to draw homology between vertebrates and chordates can be assumed to have been ancestral to all chordates. It is now believed that metameric segmentation was secondarily reduced in the tunicate lineage. Furthermore, this finding challenges previous beliefs that deuterostome evolution steadily trended towards further complexity and an increased brain size. Tunicates should not be considered primitive chordates, but instead a group that has rapidly evolved, reducing the size of its genome and gaining remarkably specialized life cycles. In light of these findings, tunicates became the preferred clade to study questions concerning vertebrate origins.

REFERENCES

Abitua, P.B., Wagner, E., Navarrete, I.A., Levine, M., 2012. Identification of a rudimentary neural crest in a non-vertebrate chordate. *Nature* 492(7427), 104–107. doi:10.1038/nature11589

Arendt, D., 2008. The evolution of cell types in animals: Emerging principles from molecular studies. *Nature Reviews: Genetics* 9(11), 868–882. doi:10.1038/nrg2416

Basch, M.L., Bronner-Fraser, M., García-Castro, M.I., 2006. Specification of the neural crest occurs during gastrulation and requires Pax7. *Nature* 441(7090), 218–222. doi:10.1038/nature04684

Belmadani, A., Tran, P.B., Ren, D., Assimacopoulos, S., Grove, E.A., Miller, R.J., 2005. The chemokine stromal cell-derived factor-1 regulates the migration of sensory neuron progenitors. *Journal of Neuroscience* 25(16), 3995–4003. doi:10.1523/JNEUROSCI.4631-04.2005

Bildsoe, H., Loebel, D.A.F., Jones, V.J., Chen, Y.-T., Behringer, R.R., Tam, P.P.L., 2009. Requirement for Twist1 in frontonasal and skull vault development in the mouse embryo. *Developmental Biology* 331(2), 176–188. doi:10.1016/j.ydbio.2009.04.034

Blair, J.E., Hedges, S.B., 2005. Molecular phylogeny and divergence times of deuterostome animals. *Molecular Biology and Evolution* 22(11), 2275–2284. doi:10.1093/molbev/msi225

Braasch, I., Brunet, F., Volff, J.-N., Schartl, M., 2009a. Pigmentation pathway evolution after whole-genome duplication in fish. *Genome Biology and Evolution* 1, 479–493. doi:10.1093/gbe/evp050

Braasch, I., Volff, J.N., Schartl, M., 2009b. The endothelin system: Evolution of vertebrate-specific ligand-receptor interactions by three rounds of genome duplication. *Molecular Biology and Evolution* 26(4), 783–799. doi:10.1093/molbev/msp015

Cañestro, C., Albalat, R., Irimia, M., Garcia-Fernàndez, J., 2013. Impact of gene gains, losses and duplication modes on the origin and diversification of vertebrates. *Seminars in Cell and Developmental Biology* 24(2), 83–94. doi:10.1016/j.semcdb.2012.12.008

Curran, K., Lister, J.A., Kunkel, G.R., Prendergast, A., Parichy, D.M., Raible, D.W., 2010. Interplay between Foxd3 and Mitf regulates cell fate plasticity in the zebrafish neural crest. *Developmental Biology* 344(1), 107–118. doi:10.1016/j.ydbio.2010.04.023

Cuvier, G., 1817. *Mémoire sur les Ascidies et sur leur anatomie.*

D'Aniello, S., D'Aniello, E., Locascio, A., Memoli, A., Corrado, M., Russo, M.T., Aniello, F., Fucci, L., Brown, E.R., Branno, M., 2006. The ascidian homolog of the vertebrate homeobox gene Rx is essential for ocellus development and function. *Differentiation* 74(5), 222–234. doi:10.1111/j.1432-0436.2006.00071.x

de Crozé, N., Maczkowiak, F., Monsoro-Burq, A.H., 2011. Reiterative Ap2A activity controls sequential steps in the neural crest gene regulatory network. *Proceedings of the National Academy of Sciences of the United States of America* 108(1), 155–160. doi:10.1073/pnas.1010740107

Dehal, P., Boore, J.L., 2005. Two rounds of whole genome duplication in the ancestral vertebrate. *PLOS Biology* 3(10), e314–319. doi:10.1371/journal.pbio.0030314

Delsuc, F., Brinkmann, H., Chourrout, D., Philippe, H., 2006. Tunicates and not cephalochordates are the closest living relatives of vertebrates. *Nature Cell Biology* 439(7079), 965–968. doi:10.1038/nature04336

Denoeud, F., Henriet, S., Mungpakdee, S., Aury, J.-M., Da Silva, C., Brinkmann, H., Mikhaleva, J., Olsen, L.C., Jubin, C., Cañestro, C., Bouquet, J.-M., Danks, G., Poulain, J., Campsteijn, C., Adamski, M., Cross, I., Yadetie, F., Muffato, M., Louis, A., Butcher, S., Tsagkogeorga, G., Konrad, A., Singh, S., Jensen, M.F., Huynh Cong, E., Eikeseth-Otteraa, H., Noel, B., Anthouard, V., Porcel, B.M., Kachouri-Lafond, R., Nishino, A., Ugolini, M., Chourrout,

P., Nishida, H., Aasland, R., Huzurbazar, S., Westhof, E., Delsuc, F., Lehrach, H., Reinhardt, R., Weissenbach, J., Roy, S.W., Artiguenave, F., Postlethwait, J.H., Manak, J.R., Thompson, E.M., Jaillon, O., Pasquier Du, L., Boudinot, P., Liberles, D.A., Volff, J.-N., Philippe, H., Lenhard, B., Roest Crollius, H., Wincker, P., Chourrout, D., 2010. Plasticity of animal genome architecture unmasked by rapid evolution of a pelagic tunicate. *Science* 330(6009), 1381–1385. doi:10.1126/science.1194167

Dorsky, R.I., Moon, R.T., Raible, D.W., 1998. Control of neural crest cell fate by the Wnt signalling pathway. *Nature* 396(6709), 370–373. doi:10.1038/24620

Fritzsch, B., Northcutt, R.G., 1993. Cranial and spinal nerve organization in amphioxus and lampreys: Evidence for an ancestral craniate pattern. *Acta Anatomica (Basel)* 148(2–3), 96–109.

Gans, C., Northcutt, R.G., 1983. Neural crest and the origin of vertebrates: A new head. *Science* 220(4594), 268–273. doi:10.1126/science.220.4594.268

García-Castro, M., Marcelle, C., Bronner-Fraser, M., 2002. Ectodermal Wnt function as a neural crest inducer. *Science* 297(5582), 848.

Gollapalli, D.R., Maiti, P., Rando, R.R., 2003. RPE65 operates in the vertebrate visual cycle by stereospecifically binding all- trans-retinyl esters. *Biochemistry* 42(40), 11824–11830. doi:10.1021/bi035227w

Green, S.A., Bronner, M.E., 2013. Gene duplications and the early evolution of neural crest development. *Seminars in Cell and Developmental Biology* 24(2), 95–100. doi:10.1016/j.semcdb.2012.12.006

Holland, N.D., Panganiban, G., Henyey, E.L., Holland, L.Z., 1996. Sequence and developmental expression of AmphiDll, an amphioxus distal-less gene transcribed in the ectoderm, epidermis and nervous system: Insights into evolution of craniate forebrain and neural crest. *Development* 122(9), 2911–2920.

Imai, K., Levine, M., Satoh, N., Satou, Y., 2006. Regulatory blueprint for a chordate embryo. *Science* 312(5777), 1183.

Incalci, M.D.R., Badri, N., Galmarini, C.M., Allavena, P., 2014. Trabectedin, a drug acting on both cancer cells and the tumour microenvironment. *British Journal of Cancer* 111(4), 646–650. doi:10.1038/bjc.2014.149

Jefferies, R.P., 1991. Two types of bilateral symmetry in the Metazoa: Chordate and bilaterian. *Ciba Foundation Symposium* 162, 94–120; discussion 121–7. doi:10.1002/9780470514160.ch7

Jeffery, W.R., 2006. Ascidian neural crest-like cells: Phylogenetic distribution, relationship to larval complexity, and pigment cell fate. *Journal of Experimental Zoology* 306B(5), 470–480. doi:10.1002/jez.b.21109

Jeffery, W.R., Chiba, T., Krajka, F.R., Deyts, C., Satoh, N., Joly, J.-S., 2008. Trunk lateral cells are neural crest-like cells in the ascidian *Ciona intestinalis*: Insights into the ancestry and evolution of the neural crest. *Developmental Biology* 324(1), 152–160. doi:10.1016/j.ydbio.2008.08.022

Jeffery, W.R., Strickler, A.G., Yamamoto, Y., 2004. Migratory neural crest-like cells form body pigmentation in a urochordate embryo. *Nature* 431(7009), 696–699. doi:10.1038/nature02975

Kaltenbach, S.L., Yu, J.-K., Holland, N.D., 2009. The origin and migration of the earliest-developing sensory neurons in the peripheral nervous system of amphioxus. *Evolution and Development* 11(2), 142–151. doi:10.1111/j.1525-142X.2009.00315.x

Kamesh, N., Aradhyam, G.K., Manoj, N., 2008. The repertoire of G protein-coupled receptors in the sea squirt *Ciona intestinalis*. *BMC Evolutionary Biology* 8, 129. doi:10.1186/1471-2148-8-129

Kotkamp, K., Kur, E., Wendik, B., Polok, B.K., Ben-Dor, S., Onichtchouk, D., Driever, W., 2014. Pou5f1/Oct4 promotes cell survival via direct activation of mych expression during zebrafish gastrulation. *PLOS ONE* 9(3), e92356–12. doi:10.1371/journal.pone.0092356

Kovalevskij, A.O., 1866. *Entwickelungsgeschichte der einfachen Ascidien.* St. Petersburg, Russia: Eggers & Schmitzdorff.

Kovalevskij, A.O., 1867. *Entwickelungsgeschichte des Amphioxus lanceolatus.* St. Petersburg, Russia: Eggers & Schmitzdorff.

Kowalevsky, A., 2009. Weitere Studien über die Entwicklungsgeschichte Des *Amphioxus lanceolatus,* nebst einem Beitrage zur Homologie des Nervensystems der Würmer und Wirbelthiere. 1–27.

Kusakabe, T., Kusakabe, R., Kawakami, I., Satou, Y., Satoh, N., Tsuda, M., 2001. CI-opsin1, a vertebrate-type opsin gene, expressed in the larval ocellus of the ascidian *Ciona intestinalis. FEBS Letters* 506(1), 69–72.

Lamb, T.D., Collin, S.P., Pugh, E.N., 2007. Evolution of the vertebrate eye: Opsins, photoreceptors, retina and eye cup. *Nature Reviews: Neuroscience* 8(12), 960–976. doi:10.1038/nrn2283

Lamborghini, J.E., 1987. Disappearance of Rohon-Beard neurons from the spinal cord of larval *Xenopus laevis. The Journal of Comparative Neurology* 264(1), 47–55. doi:10.1002/cne.902640105

Langeland, J.A., Tomsa, J.M., Jackman, W.R., Kimmel, C.B., 1998. An amphioxus snail gene: Expression in paraxial mesoderm and neural plate suggests a conserved role in patterning the chordate embryo. *Development Genes and Evolution* 208(10), 569–577.

Lynch, M., Conery, J.S., 2000. The evolutionary fate and consequences of duplicate genes. *Science* 290(5494), 1151–1155.

Maczkowiak, F., Matéos, S., Wang, E., Roche, D., Harland, R., Monsoro-Burq, A.H., 2010. The Pax3 and Pax7 paralogs cooperate in neural and neural crest patterning using distinct molecular mechanisms, in *Xenopus laevis* embryos. *Developmental Biology* 340(2), 381–396. doi:10.1016/j.ydbio.2010.01.022

Maksimovic, S., Nakatani, M., Baba, Y., Nelson, A.M., Marshall, K.L., Wellnitz, S.A., Firozi, P., Woo, S.-H., Ranade, S., Patapoutian, A., Lumpkin, E.A., 2014. Epidermal Merkel cells are mechanosensory cells that tune mammalian touch receptors. *Nature* 509(7502), 617–621. doi:10.1038/nature13250

Meulemans, D., Bronner-Fraser, M., 2002. Amphioxus and lamprey AP-2 genes: Implications for neural crest evolution and migration patterns. *Development* 129(21), 4953.

Meulemans, D., Bronner-Fraser, M., 2004. Gene-regulatory interactions in neural crest evolution and development. *Developmental Cell* 7(3), 291–299. doi:10.1016/j.devcel.2004.08.007

Nagase, T., Shimoda, Y., Sanai, Y., Nakamura, S., Harii, K., Osumi, N., 2000. Differential expression of two glucuronyltransferases synthesizing HNK-1 carbohydrate epitope in the sublineages of the rat myogenic progenitors. *Mechanisms of Development* 98(1–2), 145–149.

Nieto, M.A., 2002. The snail superfamily of zinc-finger transcription factors. *Nature Reviews: Molecular Cell Biology* 3(3), 155–166. doi:10.1038/nrm757

Noda, T., Satoh, N., 2008. A comprehensive survey of cadherin superfamily gene expression patterns in *Ciona intestinalis. Gene Expression Patterns* 8(5), 349–356. doi:10.1016/j.gep.2008.01.004

Northcutt, R.G., Gans, C., 1983. The genesis of neural crest and epidermal placodes: A reinterpretation of vertebrate origins. *The Quarterly Review of Biology* 58(1), 1–28. doi:10.1086/413055

Oda, H., Akiyama-Oda, Y., Zhang, S., 2004. Two classic cadherin-related molecules with no cadherin extracellular repeats in the cephalochordate amphioxus: Distinct adhesive specificities and possible involvement in the development of multicell-layered structures. *Journal of Cell Science* 117(13), 2757–2767. doi:10.1242/jcs.01045

Odenthal, J., Haffter, P., Vogelsang, E., Brand, M., van Eeden, F.J., Furutani-Seiki, M., Granato, M., Hammerschmidt, M., Heisenberg, C.P., Jiang, Y.J., Kane, D.A., Kelsh, R.N., Mullins, M.C., Warga, R.M., Allende, M.L., Weinberg, E.S., Nüsslein-Volhard, C., 1996. Mutations affecting the formation of the notochord in the zebrafish, *Danio rerio. Development* 123, 103–115.

Odenthal, J., Nüsslein-Volhard, C., 1998. Fork head domain genes in zebrafish. *Development Genes and Evolution* 208(5), 245–258.

Ohno, S., 1970. *Evolution by Gene Duplication.* Springer Science & Business Media, Berlin, Heidelberg. doi:10.1007/978-3-642-86659-3

Oonuma, K., Tanaka, M., Nishitsuji, K., Kato, Y., Shimai, K., Kusakabe, T.G., 2016. Revised lineage of larval photoreceptor cells in *Ciona* reveals archetypal collaboration between neural tube and neural crest in sensory organ formation. *Developmental Biology* 420(1), 178–185. doi:10.1016/j.ydbio.2016.10.014

Peel, A., Akam, M., 2003. Evolution of segmentation: Rolling back the clock. *Current Biology* 13(18), R708–710. doi:10.1016/j.cub.2003.08.045

Philippe, H., Lartillot, N., Brinkmann, H., 2005. Multigene analyses of bilaterian animals corroborate the monophyly of Ecdysozoa, Lophotrochozoa, and Protostomia. *Molecular Biology and Evolution* 22(5), 1246–1253. doi:10.1093/molbev/msi111

Rossi, C.C., Kaji, T., Artinger, K.B., 2009. Transcriptional control of Rohon-Beard sensory neuron development at the neural plate border. *Developmental Dynamics* 238(4), 931–943. doi:10.1002/dvdy.21915

Russo, M.T., Donizetti, A., Locascio, A., D'Aniello, S., Amoroso, A., Aniello, F., Fucci, L., Branno, M., 2004. Regulatory elements controlling CI-msxb tissue-specific expression during *Ciona intestinalis* embryonic development. *Developmental Biology* 267(2), 517–528. doi:10.1016/j.ydbio.2003.11.005

Santini, F., Harmon, L.J., Carnevale, G., Alfaro, M.E., 2009. Did genome duplication drive the origin of teleosts? A comparative study of diversification in ray-finned fishes. *BMC Evolutionary Biology* 9, 194–115. doi:10.1186/1471-2148-9-194

Sauka-Spengler, T., Meulemans, D., Jones, M., Bronner-Fraser, M., 2007. Ancient evolutionary origin of the neural crest gene regulatory network. *Developmental Cell* 13(3), 405–420. doi:10.1016/j.devcel.2007.08.005

Schaeffer, B., 1987. *Deuterostome Monophyly and Phylogeny, Evolutionary Biology.* Springer, Boston, MA. doi:10.1007/978-1-4615-6986-2_8

Siddall, M.E., Whiting, M.F., 1999. Long-branch abstractions. *Cladistics* 15(1), 9–24. doi:10.1111/j.1096-0031.1999.tb00391.x

Squarzoni, P., Squarzoni, P., Parveen, F., Parveen, F., Zanetti, L., Zanetti, L., Ristoratore, F., Ristoratore, F., Spagnuolo, A., Spagnuolo, A., 2011. FGF/MAPK/Ets signaling renders pigment cell precursors competent to respond to Wnt signal by directly controlling CI-Tcf transcription. *Development* 138(7), 1421–1432. doi:10.1242/dev.057323

Stolfi, A., Gainous, T.B., Young, J.J., Mori, A., Levine, M., Christiaen, L., 2010. Early chordate origins of the vertebrate second heart field. *Science* 329(5991), 565–568. doi:10.1126/science.1190181

Stolfi, A., Ryan, K., Meinertzhagen, I.A., Christiaen, L., 2015. Migratory neuronal progenitors arise from the neural plate borders in tunicates. *Nature* 527(7578), 371–374. doi:10.1038/nature15758

Tang, W.J., Chen, J.S., Zeller, R.W., 2013. Transcriptional regulation of the peripheral nervous system in *Ciona intestinalis*. *Developmental Biology* 378, 183–193. doi:10.1016/j.ydbio.2013.03.016

Tassy, O., Dauga, D., Daian, F., Sobral, D., Robin, F., Khoueiry, P., Salgado, D., Fox, V., Caillol, D., Schiappa, R., Laporte, B., Rios, A., Luxardi, G., Kusakabe, T., Joly, J.-S., Darras, S., Christiaen, L., Contensin, M., Auger, H., Lamy, C., Hudson, C., Rothbacher, U., Gilchrist, M.J., Makabe, K.W., Hotta, K., Fujiwara, S., Satoh, N., Satou, Y., Lemaire, P., 2010. The ANISEED database: Digital representation, formalization, and elucidation of a chordate developmental program. *Genome Research* 20(10), 1459–1468. doi:10.1101/gr.108175.110

Tokuoka, M., Satoh, N., Satou, Y., 2005. A bHLH transcription factor gene, Twist-like 1, is essential for the formation of mesodermal tissues of *Ciona juveniles*. *Developmental Biology* 288(2), 387–396. doi:10.1016/j.ydbio.2005.09.018

Van Otterloo, E., Li, W., Garnett, A., Cattell, M., Medeiros, D.M., Cornell, R.A., 2012. Novel Tfap2-mediated control of soxE expression facilitated the evolutionary emergence of the neural crest. *Development* 139(4), 720–730. doi:10.1242/dev.071308

Vlaeminck-Guillem, V., Carrere, S., Dewitte, F., Stehelin, D., Desbiens, X., Duterque-Coquillaud, M., 2000. The Ets family member Erg gene is expressed in mesodermal tissues and neural crests at fundamental steps during mouse embryogenesis. *Mechanisms of Development* 91(1–2), 331–335.

Wada, H., Holland, P., Sato, S., Yamamoto, H., Satoh, N., 1997. Neural tube is partially dorsalized by overexpression ofHrPax-37: The ascidian homologue ofPax-3andPax-7. *Developments in Biologicals* 187, 240–252.

Winchell, C.J., Sullivan, J., Cameron, C.B., Swalla, B.J., Mallatt, J., 2002. Evaluating hypotheses of deuterostome phylogeny and chordate evolution with new LSU and SSU ribosomal DNA data. *Molecular Biology and Evolution* 19(5), 762–776. doi:10.1093/oxfordjournals.molbev.a004134

Yu, J.-K., Meulemans, D., McKeown, S.J., Bronner-Fraser, M., 2008. Insights from the amphioxus genome on the origin of vertebrate neural crest. *Genome Research* 18(7), 1127–1132. doi:10.1101/gr.076208.108

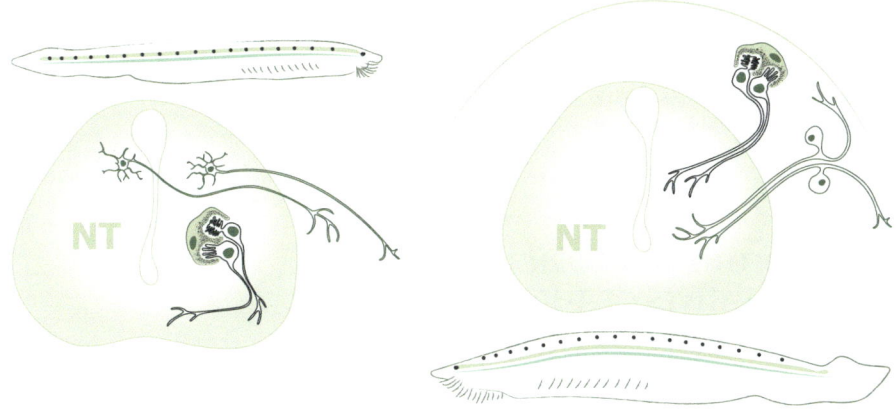

5 Elaboration of Fates in Neural Crest Lineage during Evolution

Igor Adameyko

CONTENTS

There is grandeur in this view of life, with its several powers, having been
originally breathed into a few forms or into one; and that, whilst this planet
has gone cycling on according to the fixed law of gravity, from so simple a
beginning endless forms most beautiful and most wonderful have been, and
are being, evolved.

——**Charles Darwin**, *The Origin of Species*

5.1 EVOLUTIONARY ORIGIN OF PRIMARY DERIVATIVES OF THE NEURAL CREST

Neural crest cells are incredibly multipotent and give rise to dozens of cell types
that include mesenchymal and non-mesenchymal fates essential for development of
vertebrate embryos (Dupin and Sommer, 2012). The question of how these fates
evolved and assembled into a highly coordinated lineage is of the utmost importance
for understanding the evolution of chordates.

Non-mesenchymal fates include sensory neurons (proprioceptive, mechanocep-
tive and nociceptive), autonomic neurons (sympathetic, parasympathetic, enteric),
neuroendocrine chromaffin cells, melanocytes, myelinating and non-myelinating
Schwann cells, boundary cap cells, satellite and terminal glial cells. The spectrum
of mesenchymal fates embraces chondrocytes, osteocytes, odontoblasts, dental pulp
cells, dermal fibroblasts, smooth muscle cells, pericytes, cardiomyocytes, endoneurial fibroblasts, tenocytes, adipocytes, mesenchymal cells in bone marrow or thymus
and other cell types (Hall and Gillis, 2013; Le Douarin and Kalcheim, 1999; Le
Douarin et al., 2004). Neural crest multipotency depends on the anterior-posterior
position in the embryo. In the cranial region, neural crest cells migrate in several
distinct streams and give rise to the full spectrum of mesenchymal fates in addition
to non-mesenchymal derivatives, whereas more fate-restricted trunk neural crest
cells migrate in regions posterior to the branchial arches and produce only non-
mesenchymal cell types (Espinosa-Medina et al., 2017; Simoes-Costa and Bronner,
2016). It is generally accepted that only cranial neural crest cells can generate mesenchymal fates, although there are exceptions to the rule that include mesenchymal
derivatives produced by the trunk neural crest in turtles and in cartilaginous fishes
(Cebra-Thomas et al., 2013; Gillis et al., 2017).

Neural crest cells played a key role during major evolutionary transitions that
included acquisition of cranial endoskeleton with hinged jaws, complex heart, teeth
and odontodes forming protective dermal armor, improved sensory, endocrine
and immune systems, pigmentation providing camouflage and display, myelinated
peripheral nerves and many more (Adameyko and Lallemend, 2010; Bronner, 2015;
Gans and Northcutt, 1983; Green et al., 2015). Understanding elaboration of fates in
the neural crest lineage is a key for devising evolutionary scenaria explaining these
major morphological and functional transitions. In spite of a significant progress
in an evodevo field, we still known very little about general principles of evolution
of the multipotency states and new cell fates (Arendt et al., 2016). When it comes
to the neural crest lineage, we rely on comparative analysis of existing fates within
a hierarchical phylogenetic system of evolutionary connected chordate animals.

Additionally, we take advantage of tracing the evolutionary changes of gene expression programs essential for specific functions or cell identities.

The question of elaboration of the currently known neural crest fates is intimately connected with a problem of the neural crest origin that is widely discussed in the existing literature (Donoghue et al., 2008; Nikitina et al., 2009; Van Otterloo et al., 2013; Wada, 2001; Yu, 2010). Some authors focus on the evolutionary assembly of the neural crest-specific gene regulatory networks (Donoghue et al., 2008; Yu, 2010), whereas the others discuss the evolutionary forces and transitions at cell and tissue levels (Ivashkin and Adameyko, 2013).

It is plausible that most of the modern neural crest fates were incrementally elaborated during a long period of time, which is supported by the existing literature. Thus, to understand the incremental evolution of the spectrum of fates, we should firstly consider what the proto neural crest cells could be and which fates were the "primary "or "secondary" inventions.

Since we arrived at the discussion of a cell type evolution, it might be useful to mention that evolution of organs, genes, or other functional entities often follows "the substitution of function" logic that is also sometimes called "evolution through the loss-of-function" or "neofunctionalization" (Arendt, 2008; Hottes et al., 2013). This evolutionary principle states that the main function of the structure can be gradually diminishing due to the changes in the environment (or can be lost abruptly due to a loss-of-function mutation), and this process releases the evolutionary potential of the structure, thus adapting it for something else. According to this logic, evolution of the neural crest could start from a cell type or a conglomerate of lineage-related cell types dedicated to some function that was gradually diminishing over time. Therefore, hypothetical proto neural crest cells could gain evolutionary plasticity after being released from some role inside of the CNS (central nervous system). Eventually, they developed ways to exit the neural tube and generate novel cell types mediating melanin-based pigmentation, autonomic control of organ function, development of oropharyngeal skeleton, and many more. Since we are talking about elaboration of fates in the neural crest lineage, I will not discuss the evolution of specification of the neural plate border and changes in patterning of the neural and nonneural ectoderm reviewed elsewhere (Meulemans and Bronner-Fraser, 2004).

Following the logic of neofunctionalization, we attempted to identify a structure or a cell type that is present in amphioxus (another name is *lancelet*) but lost or transformed in cyclostomes. Internal organs of the amphioxus receive innervation from the local peripheral neurons forming plexuses in the gut and atrial region. At the same time, peripheral sensory neurons of unclear origin and other sensory cell types are well represented in the lancelet skin and below (Lacalli, 2004). It is hard to envision the evolutionary scenario that would explain the forces behind redeveloping these peripheral neurons from another cellular source such as classical neural crest, given that these neurons are already successfully generated by some other progenitor cell types. Similarly, amphioxus possesses cellular cartilage inside of the larval oral cirri (Tarazona et al., 2016), and that facial cartilage does not require the existence of the canonical neural crest. However, there is something drastically different between amphioxus and currently living cyclostomes, and these are the Hesse (dorsal) ocelli—pigmented photosensory organs embedded into the neural tube along the entire length of an animal (Lacalli, 2004). These ocelli contain melanin-based pigment for efficient shading and, therefore, provide directionality of photoreception. T. Lacalli suggested that the major sensory function of these ocelli might be tracking the position of the filter-feeding animal inside of the burrow (Lacalli, 2004). At the same time, amphioxus does not possess any other melanin-containing cells participating in camouflage or photoprotection. Indeed, melanin is rather secluded to the photosensory structures and is used for photoreceptor shading in a variety of animal groups, appearing already within pigmented cells in photosensory rhopalia of jellyfishes ((Kozmik et al., 2008), discussed in (Ivashkin and Adameyko, 2013).

Cyclostomes do not have anything similar to pigmented Hesse ocelli inside of their spinal cords. However, they do possess melanin-containing pigment cells in their skin. In 2013, we suggested that the loss of the photosensory function in Hesse-like photosensory organs could drive the first steps of proto neural crest fate elaboration (see Figure 5.1). According to the current knowledge, vertebrate melanocytes and melanophores, being pigmented and serving as camouflage/display/photoprotection, are also photosensory—they express melanopsins (Provencio et al., 1998) together with components of the phototransduction pathway (Bellono et al., 2014; Wicks et al., 2011). Furthermore, melanophores and melanocytes physiologically respond to light by generating a new pigment or by dynamically redistributing pigment granules within a cell (Bellono et al., 2014; Provencio et al., 1998; Wicks et al., 2011). Thus, modern pigment cells in skin retained their photosensory properties despite losing the visual function, which is suggested as a main role of the pigmented dorsal Hesse ocelli in amphioxus. Following this logic, camouflage, photoprotection, and display could become primary functions of the former pigmented sensory cells possessing mainly the visual function in amphioxus-like animal. Yet, we did not mention what could be the driving force behind this transformation of Hesse-like ocelli into melanocytes or melanophores. One plausible explanation is based on the changes in the lifestyle of the early chordates involving the transition from sedentary filter-feeding life to a rather active predatory mode of living. Dorsal Hesse ocelli, providing the information about the animal's position in the burrow as a measure of a safety, were not much required after the transition to predation. Their major function turned nonessential, and corresponding evolutionary potential was released and transformed into light-responsive camouflage, photoprotection, and display. For the details on this hypothesis, please see Figure 5.1 and (Ivashkin and Adameyko, 2013). According to this scenario, the prototypic neural crest cell could develop from a progenitor that gained capacity to migrate outside of the neural tube towards the body periphery. This progenitor could generate scattered pigmented photosensory cells in subepithelial layers, which additionally benefited the animal with camouflage. Since the neural tube of amphioxus larva contains extremely few cells (schematized in Figure 5.1) as compared to the vertebrate analog, repositioning of an individual multipotent progenitor via a single-cell delamination or migration along the outgoing nerves could lead to the other fates forming at the periphery. This could explain how the first pigment and neuronal fates emerged from a single migratory proto neural crest cell.

This hypothesis is supported by the results obtained in a vertebrate sister group—tunicates. Abitua and coauthors reasoned that tunicate *Ciona intestinalis* has a cephalic melanocyte lineage (a9.49) that expresses Id, Snail, Ets, FoxD3—neural plate border

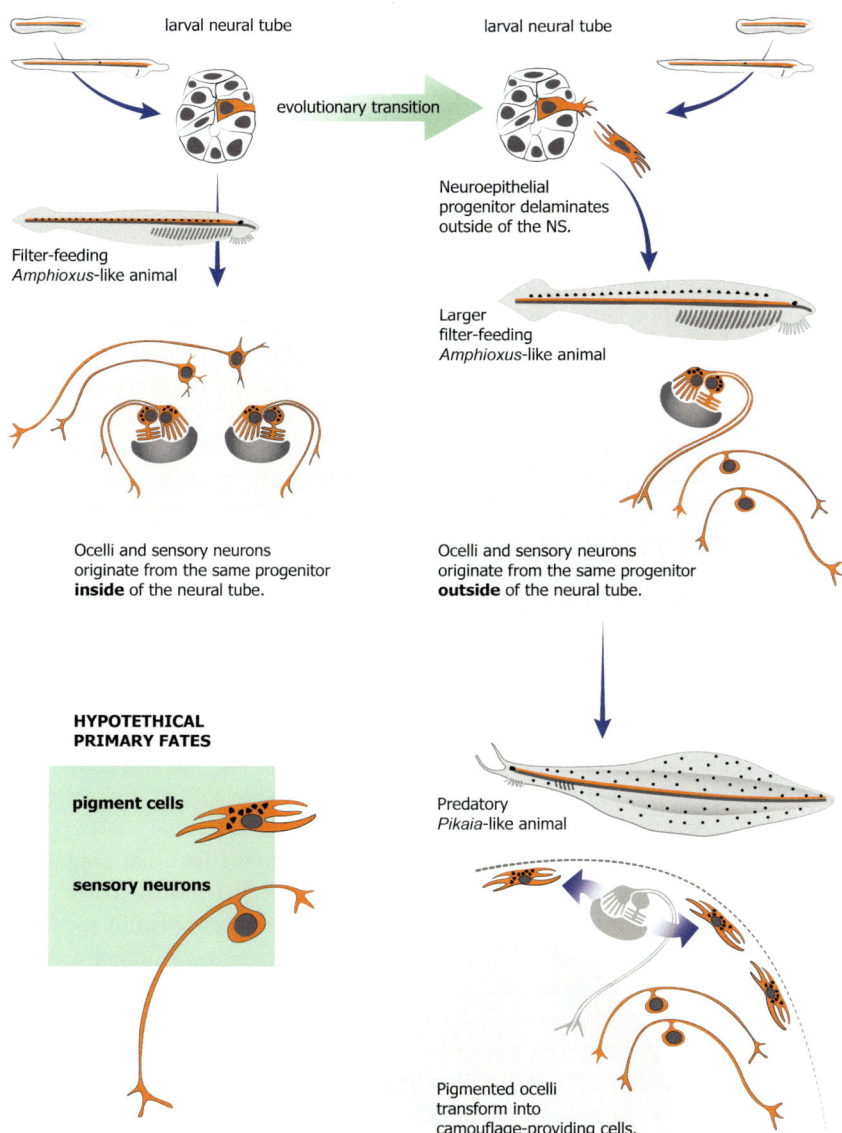

FIGURE 5.1   Hypothetical scenario of the formation of primary neural crest fates according to a principle of evolutionary neofunctionalization. (Redrawn and conceptually modified from Ivashkin and Adameyko, [2013].) Pigmented Hesse-like ocelli within the central nervous system of stem cephalochordates eventually lost their visual function and translocated to the body periphery, where they gave rise to melanocytes—photosensory pigment cells providing camouflage, photoprotection, and display. The translocation of pigmented ocelli occurred due to the delamination of the neural tube progenitor cell that could also give rise to the sensory neurons at the periphery. Similarly, the only delaminating cells observed so far in tunicate larvae turned out to be progenitors of pigment cells and sensory neurons (Abitua et al., 2012; Stolfi et al., 2015).

and neural crest–specific genes and, therefore, might be homologous to the vertebrate neural crest. Alternatively, it might represent a rudimentary crest (Abitua et al., 2012). The posterior derivatives (a10.97) of a9.49 cells arise at the neural plate border and are destined to intercalate at dorsal midline and give rise to the melanocyte of the photosensory ocellus and the gravisensing otolith (Nishida and Satoh, 1989). In a tadpole tunicate larva, the melanocyte progenitor delaminates into the neural tube lumen and stays there without demonstrating any further migratory capacity. Therefore, melanin-based pigmentation in tunicates is associated with the intra-CNS photosensory function similar to the situation in cephalochordates. Cells of a9.49 lineage express Mitf—a key conserved factor of melanocyte differentiation (Yajima et al., 2003). Their derivatives, A10.97 cells, also express Tcf/Lef transcriptional factors that mediate Wnt signaling (Squarzoni et al., 2011). Thus, it is likely that Wnt signaling plays an important role in ascidian melanogenesis similarly to its role in vertebrate neural crest. This idea gained support in experiments with targeted misexpression of Wnt7 in a9.49 cells, which led to the conversion of pigmented progenitors into ocelli (Abitua et al., 2012). Based on these results, Abitua et al. suggested that melanocyte gene regulatory network emerged before the divergence of tunicates and other chordate lineages. Furthermore, cells of a9.49 lineage were experimentally transformed into a migratory mesenchymal population of cells after the introduction of a Twist-expressing construct (Abitua et al., 2012). Based on this, authors suggested that the cooption of epithelial-to-mesenchymal transition factors and mesenchymal determinants (such as Twist) could be a key for ectomesenchyme induction in the neural crest lineage enabling the following emergence of the "new head" (Abitua et al., 2012). The capacity of the tunicate melanocyte progenitor to delaminate inside of the neural tube may also suggest that the basic delamination toolkit was also developed very early in evolution before the split of cephalochordates urochordate/euchordate ancestor.

The possible criticism of this hypothesis is related to the fact that progenitors of pigment cells in amphioxus or tunicates are not forming from the cells specified at the border of the neural epithelium. However, the concepts behind specification of neural versus nonneural ectoderm have become increasingly complex in recent years with additional insights into secondary neurulation (Shimokita and Takahashi, 2011) or the dynamics of neuromesodermal progenitors forming at the posterior end of developing embryo and producing somatic mesoderm together with the largest portion of future neural tube tissue (Henrique et al., 2015). Thus, patterning mechanisms might be flexible and evolutionarily unstable as compared to individual progenitors and adult cell type identities and functions, which provides a hope for the hypothesis depicted above.

Recently, Stolfi and coauthors discovered another cell type that delaminates and migrates from the tunicate neural plate border to form the peripherally located bipolar tail neuron (BTN)—a neuronal type that is similar and, most likely, homologous to the vertebrate neural crest–derived sensory neurons residing in dorsal root ganglia (Stolfi et al., 2015). BTNs express transcriptional factors Neurogenin and LIM-homeodomain factor Islet that are homologs to vertebrate Neurogenin 1/2 and Isl1 required for sensory neurons to develop. Authors of this study converged on the idea that the olfactorean ancestor possessed the neural plate border cells that could give rise to the peripheral neurons and pigment cells (Stolfi et al., 2015).

According to the two discussed experimental studies performed in tunicates (Abitua et al., 2012; Stolfi et al., 2015) and the evolutionary scenario focused on Hesse-like ocelli of cephalochordates (Ivashkin and Adameyko, 2013), the primary fates generated by the migratory neural plate border progenitors could include pigment cells and peripheral neuronal subtypes, whereas the mesenchymal derivatives would exemplify secondary fates. Further evolution (fueled by increasing body complexity and environmental changes) led to the increased specialization and elaboration/cooption of new fates in both non-mesenchymal and mesenchymal domains of the neural crest progeny.

KEY READING

Abitua, P.B., Wagner, E., Navarrete, I.A., Levine, M., 2012. Identification of a rudimentary neural crest in a non-vertebrate chordate. *Nature* 492, 104–107.
Ivashkin, E., Adameyko, I., 2013. Progenitors of the protochordate ocellus as an evolutionary origin of the neural crest. *Evodevo* 4, 12.
Stolfi, A., Ryan, K., Meinertzhagen, I.A., Christiaen, L., 2015. Migratory neuronal progenitors arise from the neural plate borders in tunicates. *Nature* 527, 371–374.

5.2 FURTHER ELABORATION OF THE NEURAL CREST LINEAGE COMPLEXITY: DIVERSIFICATION OF GLIAL SUBTYPES

According to the discussion above, melanocytes and peripheral neurons could be the primary derivatives of the proto neural crest. However, this notion does not help us to understand the evolutionary origin of another important neural crest derivative—peripheral glia represented by myelinating and non-myelinating Schwann cells associated with the nerve fibers. Recent insights into the development of Schwann cells in vertebrates via an intriguing stage of Schwann cell precursor (Furlan and Adameyko, 2018; Kastriti and Adameyko, 2017; Petersen and Adameyko, 2017; Zinin et al., 2014) may suggest previously unanticipated mechanisms of glial evolution. After the formation of the first afferent and efferent neurons projecting to the periphery, migrating neural crest cells associate themselves with the outgrowing nerves and transit into a state called *Schwann cell precursor* (SCP) (Dong et al., 1999; Jessen et al., 1994). During the next phases of development, most of these nerve-associated SCPs will become immature Schwann cells that, in turn, will differentiate into different subtypes of mature Schwann cells (myelinating, non-myelinating, terminal, satellite, and sensory nerve ending–associated) (Jessen and Mirsky, 2005; Kastriti and Adameyko, 2017).

However, this historical picture turned out to be incomplete, since SCPs appeared to have properties and gene expression programs that render them similar to their parental population—neural crest (Furlan and Adameyko, 2018). As discovered recently, SCPs are multipotent cells giving rise to a number of derivatives that were previously thought to develop directly and uniquely from the migratory neural crest (Figure 5.2). The spectrum of such derivatives includes melanocytes, autonomic neurons (parasympathetic and enteric), neuroendocrine chromaffin cells, dental mesenchymal stem cells, endoneurial fibroblasts, and mesenchymal cells with specialized hematopoietic stem cell niche function (Adameyko and Lallemend, 2010; Adameyko et al., 2009; Adameyko et al., 2012; Dyachuk et al., 2014; Furlan et al., 2017; Isern et al., 2014; Joseph et al., 2004; Uesaka et al., 2015). The expression of Sox10, FoxD3 and other neural crest-specific genes together with the capacity to traverse the nerves and multipotency suggest that SCP could be more of a nerve-associated neural crest subtype rather than a cell committed to a glial fate (Figure 5.2). According to this notion, Green, Uy and Bronner demonstrated that already in a lamprey, SCPs travel the thoracic nerves to populate the forming gut with enteric neurons, thus, resembling the analogous (or homologous?) mechanism of development of parasympathetic and enteric neurons in mammals (Green et al., 2017).

Furthermore, it is possible that ancient proto neural crest cells could exit the neural tube or spread after delamination via the nerve entry or exit points. Conceptually, the beginning of such spreading might not even require a delamination at the dorsal aspect of the neural tube. For example, in the case of the genetic ablation of a boundary cap population, which prevents CNS cells from crossing the CNS–PNS boundary, progenitors of the CNS glia migrate outside of the neural tube via the ventral roots and myelinate proximal regions of the outgoing peripheral nerves (Coulpier et al., 2010). Taken together, SCPs represent a special nerve-dependent and disseminating subtype of the multipotent neural crest rather than a cell that in the future will necessarily commit to become glia and serve the neurons (see Figure 5.2).

Hypothetically, ancient proto neural crest cells could resemble modern SCPs much more than we expect. In line with this, the lancelets develop highly specialized

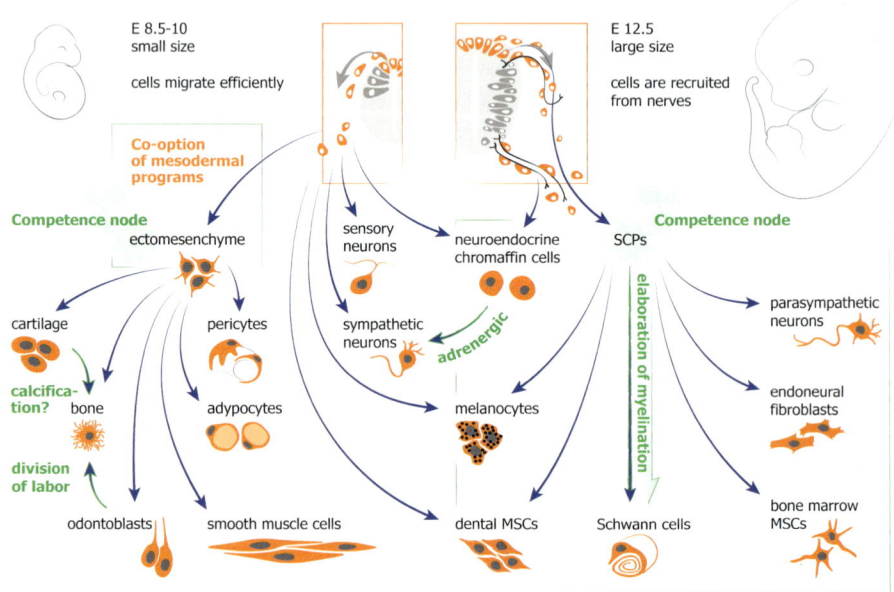

FIGURE 5.2 Neural crest lineage tree and evolutionary connections between different neural crest–derived cell types. The scheme highlights the evolutionary innovations (arrows highlighted with green) that led to the acquisition of new cells types within the lineage.

motor nerve–like synaptic structures directly on the ventral neural tube surface instead of growing ventral roots with motor axons extending towards the periphery (Fritzsch and Northcutt, 1993). The absence of the outgoing ventral motor axons combined with the presence of developed dorsal roots could predefine the benefits of the dorsal location, where the exit and entry roots could provide for proto neural crest navigation, either after delamination or via the direct nerve-guided exit of CNS-resident progenitors.

The capacity of the crest and SCPs to populate the peripheral nerves could enable another innovation next in line: elaborating true peripheral glial cells serving the neurons and developing insulation strategies for the faster propagation of the action potentials. There are very few scattered peripheral glial cells identified in amphioxus, and the origin or properties of these glia-like cells are currently unknown (Flood, 1966). In cyclostomes possessing almost all known neural crest derivatives, the nerves are nonmyelinated despite the fact that they are densely covered by peripheral glial cells with unknown functions (Bullock et al., 1984). Nevertheless, the analysis of a lamprey genome revealed the presence of numerous genes involved in myelination in gnathostomes, including major components of myelin sheaths such as Mpz, Plp, Pmp22, Mal, Mbp, enzyme CNP, and transcription factor Mytll (Smith et al., 2013). This could be seen as a preadaptation to myelination or, alternatively, it may suggest that myelination capacity was a property of the last common vertebrate ancestor that somehow got lost in cyclostomes. leaving only the traces of genetic program formerly involved in myelination. This last scenario does not sound convincing because tunicates (*Ciona intestinalis*) also possess homologs of Plp1 and Mytll and lack myelination at the same time (Smith et al., 2013).

First myelinated axons in the lineage of chordates are identified in cartilaginous and bony fish. Elaboration of myelination is a powerful trend also observed in different invertebrate lineages including Ecdysozoa (megacalanoid copepods, euphausiid and mysid shrimp, decapod shrimp) and Lophotrochozoa (earthworms, bamboo worms, parchment worms) (Hartline and Colman, 2007; Schweigreiter et al., 2006). Some authors have suggested that the primary role of myelin was energy conservation, and that only later was myelin recruited into the nervous system for insulating the axons (Stiefel et al., 2013). Composition of the molecular toolkit involved in myelination supports independent elaboration of myelinating glial cells in the line of chordate animals (Mobius et al., 2008).

The peripheral myelinating glial cells share a striking similarity with myelinating cells residing within the CNS—oligodendrocytes. Gene expression programs directing the development of oligodendrocytes and Schwann cells, as well as many of their properties, are similar between these cell types (Jacob, 2015; Kastriti and Adameyko, 2017). In case of boundary cap ablation, progenitors of oligodendrocytes leak through the motor nerve exit point and myelinate nerve fibers extending to the periphery (Coulpier et al., 2010). In evolution, myelination appeared around the same time in both central and peripheral nervous systems during transition from agnathans to gnathostomes. However, we can only speculate regarding whether the myelination program was innovated by oligodendrocyte-like cells and then coopted to the neural crest lineage or the other way around.

Increase in the body size of predatory and diversifying chordates could require a fast system of muscle control and sensory information transfer. This could stimulate the elaboration of peripheral myelination providing fast action potential propagation along the long axons targeting muscles. This argument rather supports the assembly of a myelination program in peripheral nerve–associated cells with following cooption of this program to the CNS glial populations.

On the other hand, cyclostomes are capable of fast and precise movements controlled by large-diameter axons that are conceptually similar to the giant axons of cephalopods (Clay, 1985; Hill et al., 2008). Thus, the motor stimuli in these animals are efficiently relayed via a mechanism that does not require myelination but rather employs large-diameter axons to hasten action potential propagation, given sufficient space inside of the brain and spinal cord. Competition for the space that is necessary to host large numbers of neurons generating an increase in computational power could favor miniaturization of the large-diameter nerve tracks during evolution. Thus, the elaboration of a myelination program could take place primarily inside of the CNS for the sake of such miniaturization, and then could be exported to the peripheral nerves.

Production of a myelin is not the only requirement for such transition from large to small-diameter conducting fibers. Clustering of voltage-gated ion channels at gaps within a myelin sheath, the Ranvier nodes, is another key property required to build an efficient small-diameter insulated fiber. Surprisingly, nonmyelinated axons of a lamprey demonstrate the presence of narrow segments enriched with ion channels similarly to the currently living gnathostomes (Hill et al., 2008). Such clustering is achieved because of the linkage of ion channels to the cytoskeleton via the adaptor protein ankyrin-G. The anchor motif for clustering the sodium channels evolved before the split of cephalochordates, whereas the anchoring motif for potassium channels was acquired later, at the onset of myelination (Hill et al., 2008). Hence, clustering of sodium voltage-gated ion channels in noninsulated fibers might be considered as a key preadaptation to the development of myelination by vertebrate glial cells. Coclustering of potassium channels and synthesis of myelin could be the next-in-line innovations required to build the modern fast conductive system.

Very little is currently known or suggested about the evolution of peripheral non-myelinating Remak cells, terminal glial cells at the neuromuscular junctions, or satellite glial cells inhabiting ganglia of PNS (Kastriti and Adameyko, 2017). Motor axons in cyclostomes are already insulated by non-myelinating Schwann cells, and, when the axons reach the neuromuscular junction, glial cells demonstrate a plug-like shape and completely cover terminal end plates on white and intermediate muscle fibers (Korneliussen, 1973). Thus, terminal glial cells and possibly other specialized peripheral glial subtypes could evolve before the myelinating cells.

KEY READING

Kastriti, M.E., Adameyko, I., 2017. Specification, plasticity and evolutionary origin of peripheral glial cells. *Curr Opin Neurobiol* 47, 196–202.

Green, S.A., Uy, B.R., Bronner, M.E., 2017. Ancient evolutionary origin of vertebrate enteric neurons from trunk-derived neural crest. *Nature* 544, 88–91.

Furlan, A., Adameyko, I., 2018. Schwann cell precursor: a neural crest cell in disguise? *Dev Biol* 4.

Hartline, D.K., Colman, D.R., 2007. Rapid conduction and the evolution of giant axons and myelinated fibers. *Curr Biol* 17, R29–R35.

5.3 DIVERSIFICATION OF SENSORY AND AUTONOMIC NEURONAL SUBTYPES

Sensory neurons projecting to the body periphery are present already in cephalochordates and tunicates (Lacalli, 2004), although their functional subtypes accommodating different sensory modalities are not well investigated. Dorsal Retzius bipolar cells are the intramedullary neurons residing inside of the amphioxus CNS. Being inside of the neural tube, these sensory neurons extend their processes to the periphery of the body, presumably ending up in skin (Lacalli, 2004), which would render them mechanosensory or nociceptive. Amphioxus is not an easily tractable system, whereas tunicate development is amenable for lineage tracing and genetic manipulations, which has resulted in identification of delaminating sensory bipolar tail neurons that originate from the neural plate border, thus resembling neural crest (Stolfi et al., 2015). At this point, we do not know if there are different subtypes of bipolar tail neurons or dorsal bipolar Retzius cells that could correspond to proprioceptive, mechanoreceptive, and nociceptive neurons of vertebrate dorsal root ganglia.

In teleost fish, Rohon-Beard cells represent transient mechanosensory neurons that originate inside of the dorsal neural tube from the lateral neural ectoderm, thus being somewhat similar to dorsal Retzius bipolar cells from amphioxus in terms of their origin and localization. During later development, Rohon-Beard neurons degenerate and are replaced by more diverse neural crest–derived sensory neurons in dorsal root ganglia outside of the neural tube (Rossi et al., 2009). Hence, functionally similar cell types can be reproduced twice during the development of an individual animal from two different cell sources. Cyclostomes also possess both transient Rohon-Beard sensory neurons and dorsal root ganglia neurons (Nakao and Ishizawa, 1987a, b), which strengthens the plausibility of cooption of a sensory neuronal fate from an intra-CNS progenitor into migratory proto neural crest lineage, although the driving forces of such evolutionary transformation are not clear. This is quite probable, given that major populations of peripheral sensory neurons including Rohon-Beard cells, neural crest-derived sensory neurons, and even the intra-CNS neurons of mesencephalic trigeminal nucleus (collecting proprioceptive information from jaws) express Islet1, Runx, Brn3a, Drg11, and Trk receptors showing a large degree of commonality between gene expression programs (Dyer et al., 2014). This commonality may reflect the interlinked evolutionary processes that occurred during elaboration of the origin mechanisms providing generation of these related cell types. Common gene expression programs observed in neural crest–derived and intra-CNS neurons suggest the assembly of sensory neuronal fate before the elaboration of the neural crest. Since the neurons of mesencephalic trigeminal nucleus contain neurons functionally similar to proprioceptors residing in dorsal root ganglia (Lipovsek et al., 2017), it is possible to hypothesize that different sensory subtypes corresponding to mechanoreception, pain reception, and proprioception could evolve before the origin of organisms possessing the neural crest.

Initial evolution of these neuronal subtypes could proceed according to the "division of labor" model previously established for the evolution of photoreceptors (Arendt et al., 2009; Arendt et al., 2016). The corresponding evolutionary transformations include changes in the so-called "core regulatory complex" (CoRC) that is similar to a gene regulatory network consisting of transcription factors. Changes in this complex render emerging cell types distinct and further facilitate their independent evolution via elaboration of apomeres—independent functional modules responsible for unique properties or functions (Arendt et al., 2016).

The evolution of fates within the autonomic nervous system of chordates is also enigmatic. Amphioxus possesses autonomic neurons organized into local oral, atrial, and enteric plexuses. Functional organization of these plexuses is not well understood in terms of the modalities and subtypes of the participating neurons that might include integrated sensory, motor, and interneuron populations (Lacalli, 2004). Subdivision of plexuses and other autonomic peripheral neurons into sympathetic and parasympathetic compartments based on neuromediatory properties and gene expression programs is not sufficiently studied in cephalochordates, tunicates, and cyclostomes. Experiments on tunicate peripheral plexuses confirmed its autonomic nature since the surgical separation of the plexus from the cerebral ganglion did not significantly change its activity and general function (Mackie and Wyeth, 2000). The developmental origin of these autonomic neurons is not understood and requires further investigation. The evolutionary scenario that explains how different autonomic fates were coopted into the neural crest lineage is also missing.

Cyclostomes possess scattered autonomic neurons (Green and Bronner, 2014; Nilsson, 2011) and chromaffin cells (Paiement and McMillan, 1975). Despite the fact that an organized sympathetic chain was not found in lampreys and hagfishes as reviewed by (Haming et al., 2011), cyclostomes hypothetically might have developed scattered neural crest–derived sympathetic and possibly parasympathetic neuronal subtypes. In cyclostomes, chromaffin cells are found to form the irregular layer in the tunica adventitia of the dorsal aorta and the segmental arteries of the ammocoete (Green and Bronner, 2014; Paiement and McMillan, 1975) instead of being grouped into a cluster similar to the adrenal gland medulla or Zuckerkandl organ observed in mammals.

Neuroendocrine chromaffin cells could be the pioneering sympathetic components at the body periphery since they are capable of releasing adrenalin and noradrenalin into the systemic circulation or nearby tissues to act as endocrine and paracrine signals (Eiden and Jiang, 2018). Such a neuroendocrine module does not require precise connections with target organs via nerve terminals to regulate the operation of multiple tissues. Thus, fine control of sympathetic regulation could evolve later at the level of protosympathetic neurons that coopted the adrenergic phenotype from potentially preexisting neuroendocrine chromaffin cells as suggested by the similar differentiation and effector genetic programs. Indeed, the developmental program of chromaffin cells and sympathetic neurons is very similar and embraces key gene regulatory network components such as Ascl1, Phox2b, Phox2a, and other genes (Furlan et al., 2017). However, transcriptional factors Phox2, Hand,

and Ash/Ascl were found not to be expressed in the same cells during early cyclostome development (Haming et al., 2011). Still, these master regulators may switch on and off sequentially depending on their positions in the hierarchy of molecular events leading to chromaffin and sympathetic phenotypes in cyclostomes. The order of analogous molecular events might be different in gnathostomes. This will definitely require further investigation.

KEY READING

Arendt, D., Musser, J.M., Baker, C.V.H., Bergman, A., Cepko, C., Erwin, D.H., Pavlicev, M., Schlosser, G., Widder, S., Laubichler, M.D., Wagner, G.P., 2016. The origin and evolution of cell types. *Nat Rev Genet* 17, 744–757.
Green, S.A., Bronner, M.E., 2014. The lamprey: A jawless vertebrate model system for examining origin of the neural crest and other vertebrate traits. *Differentiation* 87, 44–51.
Lacalli, T.C., 2004. Sensory systems in amphioxus: A window on the ancestral chordate condition. *Brain Behav Evol* 64, 148–162.

5.4 ACQUISITION AND DIVERSIFICATION OF MESENCHYMAL FATES

5.4.1 ACQUISITION OF ECTOMESENCHYME

We are not aware of animals that possess neural crest cells generating only non-mesenchymal derivatives such as neurons, Schwann cells, and melanocytes. Yet, in the absence of these examples, the idea about the primary elaboration of pigment and neuronal fates remains a barely supported hypothesis. In fact, acquisition of mesenchymal fates could occur at the same time as neuromelanogenic derivatives. Strictly speaking, it is not clear which fates could precede the others. For instance, mesenchymal fates could be elaborated in a concomitant way with neuronal and pigmented derivatives as a result of improving proto-neural crest migration strategies. This could involve a recruitment of mesenchymal gene expression programs necessary to promote epithelia-to-mesenchymal transition and to generate actively moving migratory cells.

The spectrum of mesenchymal fates generated by the cranial neural crest lineage includes chondrocytes, osteocytes, adipocytes, dermal fibroblasts, odontoblasts and pulp cells, cells of dermal papilla, pericytes, smooth muscle cells of a vascular wall, smooth muscle cells and cardiomyocytes of the heart, some specialized cell types in the bone marrow and thymus, and many more (Hall and Gillis, 2013; Le Douarin and Kalcheim, 1999; Le Douarin et al., 2004). Primary anterior–posterior subdivision of the neural crest into cranial, vagal, and trunk regions with different competences and patterning mechanisms could potentially evolve before the elaboration of first mesenchymal derivatives. As I mentioned in the beginning of this chapter, trunk neural crest cells are not competent to generate mesenchymal cell types *in vivo* in the majority of cases (Le Douarin and Kalcheim, 1999), although there are exceptions to this rule. In turtles, trunk neural crest cells generate skeletogenic tissues of the forming plastron bones (Cebra-Thomas et al., 2007; Cebra-Thomas et al., 2013; Clark et al., 2001; Gilbert et al., 2007). Similarly, odontoblasts of the dermal armor (odontodes) in skates are derived from the trunk neural crest according to the recent DiI tracing experiments (Gillis et al., 2017). Finally, trunk neural crest cells give rise to populations of mesenchymal cells inside of the peripheral nerves, and these crest-derived mesenchymal cells actively participate in limb regeneration contributing to the blastema (Carr et al., 2019). These exceptions could hypothetically highlight the evolutionary plasticity of non-mesenchymogenic neural crest populations and show an example of how mesenchymal fates could be introduced among the spectrum of nonmesenchymal derivatives. Thus, it might be tempting to speculate that trunk neural crest always has some sort of potential to form mesenchymal progenies.

Mesenchymogenic cranial neural crest cells migrate in distinct streams (Kulesa et al., 2010) and generate nonmigratory primary ectomesenchyme that later undergoes multi-lineage differentiation (Kaucka et al., 2016) in the head and neck. Ectomesenchyme is a multipotent tissue that undergoes spatial patterning and generates clones that become gradually fate restricted, depending on their position and combinatorial extrinsic signaling code (Kaucka et al., 2016). Discussions about how neural crest could gain the capacity to generate cartilage or bone are often found in the literature (Hall and Gillis, 2013). Yet, the neural crest cells do not generate cartilage and bone immediately after migrating to their target destinations, but require the intermediate stage of ectomesenchyme. The majority, if not all, terminal mesenchymal fates are ectomesenchyme-derived, and only via ectomesenchyme connect their lineages with the ancestral neural crest root. This multipotency range relates neural crest-derived ectomesenchyme to mesoderm that generates similar cell types in the rest of the body (Loebel et al., 2003). Mesoderm-derived cell types are conserved among many invertebrate groups, and the capacity of mesoderm to generate muscle and connective tissues was likely present in the common bilaterian ancestor, thus preceding the elaboration of the neural crest (Technau and Scholz, 2003). Hence, cranial neural crest–derived ectomesenchyme can be considered a mesoderm-like tissue. Differentiation of cartilage and bone from the ectomesenchyme follows the same molecular principles as specification of these skeletogenic tissues from the mesodermal progenitors.

Novelties at a cellular level can be either innovated or coopted from elsewhere as genetic programs. In agreement with provided examples, multipotency and competence spectrum of the ectomesenchyme could be indeed co-opted from the mesoderm (Figure 5.3). If true, after such cooption, neural crest cells gained the capacity to generate a tissue that is competent to produce almost everything that mesoderm was competent to produce, including skeletogenic elements. This logic suggests the evolutionary mechanism whereby the transformation of a cell lineage could rely on co-options of different oligo- or multipotent progenitor fates and states (those we will call *competence nodes*, see Figure 5.2) from other lineages in addition to the division of labor and neofunctionalization (Figure 5.3).

Division of labor is an evolutionary strategy that enables partitioning multiple functions performed by an ancestral multifunctional single cell type into derived cell subtypes, which diversify from what is rather a functionally universal cell over a period of time (Arendt et al., 2009). In case of the neural crest, the division of labor could generate the diversity of mesenchymal derivatives supplementing cardiac and

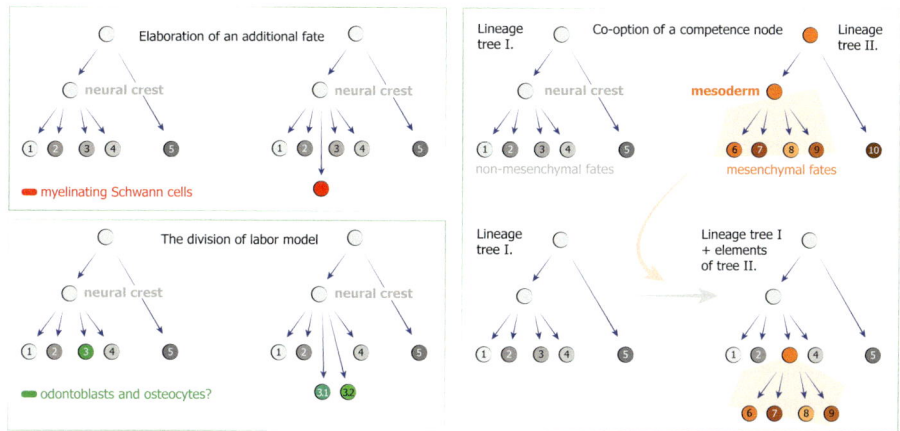

FIGURE 5.3 Different ways of evolving the lineage tree structure based on elaboration or recruitment of new terminal fates and cooption of multipotent progenitor competence nodes.

craniofacial development and could play a role in ensuring the diversity of peripheral sensory and autonomic neuronal subtypes. At the same time, neofunctionalization brings innovations based on the evolution of molecular properties (such as dentin structure) and circuits (Assis and Bachtrog, 2013) (Figure 5.3).

Overall, the combination of different strategies enables acquisition of an additional and ectopic competence spectrum within any given cell lineage after the successful cooption of a progenitor or a stem cell fate from elsewhere. This logic is supported by cases of convergent generation of various cell types within an individual animal as exemplified by the mesoderm- and neural crest–derived chondrocytes, osteocytes, adipocytes, dermal fibroblasts, or the neurons of mesencephalic trigeminal nucleus and proprioceptors from the neural crest–derived dorsal root ganglia. Acquisition of the oligopotent mesoderm-like ectomesenchyme during evolution via the cooption of the mesodermal progenitor state into the neural crest lineage could evoke a new momentum and became a key landmark in the rise of chordates as explained by the "new head" hypothesis (Gans and Northcutt 1983).

5.4.2 Acquisition of Cartilage

In vertebrates, cartilage tissue was essential for the evolution of the cranial skeleton, biting jaws, and endoskeleton of the whole body. For a long time, the idea about vertebrate cartilage being elaborated in the lineage of chordates *de novo* was quite popular. At the same time, numerous invertebrate protostome and deuterostome animals such as hemichordates, horseshoe crabs, cephalopods, and sabellid polychaete worms demonstrated the presence of cartilaginous tissues morphologically similar to the cellular or acellular cartilages found in vertebrates (Cole and Hall, 2004). Until recently, it was still unclear whether these invertebrate cartilages are homologs to the vertebrate cartilage and develop according to the similar molecular and cellular mechanisms or were independently elaborated with recruitment of diverse and independent molecular tools.

Tarazona and coauthors (Tarazona et al., 2016) resolved this puzzle by analyzing molecular properties of cartilage development in *Sepia* (cephalopod mollusk) and *Limulus* (horseshoe crab). The authors revealed that induction of chondroprogenitor cells is controlled by the concerted action of Shh and β-catenin signaling, similarly to the situation in developing vertebrate embryos (Snowball et al., 2015). Furthermore, the expression of well-established chondrogenic genes (proto-orthologues of the vertebrate SoxE and collagen2α1) in developing chondrocytes turned out to be conserved in mollusk and arthropod model systems. Taken together, these results support the notion that chondrogenic gene regulatory network and underlying effector programs evolved in a common bilateralian ancestor and were not elaborated from scratch by the emerging neural crest or in chordates in general (Tarazona et al., 2016).

Tunicates so far have not shown any signs of cartilage-like tissue, whereas amphioxus possesses acellular stiff connective tissue enriched with extracellular matrix and supporting pharyngeal bars (Mansfield et al., 2015). This stiff connective tissue matrix contains collagen type II and resembles cartilage-like tissue found in enteropneust hemichordates or the acellular cartilage of cyclostomes (Hall and Gillis, 2013; Rychel et al., 2006; Rychel and Swalla, 2007).

Recently, a groundbreaking discovery by Jandzik and coauthors revealed the existence of a transient cellular cartilage in the oral cirri from a larval stage of a Florida amphioxus (*Branchiostoma floridae*)—an animal that does not possess the canonical neural crest (Jandzik et al., 2015). Similarly, some extinct presumably chordate organisms (*Haikouella lanceolate*, *Pikaia gracilens*) also demonstrated the presence of oral cirri that potentially could be supported by the cartilage (Chen et al., 1999; Morris and Caron, 2012). Jandzik et al. clearly suggested that chordate-type cartilage was elaborated as a tissue or a cell type before neural crest acquisition. Later, this cellular cartilage might have adapted different ways of spreading through the body via the evolution of genomic regulatory regions defining tissue-specific expression of chondrogenic programs. Evolution of such regulatory elements could indeed shift the cell lineage–related specificity of cartilage induction or, in other words, could create conditions for coopting the cartilage program into the ectopic locations such as the neural crest. Jandzik et al. provided a coherent explanation of how SoxE, a transcription factor critical for cartilage induction, gained a new cis-regulatory sequence directing its expression in the neural lineage of vertebrate animals. Authors of this study favored this way of enabling cartilage formation in neural crest lineage instead of other alternative mechanisms that could involve changes in the DNA-binding properties of SoxE and other transcription factors that are hierarchically interconnected and can drive chondrogenic differentiation. At the same time, Jandzik et al. suggested that amphioxus might form its head cartilage from the coelomic mesothelium, which would render mesendoderm as the original structure competent to produce cartilage in cephalochordates (Jandzik et al., 2015). Thus, the chondrocyte fate was likely coopted by the proto neural crest cells from mesodermal or mesendodermal lineages. As we mentioned above, the cooption of a chondrocyte fate could happen as a part of a more general co-option of an upstream mesoderm-like state (*competence node*) leading to formation of the neural crest-derived ectomesenchyme generating all types of skeletogenic tissues.

5.4.3 ACQUISITION OF DENTIN AND BONE

Of course, the co-option of competence nodes is not the only way to drive the evolution of lineage. The elaboration of specific properties and cell fates from scratch must also take place in nature. It seems that odontoblasts producing dentin might be one of the few original inventions in the neural crest lineage in addition to, hypothetically, myelinating glia and bone.

Evolution of an odontoblast as a cell type is closely connected to the evolution of teeth and odontodes since those represent the earliest mineralized tissue found within the dermal armor (Gillis et al., 2017). At this point, we do not have a good evolutionary scenario explaining the elaboration of the first mineralization program from scratch. The analysis of the evolution of the genetic toolkit necessary to secrete mineralized matrix revealed that one set of the important components, calcium binding phosphoprotein family (SCPP) members, emerged from the 5' region of the ancestral Sparcl1 (Sparc-like 1) gene (Kawasaki et al., 2004). Furthermore, elaboration of Scpp genes occurred already after the divergence of cartilaginous and bony fishes. Consistently, only two Scpp-related genes (Sparc and Sparcl1) were identified in a genome of an elephant shark, which deviated early from the lineage leading to bony fishes (Venkatesh et al., 2014), discussed in (Lv et al., 2017). Thus, mineralized tissues existing before this split and including those of the neural crest origin utilized some other means of mineralization, for instance, SPARC proteins (Kawasaki et al., 2004). The authors of this study suggested that evolution of mineralization could begin with massively secreted collagens acting to reinforce the dermis (Kawasaki et al., 2004), which could somehow resemble the cases of calcified tendons (Merolla et al., 2015). Still, the questions of how the biomineralization gene expression program was introduced into the neural crest and how the prospective odontoblast identity became encoded within the hierarchy of the neural crest lineage remain unexplained.

In extinct agnathans and gnathostomes, the dermal armor was built of odontodes and often covered the entire body of an animal, including the trunk region. Later, during evolution, the dermal armor regressed to the cranial part of the body (Shimada et al., 2013). Since dentin-producing cells—odontoblasts—are exclusively neural crest derived, it might be possible that the trunk neural crest could give rise to odontodes in a trunk region as well, thus, showing skeletogenic properties similar to the cranial neural crest. Modern teleost fish scales are built of rather acellular bone and are generated by osteoblasts and osteocytes originating from the mesoderm, not the neural crest (Shimada et al., 2013). However, the situation is different when it comes to scales in cartilaginous fishes. Those scales develop in all regions of the body, including postcranial areas, and are composed of dentin and bone resembling odontodes of extinct species. Lineage tracing of the neural crest progeny with the DiI labelling technique revealed that trunk neural crest of the cartilaginous fishes (as shown in a skate *Leucoraja erinacea*) gives rise to the dentin-producing cells (*bona fide* odontoblasts) in the postcranial region. The authors of this study suggested that the ancestral body armor consisted of two parts: osteogenic (mesoderm-derived) and odontogenic (neural crest-derived) (Gillis et al., 2017). During further evolution, different combinations of losses and conservations of the bony and dentinal components led to the diversity of scales and odontodes in currently living vertebrate lineages (Gillis et al., 2017; Sire et al., 2009).

If trunk neural crest cells can give rise to odontoblasts in the postcranial region in some animals, it might be possible that other mesenchymal fates are also enabled in the trunk neural crest of selected species. Due to evolutionary plasticity or some degree of conserved competence in the trunk neural crest, turtles also show the presence of mesenchymal derivatives of the trunk neural crest. The osteocytes that build developing plastron and nuchal bones are neural crest–derived, and the trunk neural crest cells firstly migrate to the carapacial staging area and only then to the future sites of the bone induction within the trunk (Gilbert et al., 2007).

Odontoblasts and osteocytes are similar cell types since both produce organized mineralized matrix (Kawasaki et al., 2004) and are generated within the neural crest lineage. As compared to osteocytes, odontoblasts perform a broader set of functions since they are sensory cells (Magloire et al., 2009) innervated by pain fibers. They express mechanosensory and thermosensory ion channels (Allard et al., 2006; Khatibi Shahidi et al., 2015; Okumura et al., 2005) and play a role in dental pain transmission. Odontoblast also project into the dentin with long perfectly oriented processes going inside of the dentinal tubules (Khatibi Shahidi et al., 2015). Bone, as a tissue, and an osteocyte, as a fate, could be a result of a transformation that started with the odontoblast program according to the evolutionary scenario based on a spread of mineralization capacity from the skin basal membrane deeper into the dermis, where the acellular and cellular bone would form at the odontode base (Donoghue et al., 2006; Wagner and Aspenberg, 2011). Based on this, bone could be innovated within the neural crest lineage by transforming odontogenic program and only then coopted by the mesoderm. Although this scenario could take place, it is more likely that the mineralization program was co-opted from neural crest–derived odontoblasts into the mesodermal bone because the composite scales of skates are made of neural crest–derived dentin and mesodermal bone (Gillis et al., 2017).

After the elaboration of dentin and bone, the genetic programs for mineralized matrix deposition could be further exported to other cell lineages and cell types including, for instance, the case of recently discovered osteogenic chondrocytes participating in ossification processes discovered in teleost fish and mammals (Paul et al., 2016; Yang et al., 2014). Indeed, Paul et al. discovered that during jawbone regeneration, periosteum-derived chondrocytes express genes typical for osteoblast differentiation and participate in extensive mineralization of the regenerating jawbone cartilage (Paul et al., 2016). At the same time, Yang et al. revealed that hypertrophic chondrocytes give rise to osteoblasts during normal developmental endochondral ossification in mammals (Yang et al., 2014). In line with this reasoning, bone preceded mineralized cartilage as evident from the fossil record (Gomez-Picos and Eames, 2015), although it does not completely exclude the possibility that eventual mineralization of cartilage could lead to the elaboration of bone.

5.4.4   OTHER ECTOMESENCHYME-DERIVED CELL TYPES AND CRANIAL NEURAL CREST

The spectrum of mesenchymal fates that are derivatives of the neural crest includes a number of nonskeletogenic fates, for instance, adipocytes and perivascular and muscle cells. It is hard to reason about the evolution of, for instance, perivascular cells since they leave no traces in the fossil record and were never sufficiently analyzed in extant cephalochordates, tunicates, and agnathans. Conglomerates of cells resembling white and brown adipose tissue were detected in the perimeningeal tissue of lampreys (Muller, 1968). The embryonic origin of those putative adipocytes is unknown and requires further investigation. Despite the fact that invertebrate and vertebrate taxa share a fair portion of molecular toolkit that is related to lipid metabolism and storage, the divergence is also significant, which, together, does not help to clarify whether lipid-storing cells in vertebrates and invertebrates are homologous cell types (Ottaviani et al., 2011).

In amniotes, neural crest cells give rise to the mesenchymal component of the thyroid and parathyroid glands, the thymus, perivascular smooth muscle cells in the face and neck regions, and the muscle-type derivatives within the heart (Keyte et al., 2014). Cardiac neural crest cells migrate as a stream following specific navigation signals (Toyofuku et al., 2008), guiding cells ventrally to the developing heart where they contribute to the structures of spiraculum, cushions, valves, outflow septum of the heart, and the smooth muscle part of the arterial walls of the heart outflow tract (Kirby et al., 1983; Snider et al., 2007). Soon after the first discoveries of cardiac neural crest in chick and mouse embryos, it became clear that cardiac neural crest cells are key not only for correct heart development but also for the evolution of its complexity (Keyte et al., 2014). Neural crest cells contributed to the evolutionary reinforcement of aortic arch arteries that continued to evolve into the great arteries of the heart and assisted the septation of the outflow tract in amniotes (Keyte et al., 2014). In amphibian embryos, cardiac neural crest cells do not contribute to the development of the outflow septum that separates two circles of blood circulation, whereas in birds and mammals, the contribution of the crest to the septation is important for correct heart development and efficient blood oxygenation (Lee and Saint-Jeannet, 2011). Elaboration and tuning of cardiac neural crest and its derivatives required extensive coevolution with other cardiac progenitors including second heart field (Keyte et al., 2014).

In chick and mouse heart, the neural crest cells give rise to smooth muscles, fibroblasts, pigment cells, and components of the local autonomic nervous system, but not to cardiomyocytes. Surprisingly, a series of studies reveled that in zebrafish, cardiac neural crest cells give rise to a significant proportion of cardiomyocytes in all parts of the developing fish heart (Cavanaugh et al., 2015; Li et al., 2003). This example may suggest that some fates and competences might be lost in the neural crest lineage, rather than just acquired. Some unique types of a cyclostome cartilage (Kaucka and Adameyko, 2017; Zhang et al., 2009) were also probably lost during the later steps in evolutionary transition from agnathans to gnathostomes. Alternatively, it is possible that some fates developed only after the split of the specific animal group (teleost fish or cyclostomes) from the stem connecting this group with the common ancestor. Thus, such fates were never acquired in our evolutionary lineage.

Early evolution of cardiac neural crest and their derivatives, according to Keyte at al., could start from the reinforcement of large pharyngeal arch arteries with a smooth muscle layer. Later, this layer was recruited into the developing heart to aid the outflow tract septation (Keyte et al., 2014). If true, the cooption of mesodermal competences including smooth muscle fate into cranial neural crest could be the starting point of shaping the future of cardiac neural crest stream.

The analysis of cardiac development–related gene regulatory networks showed that Tbx1 gene is necessary for the proliferation of second heart field progenitors in teleost fish (Nevis et al., 2013). At the same time, Tbx1 is expressed in pharyngeal arches of lampreys and other vertebrates (Sauka-Spengler et al., 2002). Knockout of Tbx1 in fish resulted in abnormal development of aortic arch arteries and neural crest–derived cartilages (Piotrowski et al., 2003), whereas in *Xenopus*, depletion of Tbx1 level caused abnormal heart looping, aortic arch abnormalities, and affected craniofacial skeleton (Tazumi et al., 2010; Tran et al., 2011). Based on this, Keyte et al. suggested that Tbx1 was coopted into the neural crest lineage to control the third stream of neural crest migration towards the heart (Keyte et al., 2014).

KEY READING

Tarazona, O.A., Slota, L.A., Lopez, D.H., Zhang, G., Cohn, M.J., 2016. The genetic program for cartilage development has deep homology within Bilateria. *Nature* 533, 86–89.

Jandzik, D., Garnett, A.T., Square, T.A., Cattell, M.V., Yu, J.K., Medeiros, D.M., 2015. Evolution of the new vertebrate head by co-option of an ancient chordate skeletal tissue. *Nature* 518, 534–537.

Kaucka, M., Adameyko, I., 2017. Evolution and development of the cartilaginous skull: From a lancelet towards a human face. *Semin Cell Dev Biol* 91, 2–12.

Keyte, A.L., Alonzo-Johnsen, M., Hutson, M.R., 2014. Evolutionary and developmental origins of the cardiac neural crest: Building a divided outflow tract. *Birth Defects Res C Embryo Today* 102, 309–323.

Arendt, D., Musser, J.M., Baker, C.V.H., Bergman, A., Cepko, C., Erwin, D.H., Pavlicev, M., Schlosser, G., Widder, S., Laubichler, M.D., Wagner, G.P., 2016. The origin and evolution of cell types. *Nat Rev Genet* 17, 744–757.

5.5   CONCLUDING REMARKS

Evolution of the neural crest lineage likely proceeded in steps starting with primary elaboration of the early crest generating only few fates such as pigment cells and neurons. This set of primary fates underwent further expansion during evolutionary modifications of the neural crest lineage tree. At the fundamental level, incremental evolution of fates in the neural crest lineage required mechanisms enabling acquisition of new potentials via coopting the competence nodes and individual gene expression programs from other lineages or elaborating new fates and functions from scratch. Neofunctionalization and division of labor could be potent driving forces preparing neural crest for such co-options and *de novo* developments. The evolution of the lineage tree structure including tuning the hierarchy of progenitor fates redefined the overall multipotency and provided for an acquisition of a major mesoderm-like ectomesenchymal competence node. This, in turn, permitted the development of a "new head" with complex articulated endoskeleton, improved sensory control, and advanced dentition. Soon after the rise, neural crest cells began to function as a new germ layer that gradually became more and more potent in bringing further innovations at the cellular and tissue levels. Neural crest–derived cell types occupied strategic positions in the body and catalyzed key transformation of the oral apparatus, cardio-vascular system, peripheral innervation, and many more structures. Through this, the incremental evolution of fates and neural crest derivatives within multiple tissues and organs shaped the future story of the competitiveness and success of our chordate ancestors.

REFERENCES

Abitua, P.B., Wagner, E., Navarrete, I.A., Levine, M., 2012. Identification of a rudimentary neural crest in a non-vertebrate chordate. *Nature* 492(7427), 104–107.

Adameyko, I., Lallemend, F., 2010. Glial versus melanocyte cell fate choice: Schwann cell precursors as a cellular origin of melanocytes. *Cell Mol Life Sci* 67(18), 3037–3055.

Adameyko, I., Lallemend, F., Aquino, J.B., Pereira, J.A., Topilko, P., Muller, T., Fritz, N., Beljajeva, A., Mochii, M., Liste, I., Usoskin, D., Suter, U., Birchmeier, C., Ernfors, P., 2009. Schwann cell precursors from nerve innervation are a cellular origin of melanocytes in skin. *Cell* 139(2), 366–379.

Adameyko, I., Lallemend, F., Furlan, A., Zinin, N., Aranda, S., Kitambi, S.S., Blanchart, A., Favaro, R., Nicolis, S., Lubke, M., Muller, T., Birchmeier, C., Suter, U., Zaitoun, I., Takahashi, Y., Ernfors, P., 2012. Sox2 and Mitf cross-regulatory interactions consolidate progenitor and melanocyte lineages in the cranial neural crest. *Development* 139(2), 397–410.

Allard, B., Magloire, H., Couble, M.L., Maurin, J.C., Bleicher, F., 2006. Voltage-gated sodium channels confer excitability to human odontoblasts: Possible role in tooth pain transmission. *J Biol Chem* 281(39), 29002–29010.

Arendt, D., 2008. The evolution of cell types in animals: Emerging principles from molecular studies. *Nat Rev Genet* 9(11), 868–882.

Arendt, D., Hausen, H., Purschke, G., 2009. The 'division of labour' model of eye evolution. *Philos Trans R Soc Lond B Biol Sci* 364(1531), 2809–2817.

Arendt, D., Musser, J.M., Baker, C.V.H., Bergman, A., Cepko, C., Erwin, D.H., Pavlicev, M., Schlosser, G., Widder, S., Laubichler, M.D., Wagner, G.P., 2016. The origin and evolution of cell types. *Nat Rev Genet* 17(12), 744–757.

Assis, R., Bachtrog, D., 2013. Neofunctionalization of young duplicate genes in Drosophila. *Proc Natl Acad Sci U S A* 110(43), 17409–17414.

Bellono, N.W., Najera, J.A., Oancea, E., 2014. UV light activates a Galphaq/11-coupled phototransduction pathway in human melanocytes. *J Gen Physiol* 143(2), 203–214.

Bronner, M.E., 2015. Evolution: On the crest of becoming vertebrate. *Nature* 527(7578), 311–312.

Bullock, T.H., Moore, J.K., Fields, R.D., 1984. Evolution of myelin sheaths: Both lamprey and hagfish lack myelin. *Neurosci Lett* 48(2), 145–148.

Carr, M.J., Toma, J.S., Johnston, A.P.W., Steadman, P.E., Yuzwa, S.A., Mahmud, N., Frankland, P.W., Kaplan, D.R., Miller, F.D., 2019. Mesenchymal precursor cells in adult nerves contribute to mammalian tissue repair and regeneration. *Cell Stem Cell* 24(2), 240–256, e249.

Cavanaugh, A.M., Huang, J., Chen, J.N., 2015. Two developmentally distinct populations of neural crest cells contribute to the zebrafish heart. *Dev Biol* 404(2), 103–112.

Cebra-Thomas, J.A., Betters, E., Yin, M., Plafkin, C., McDow, K., Gilbert, S.F., 2007. Evidence that a late-emerging population of trunk neural crest cells forms the plastron bones in the turtle *Trachemys scripta*. *Evol Dev* 9(3), 267–277.

Cebra-Thomas, J.A., Terrell, A., Branyan, K., Shah, S., Rice, R., Gyi, L., Yin, M., Hu, Y., Mangat, G., Simonet, J., Betters, E., Gilbert, S.F., 2013. Late-emigrating trunk neural crest cells in turtle embryos generate an osteogenic ectomesenchyme in the plastron. *Dev Dyn* 242(11), 1223–1235.

Chen, J., Huang, J., Li, C., 1999. An early Cambrian craniate-like chordate. *Nature* 402(6761), 518–522.

Clark, K., Bender, G., Murray, B.P., Panfilio, K., Cook, S., Davis, R., Murnen, K., Tuan, R.S., Gilbert, S.F., 2001. Evidence for the neural crest origin of turtle plastron bones. *Genesis* 31(3), 111–117.

Clay, J.R., 1985. Potassium current in the squid giant axon. *Int Rev Neurobiol* 27, 363–384.

Cole, A.G., Hall, B.K., 2004. The nature and significance of invertebrate cartilages revisited: Distribution and histology of cartilage and cartilage-like tissues within the Metazoa. *Zoology (Jena)* 107(4), 261–273.

Coulpier, F., Decker, L., Funalot, B., Vallat, J.M., Garcia-Bragado, F., Charnay, P., Topilko, P., 2010. CNS/PNS boundary transgression by central glia in the absence of Schwann cells or Krox20/Egr2 function. *J Neurosci* 30(17), 5958–5967.

Dong, Z., Sinanan, A., Parkinson, D., Parmantier, E., Mirsky, R., Jessen, K.R., 1999. Schwann cell development in embryonic mouse nerves. *J Neurosci Res* 56(4), 334–348.

Donoghue, P.C., Graham, A., Kelsh, R.N., 2008. The origin and evolution of the neural crest. *BioEssays* 30(6), 530–541.

Donoghue, P.C., Sansom, I.J., Downs, J.P., 2006. Early evolution of vertebrate skeletal tissues and cellular interactions, and the canalization of skeletal development. *J Exp Zool B Mol Dev Evol* 306(3), 278–294.

Dupin, E., Sommer, L., 2012. Neural crest progenitors and stem cells: From early development to adulthood. *Dev Biol* 366(1), 83–95.

Dyachuk, V., Furlan, A., Shahidi, M.K., Giovenco, M., Kaukua, N., Konstantinidou, C., Pachnis, V., Memic, F., Marklund, U., Muller, T., Birchmeier, C., Fried, K., Ernfors, P., Adameyko, I., 2014. Neurodevelopment. Parasympathetic neurons originate from nerve-associated peripheral glial progenitors. *Science* 345(6192), 82–87.

Dyer, C., Linker, C., Graham, A., Knight, R., 2014. Specification of sensory neurons occurs through diverse developmental programs functioning in the brain and spinal cord. *Dev Dyn* 243(11), 1429–1439.

Eiden, L.E., Jiang, S.Z., 2018. What's new in endocrinology: The chromaffin cell. *Front Endocrinol (Lausanne)* 9, 711.

Espinosa-Medina, I., Jevans, B., Boismoreau, F., Chettouh, Z., Enomoto, H., Muller, T., Birchmeier, C., Burns, A.J., Brunet, J.F., 2017. Dual origin of enteric neurons in vagal Schwann cell precursors and the sympathetic neural crest. *Proc Natl Acad Sci U S A* 114(45), 11980–11985.

Flood, P.R., 1966. A peculiar mode of muscular innervation in amphioxus. Light and electron microscopic studies of the so-called ventral roots. *J Comp Neurol* 126(2), 181–217.

Fritzsch, B., Northcutt, R.G., 1993. Cranial and spinal nerve organization in amphioxus and lampreys: Evidence for an ancestral craniate pattern. *Acta Anat (Basel)* 148(2–3), 96–109.

Furlan, A., Adameyko, I., 2018. Schwann cell precursor: A neural crest cell in disguise? *Dev Biol* 444(1), 25–35.

Furlan, A., Dyachuk, V., Kastriti, M.E., Calvo-Enrique, L., Abdo, H., Hadjab, S., Chontorotzea, T., Akkuratova, N., Usoskin, D., Kamenev, D., Petersen, J., Sunadome, K., Memic, F., Marklund, U., Fried, K., Topilko, P., Lallemend, F., Kharchenko, P.V., Ernfors, P., Adameyko, I., 2017. Multipotent peripheral glial cells generate neuroendocrine cells of the adrenal medulla. *Science* 357(6346).

Gans, C., Northcutt, R.G., 1983. Neural crest and the origin of vertebrates: A new head. *Science* 220(4594), 268–273.

Gilbert, S.F., Bender, G., Betters, E., Yin, M., Cebra-Thomas, J.A., 2007. The contribution of neural crest cells to the nuchal bone and plastron of the turtle shell. *Integr Comp Biol* 47(3), 401–408.

Gillis, J.A., Alsema, E.C., Criswell, K.E., 2017. Trunk neural crest origin of dermal denticles in a cartilaginous fish. *Proc Natl Acad Sci U S A* 114(50), 13200–13205.

Gomez-Picos, P., Eames, B.F., 2015. On the evolutionary relationship between chondrocytes and osteoblasts. *Front Genet* 6, 297.

Green, S.A., Bronner, M.E., 2014. The lamprey: A jawless vertebrate model system for examining origin of the neural crest and other vertebrate traits. *Differentiation* 87(1–2), 44–51.

Green, S.A., Simoes-Costa, M., Bronner, M.E., 2015. Evolution of vertebrates as viewed from the crest. *Nature* 520(7548), 474–482.

Green, S.A., Uy, B.R., Bronner, M.E., 2017. Ancient evolutionary origin of vertebrate enteric neurons from trunk-derived neural crest. *Nature* 544(7648), 88–91.

Hall, B.K., Gillis, J.A., 2013. Incremental evolution of the neural crest, neural crest cells and neural crest-derived skeletal tissues. *J Anat* 222(1), 19–31.

Haming, D., Simoes-Costa, M., Uy, B., Valencia, J., Sauka-Spengler, T., Bronner-Fraser, M., 2011. Expression of sympathetic nervous system genes in Lamprey suggests their recruitment for specification of a new vertebrate feature. *PLOS ONE* 6(10), e26543.

Hartline, D.K., Colman, D.R., 2007. Rapid conduction and the evolution of giant axons and myelinated fibers. *Curr Biol* 17(1), R29–35.

Henrique, D., Abranches, E., Verrier, L., Storey, K.G., 2015. Neuromesodermal progenitors and the making of the spinal cord. *Development* 142(17), 2864–2875.

Hill, A.S., Nishino, A., Nakajo, K., Zhang, G., Fineman, J.R., Selzer, M.E., Okamura, Y., Cooper, E.C., 2008. Ion channel clustering at the axon initial segment and node of Ranvier evolved sequentially in early chordates. *PLOS Genet* 4(12), e1000317.

Hottes, A.K., Freddolino, P.L., Khare, A., Donnell, Z.N., Liu, J.C., Tavazoie, S., 2013. Bacterial adaptation through loss of function. *PLOS Genet* 9(7), e1003617.

Isern, J., Garcia-Garcia, A., Martin, A.M., Arranz, L., Martin-Perez, D., Torroja, C., Sanchez-Cabo, F., Mendez-Ferrer, S., 2014. The neural crest is a source of mesenchymal stem cells with specialized hematopoietic stem cell niche function. *eLife* 3, e03696.

Ivashkin, E., Adameyko, I., 2013. Progenitors of the protochordate ocellus as an evolutionary origin of the neural crest. *EvoDevo* 4(1), 12.

Jacob, C., 2015. Transcriptional control of neural crest specification into peripheral glia. *Glia* 63(11), 1883–1896.

Jandzik, D., Garnett, A.T., Square, T.A., Cattell, M.V., Yu, J.K., Medeiros, D.M., 2015. Evolution of the new vertebrate head by co-option of an ancient chordate skeletal tissue. *Nature* 518(7540), 534–537.

Jessen, K.R., Brennan, A., Morgan, L., Mirsky, R., Kent, A., Hashimoto, Y., Gavrilovic, J., 1994. The Schwann cell precursor and its fate: A study of cell death and differentiation during gliogenesis in rat embryonic nerves. *Neuron* 12(3), 509–527.

Jessen, K.R., Mirsky, R., 2005. The origin and development of glial cells in peripheral nerves. *Nat Rev Neurosci* 6(9), 671–682.

Joseph, N.M., Mukouyama, Y.S., Mosher, J.T., Jaegle, M., Crone, S.A., Dormand, E.L., Lee, K.F., Meijer, D., Anderson, D.J., Morrison, S.J., 2004. Neural crest stem cells undergo multilineage differentiation in developing peripheral nerves to generate endoneurial fibroblasts in addition to Schwann cells. *Development* 131(22), 5599–5612.

Kastriti, M.E., Adameyko, I., 2017. Specification, plasticity and evolutionary origin of peripheral glial cells. *Curr Opin Neurobiol* 47, 196–202.

Kaucka, M., Adameyko, I., 2017. Evolution and development of the cartilaginous skull: From a lancelet towards a human face. *Semin Cell Dev Biol* 91, 2–12.

Kaucka, M., Ivashkin, E., Gyllborg, D., Zikmund, T., Tesarova, M., Kaiser, J., Xie, M., Petersen, J., Pachnis, V., Nicolis, S.K., Yu, T., Sharpe, P., Arenas, E., Brismar, H., Blom, H., Clevers, H., Suter, U., Chagin, A.S., Fried, K., Hellander, A., Adameyko, I., 2016. Analysis of neural crest-derived clones reveals novel aspects of facial development. *Sci Adv* 2(8), e1600060.

Kawasaki, K., Suzuki, T., Weiss, K.M., 2004. Genetic basis for the evolution of vertebrate mineralized tissue. *Proc Natl Acad Sci U S A* 101(31), 11356–11361.

Keyte, A.L., Alonzo-Johnsen, M., Hutson, M.R., 2014. Evolutionary and developmental origins of the cardiac neural crest: Building a divided outflow tract. *Birth Defects Res C Embryo Today* 102(3), 309–323.

Khatibi Shahidi, M., Krivanek, J., Kaukua, N., Ernfors, P., Hladik, L., Kostal, V., Masich, S., Hampl, A., Chubanov, V., Gudermann, T., Romanov, R.A., Harkany, T., Adameyko, I., Fried, K., 2015. Three-dimensional imaging reveals new compartments and structural adaptations in odontoblasts. *J Dent Res* 94(7), 945–954.

Kirby, M.L., Gale, T.F., Stewart, D.E., 1983. Neural crest cells contribute to normal aortico-pulmonary septation. *Science* 220(4601), 1059–1061.

Korneliussen, H., 1973. Ultrastructure of motor nerve terminals on different types of muscle fibers in the Atlantic hagfish (*Myxine glutinosa*, L.). Occurrence of round and elongated profiles of synaptic vesicles and dense-core vesicles. *Z Zellforsch Mikrosk Anat* 147(1), 87–105.

Kozmik, Z., Ruzickova, J., Jonasova, K., Matsumoto, Y., Vopalensky, P., Kozmikova, I., Strnad, H., Kawamura, S., Piatigorsky, J., Paces, V., Vlcek, C., 2008. Assembly of the cnidarian camera-type eye from vertebrate-like components. *Proc Natl Acad Sci U S A* 105(26), 8989–8993.

Kulesa, P.M., Bailey, C.M., Kasemeier-Kulesa, J.C., McLennan, R., 2010. Cranial neural crest migration: New rules for an old road. *Dev Biol* 344(2), 543–554.

Lacalli, T.C., 2004. Sensory systems in amphioxus: A window on the ancestral chordate condition. *Brain Behav Evol* 64(3), 148–162.

Le Douarin, N., Kalcheim, C., 1999. *The Neural Crest*, 2nd ed. Cambridge University Press, Cambridge, UK/New York, NY.

Le Douarin, N.M., Creuzet, S., Couly, G., Dupin, E., 2004. Neural crest cell plasticity and its limits. *Development* 131(19), 4637–4650.

Lee, Y.H., Saint-Jeannet, J.P., 2011. Cardiac neural crest is dispensable for outflow tract septation in Xenopus. *Development* 138(10), 2025–2034.

Li, Y.X., Zdanowicz, M., Young, L., Kumiski, D., Leatherbury, L., Kirby, M.L., 2003. Cardiac neural crest in zebrafish embryos contributes to myocardial cell lineage and early heart function. *Dev Dyn* 226(3), 540–550.

Lipovsek, M., Ledderose, J., Butts, T., Lafont, T., Kiecker, C., Wizenmann, A., Graham, A., 2017. The emergence of mesencephalic trigeminal neurons. *Neural Dev* 12(1), 11.

Loebel, D.A., Watson, C.M., De Young, R.A., Tam, P.P., 2003. Lineage choice and differentiation in mouse embryos and embryonic stem cells. *Dev Biol* 264(1), 1–14.

Lv, Y., Kawasaki, K., Li, J., Li, Y., Bian, C., Huang, Y., You, X., Shi, Q., 2017. A genomic survey of SCPP family genes in fishes provides novel insights into the evolution of fish scales. *Int J Mol Sci* 18(11).

Mackie, G., Wyeth, R., 2000. Conduction and coordination in deganglionated ascidians. *Can J Zool* 78(9), 1626–1639.

Magloire, H., Couble, M.L., Thivichon-Prince, B., Maurin, J.C., Bleicher, F., 2009. Odontoblast: A mechano-sensory cell. *J Exp Zool B Mol Dev Evol* 312B(5), 416–424.

Mansfield, J.H., Haller, E., Holland, N.D., Brent, A.E., 2015. Development of somites and their derivatives in amphioxus, and implications for the evolution of vertebrate somites. *EvoDevo* 6, 21.

Merolla, G., Bhat, M.G., Paladini, P., Porcellini, G., 2015. Complications of calcific tendinitis of the shoulder: A concise review. *J Orthop Traumatol* 16(3), 175–183.

Meulemans, D., Bronner-Fraser, M., 2004. Gene-regulatory interactions in neural crest evolution and development. *Dev Cell* 7(3), 291–299.

Mobius, W., Patzig, J., Nave, K.A., Werner, H.B., 2008. Phylogeny of proteolipid proteins: Divergence, constraints, and the evolution of novel functions in myelination and neuroprotection. *Neuron Glia Biol* 4(2), 111–127.

Morris, S.C., Caron, J.B., 2012. Pikaia gracilens Walcott, a stem-group chordate from the Middle Cambrian of British Columbia. *Biol Rev Camb Philos Soc* 87(2), 480–512.

Muller, H., 1968. [Fine structure and lipid formation in fat cells of the perimeningeal tissue of lampreys under normal and experimental conditions]. *Z Zellforsch Mikrosk Anat* 84(4), 585–608.

Nakao, T., Ishizawa, A., 1987a. Development of the spinal nerves in the lamprey: I. Rohon-Beard cells and interneurons. *J Comp Neurol* 256(3), 342–355.

Nakao, T., Ishizawa, A., 1987b. Development of the spinal nerves in the lamprey: III. Spinal ganglia and dorsal roots in 26-day (13 mm) larvae. *J Comp Neurol* 256(3), 369–385.

Nevis, K., Obregon, P., Walsh, C., Guner-Ataman, B., Burns, C.G., Burns, C.E., 2013. Tbx1 is required for second heart field proliferation in zebrafish. *Dev Dyn* 242(5), 550–559.

Nikitina, N., Sauka-Spengler, T., Bronner-Fraser, M., 2009. Chapter 1. Gene regulatory networks in neural crest development and evolution. *Curr Top Dev Biol* 86, 1–14. Chapter 1.

Nilsson, S., 2011. Comparative anatomy of the autonomic nervous system. *Auton Neurosci* 165(1), 3–9.

Nishida, H., Satoh, N., 1989. Determination and regulation in the pigment cell lineage of the ascidian embryo. *Dev Biol* 132(2), 355–367.

Okumura, R., Shima, K., Muramatsu, T., Nakagawa, K., Shimono, M., Suzuki, T., Magloire, H., Shibukawa, Y., 2005. The odontoblast as a sensory receptor cell? The expression of TRPV1 (VR-1) channels. *Arch Histol Cytol* 68, 251–257.

Ottaviani, E., Malagoli, D., Franceschi, C., 2011. The evolution of the adipose tissue: A neglected enigma. *Gen Comp Endocrinol* 174(1), 1–4.

Paiement, J.M., McMillan, D.B., 1975. The extracardiac chromaffin cells of larval lampreys. *Gen Comp Endocrinol* 27(4), 495–508.

Paul, S., Schindler, S., Giovannone, D., de Millo Terrazzani, A., Mariani, F.V., Crump, J.G., 2016. Ihha induces hybrid cartilage-bone cells during zebrafish jawbone regeneration. *Development* 143(12), 2066–2076.

Petersen, J., Adameyko, I., 2017. Nerve-associated neural crest: Peripheral glial cells generate multiple fates in the body. *Curr Opin Genet Dev* 45, 10–14.

Piotrowski, T., Ahn, D.G., Schilling, T.F., Nair, S., Ruvinsky, I., Geisler, R., Rauch, G.J., Haffter, P., Zon, L.I., Zhou, Y., Foott, H., Dawid, I.B., Ho, R.K., 2003. The zebrafish van gogh mutation disrupts tbx1, which is involved in the DiGeorge deletion syndrome in humans. *Development* 130(20), 5043–5052.

Provencio, I., Jiang, G., De Grip, W.J., Hayes, W.P., Rollag, M.D., 1998. Melanopsin: An opsin in melanophores, brain, and eye. *Proc Natl Acad Sci U S A* 95(1), 340–345.

Rossi, C.C., Kaji, T., Artinger, K.B., 2009. Transcriptional control of Rohon-Beard sensory neuron development at the neural plate border. *Dev Dyn* 238(4), 931–943.

Rychel, A.L., Smith, S.E., Shimamoto, H.T., Swalla, B.J., 2006. Evolution and development of the chordates: Collagen and pharyngeal cartilage. *Mol Biol Evol* 23(3), 541–549.

Rychel, A.L., Swalla, B.J., 2007. Development and evolution of chordate cartilage. *J Exp Zool B Mol Dev Evol* 308(3), 325–335.

Sauka-Spengler, T., Le Mentec, C., Lepage, M., Mazan, S., 2002. Embryonic expression of Tbx1, a DiGeorge syndrome candidate gene, in the lamprey *Lampetra fluviatilis*. *Gene Expr Patterns* 2(1–2), 99–103.

Schweigreiter, R., Roots, B.I., Bandtlow, C.E., Gould, R.M., 2006. Understanding myelination through studying its evolution. *Int Rev Neurobiol* 73, 219–273.

Shimada, A., Kawanishi, T., Kaneko, T., Yoshihara, H., Yano, T., Inohaya, K., Kinoshita, M., Kamei, Y., Tamura, K., Takeda, H., 2013. Trunk exoskeleton in teleosts is mesodermal in origin. *Nat Commun* 4, 1639.

Shimokita, E., Takahashi, Y., 2011. Secondary neurulation: Fate-mapping and gene manipulation of the neural tube in tail bud. *Dev Growth Differ* 53(3), 401–410.

Simoes-Costa, M., Bronner, M.E., 2016. Reprogramming of avian neural crest axial identity and cell fate. *Science* 352(6293), 1570–1573.

Sire, J.Y., Donoghue, P.C., Vickaryous, M.K., 2009. Origin and evolution of the integumentary skeleton in non-tetrapod vertebrates. *J Anat* 214(4), 409–440.

Smith, J.J., Kuraku, S., Holt, C., Sauka-Spengler, T., Jiang, N., Campbell, M.S., Yandell, M.D., Manousaki, T., Meyer, A., Bloom, O.E., Morgan, J.R., Buxbaum, J.D., Sachidanandam, R., Sims, C., Garruss, A.S., Cook, M., Krumlauf, R., Wiedemann, L.M., Sower, S.A., Decatur, W.A., Hall, J.A., Amemiya, C.T., Saha, N.R., Buckley, K.M., Rast, J.P., Das, S., Hirano, M., McCurley, N., Guo, P., Rohner, N., Tabin, C.J., Piccinelli, P., Elgar, G., Ruffier, M., Aken, B.L., Searle, S.M., Muffato, M., Pignatelli, M., Herrero, J., Jones, M., Brown, C.T., Chung-Davidson, Y.W., Nanlohy, K.G., Libants, S.V., Yeh, C.Y., McCauley, D.W., Langeland, J.A., Pancer, Z., Fritzsch, B., de Jong, P.J., Zhu, B., Fulton, L.L., Theising, B., Flicek, P., Bronner, M.E., Warren, W.C., Clifton, S.W., Wilson, R.K., Li, W., 2013. Sequencing of the sea lamprey (*Petromyzon marinus*) genome provides insights into vertebrate evolution. *Nat Genet* 45(4), 415–421.

Snider, P., Olaopa, M., Firulli, A.B., Conway, S.J., 2007. Cardiovascular development and the colonizing cardiac neural crest lineage. *Sci World J* 7, 1090–1113.

Snowball, J., Ambalavanan, M., Whitsett, J., Sinner, D., 2015. Endodermal Wnt signaling is required for tracheal cartilage formation. *Dev Biol* 405(1), 56–70.

Squarzoni, P., Parveen, F., Zanetti, L., Ristoratore, F., Spagnuolo, A., 2011. FGF/MAPK/Ets signaling renders pigment cell precursors competent to respond to Wnt signal by directly controlling CI-Tcf transcription. *Development* 138(7), 1421–1432.

Stiefel, K.M., Torben-Nielsen, B., Coggan, J.S., 2013. Proposed evolutionary changes in the role of myelin. *Front Neurosci* 7, 202.

Stolfi, A., Ryan, K., Meinertzhagen, I.A., Christiaen, L., 2015. Migratory neuronal progenitors arise from the neural plate borders in tunicates. *Nature* 527(7578), 371–374.

Tarazona, O.A., Slota, L.A., Lopez, D.H., Zhang, G., Cohn, M.J., 2016. The genetic program for cartilage development has deep homology within Bilateria. *Nature* 533(7601), 86–89.

Tazumi, S., Yabe, S., Uchiyama, H., 2010. Paraxial T-box genes, Tbx6 and Tbx1, are required for cranial chondrogenesis and myogenesis. *Dev Biol* 346(2), 170–180.

Technau, U., Scholz, C.B., 2003. Origin and evolution of endoderm and mesoderm. *Int J Dev Biol* 47(7–8), 531–539.

Toyofuku, T., Yoshida, J., Sugimoto, T., Yamamoto, M., Makino, N., Takamatsu, H., Takegahara, N., Suto, F., Hori, M., Fujisawa, H., Kumanogoh, A., Kikutani, H., 2008. Repulsive and attractive semaphorins cooperate to direct the navigation of cardiac neural crest cells. *Dev Biol* 321(1), 251–262.

Tran, H.T., Delvaeye, M., Verschuere, V., Descamps, E., Crabbe, E., Van Hoorebeke, L., McCrea, P., Adriaens, D., Van Roy, F., Vleminckx, K., 2011. ARVCF depletion cooperates with Tbx1 deficiency in the development of 22q11.2DS-like phenotypes in Xenopus. *Dev Dyn* 240(12), 2680–2687.

Uesaka, T., Nagashimada, M., Enomoto, H., 2015. Neuronal differentiation in Schwann cell lineage underlies postnatal neurogenesis in the enteric nervous system. *J Neurosci* 35(27), 9879–9888.

Van Otterloo, E., Cornell, R.A., Medeiros, D.M., Garnett, A.T., 2013. Gene regulatory evolution and the origin of macroevolutionary novelties: Insights from the neural crest. *Genesis* 51(7), 457–470.

Venkatesh, B., Lee, A.P., Ravi, V., Maurya, A.K., Lian, M.M., Swann, J.B., Ohta, Y., Flajnik, M.F., Sutoh, Y., Kasahara, M., Hoon, S., Gangu, V., Roy, S.W., Irimia, M., Korzh, V., Kondrychyn, I., Lim, Z.W., Tay, B.H., Tohari, S., Kong, K.W., Ho, S., Lorente-Galdos, B., Quilez, J., Marques-Bonet, T., Raney, B.J., Ingham, P.W., Tay, A., Hillier, L.W., Minx, P., Boehm, T., Wilson, R.K., Brenner, S., Warren, W.C., 2014. Elephant shark genome provides unique insights into gnathostome evolution. *Nature* 505(7482), 174–179.

Wada, H., 2001. Origin and evolution of the neural crest: A hypothetical reconstruction of its evolutionary history. *Dev Growth Differ* 43(5), 509–520.

Wagner, D.O., Aspenberg, P., 2011. Where did bone come from? *Acta Orthop* 82(4), 393–398.

Wicks, N.L., Chan, J.W., Najera, J.A., Ciriello, J.M., Oancea, E., 2011. UVA phototransduction drives early melanin synthesis in human melanocytes. *Curr Biol* 21(22), 1906–1911.

Yajima, I., Endo, K., Sato, S., Toyoda, R., Wada, H., Shibahara, S., Numakunai, T., Ikeo, K., Gojobori, T., Goding, C.R., Yamamoto, H., 2003. Cloning and functional analysis of ascidian Mitf in vivo: Insights into the origin of vertebrate pigment cells. *Mech Dev* 120(12), 1489–1504.

Yang, L., Tsang, K.Y., Tang, H.C., Chan, D., Cheah, K.S., 2014. Hypertrophic chondrocytes can become osteoblasts and osteocytes in endochondral bone formation. *Proc Natl Acad Sci U S A* 111(33), 12097–12102.

Yu, J.K., 2010. The evolutionary origin of the vertebrate neural crest and its developmental gene regulatory network--Insights from amphioxus. *Zoology (Jena)* 113(1), 1–9.

Zhang, G., Eames, B.F., Cohn, M.J., 2009. Chapter 2. Evolution of vertebrate cartilage development. *Curr Top Dev Biol* 86, 15–42.

Zinin, N., Adameyko, I., Wilhelm, M., Fritz, N., Uhlen, P., Ernfors, P., Henriksson, M.A., 2014. MYC proteins promote neuronal differentiation by controlling the mode of progenitor cell division. *EMBO Rep* 15(4), 383–391.

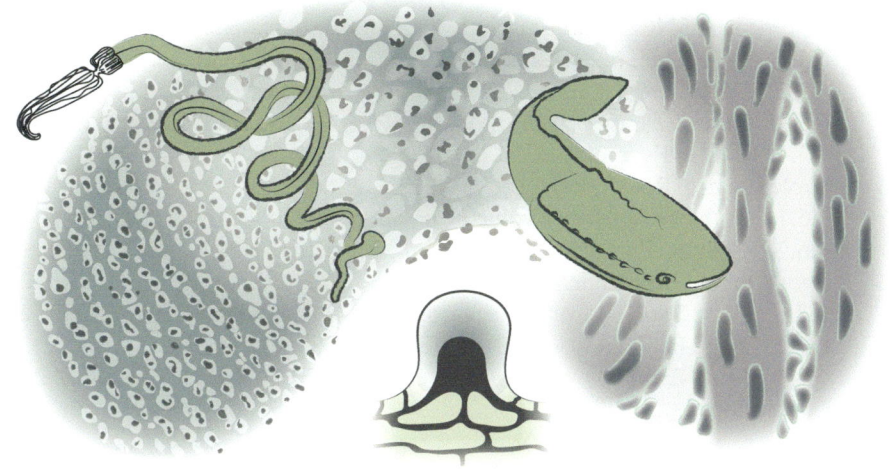

6 On the Evolution of Skeletal Cells before and after Neural Crest

*Brian F. Eames, Patsy Gómez-Picos,
and David Jandzik*

CONTENTS

6.1   ONE OF THESE THREE IS NOT LIKE THE OTHERS: NEURAL CREST, CARTILAGE, OR BONE AS A VERTEBRATE SYNAPOMORPHY?

Our understanding of how skeletal tissues evolved relative to neural crest cells has changed a lot in the past few decades. Before then, both cartilage and bone were considered synapomorphies, or defining features, of vertebrates (Gans and Northcutt, 1983). From a comparative anatomist's point of view, the distribution of cartilage among extant vertebrates spanned "down" the phylogenetic tree from bony fish (osteichthyans) to cartilaginous fish (chondrichthyans) to even jawless fish (agnathan), while bone was observed only in osteichthyans (Figure 6.1). These observations led to a traditional phylogenetic model whereby cartilage evolved within the ancestors to all vertebrates, and bone appeared later in the ancestor to osteichthyans. Discovery of a sufficient fossil record, however, rewrote the false story that bone evolved in the ancestral osteichthyans. Not only did primitive chondrichthyans have abundant bone, but primitive agnathans did also (Coates et al., 2018; Coates et al., 1998; Janvier, 1996). Therefore, the revised model was that cartilage and bone evolved in the ancestral vertebrates.

Embryologists recently added neural crest cells to the list of vertebrate synapomorphies. These migratory cells were identified over 150 years ago (His, 1868), but the demonstration of neural crest in extant agnathans in the past 20 years solidified the understanding that neural crest cells are a vertebrate synapomorphy (Horigome et al., 1999; McCauley and Bronner-Fraser, 2003; Ota et al., 2007). The contributions of neural crest to skeletal tissues of the vertebrate head were demonstrated through cell labelling and transplant experiments (Couly et al., 1992; Nikitina et al., 2009; Noden, 1988). Thus, it was tempting to speculate that cartilage and bone evolved along with the appearance of neural crest in the ancestral vertebrates, but recent data refute this hypothesis.

In this chapter, we clarify the evolutionary relationship between the appearance of neural crest and the ability to form cartilage and bone. Recent studies confirm that the ability to make cartilage (and a related tissue, the notochord) preceded the

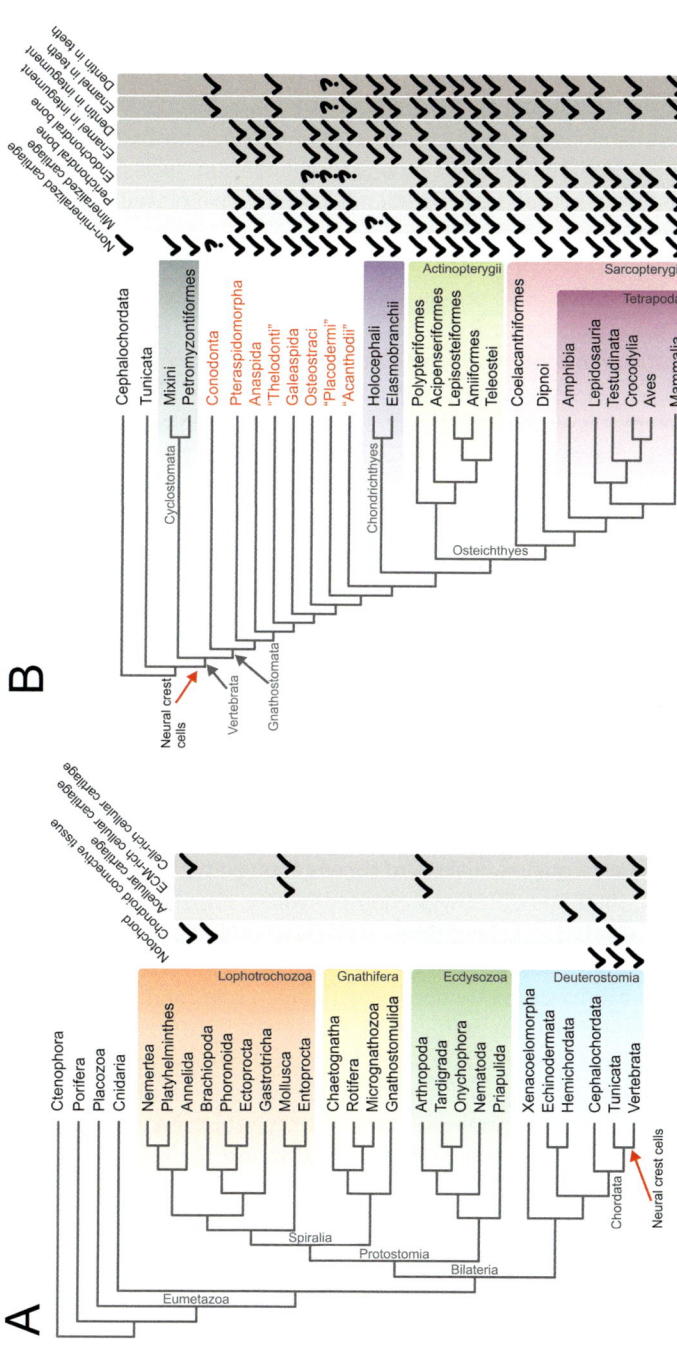

FIGURE 6.1 **Phylogenetic distribution of skeletal tissues in metazoans (A) and chordates (B).** Names of the fossil clades are in red, while clades with names in "" are likely paraphyletic, though traditionally recognized, fossil groups. The term *enamel* is defined broadly here and includes enameloid as well as ganoine, while the term *dentin* includes both orthodentin and mesodentin. Characters in extant lineages do not reflect occurrence in fossil members of those lineages. "?" represents a character with contradictory evidence in the literature. The phylogenetic trees are based on (Donoghue and Rucklin, 2016; Dunn et al., 2014; Irisarri et al., 2017; Marletaz et al., 2019; Whelan et al., 2017). The characters mapped on the phylogenetic trees are based on (Cole and Hall, 2004; Donoghue and Rucklin, 2016; Donoghue et al., 2006; Goudemand et al., 2011; Hall, 2005; Keating et al., 2015; Leprevost et al., 2017; Sire et al., 2009).

appearance of neural crest (Annona et al., 2015; Cole and Hall, 2004; Rychel et al., 2006). Neural crest might have increased the number of cartilage types in vertebrates. Interestingly, the ability to make bone might have evolved slightly after the appearance of neural crest, if cladistics supports the notion that cyclostomes (such as lamprey and hagfish, whose ancestors have never shown evidence of bone formation) are one of the earliest clades of agnathans (Janvier, 1996). Since bone appears to have formed first in a region of the skeleton (exoskeleton) that was demonstrated recently to be derived from neural crest, neural crest might have played a key role in the evolution of bone (Gillis et al., 2017).

Given that cartilage was recently confirmed to precede the appearance of neural crest, which, in turn, preceded bone during chordate evolution (Annona et al., 2015; Cole and Hall, 2004; Janvier, 1996; Rychel et al., 2006), this chapter also presents likely molecular genetic scenarios by which neural crest coopted or otherwise evolved the ability to form skeletal tissues. Gene regulatory networks (GRNs) are discrete sets of genes that interact with environmental cues to impart biological traits, such as the formation of cartilage and bone (Davidson and Levine, 2008; Gomez-Picos and Eames, 2015; Levine and Davidson, 2005). Here, we argue that neural crest first coopted a GRN that was used to make cartilage in other cell lineages in the ancestors to vertebrates, and then neural crest played a significant role in evolving various vertebrate-specific cartilage types through modifications to this ancestral cartilage GRN. The appearance of a bone GRN within neural crest turns out to be more complicated to model with current data. The fossil record suggests that bone evolved from dermal armor in primitive jawless vertebrates (Janvier, 1996; Smith and Hall, 1990). Since there is little molecular genetic data to support this, however, and the current fossil record might not be representative of all early mineralized tissues, we present three hypotheses for the evolution of a GRN underlying bone formation.

In the first hypothesis, neural crest cells coopted a mineralization GRN that was present in the last common ancestor of all deuterostomes. In the latter two hypotheses, a GRN driving mineralization of skeletal tissues evolved within neural crest. In the second hypothesis, the bone GRN was coopted from a mineralization program underlying formation of related skeletal tissues—enamel and dentin—in the ancestor to non-cyclostome vertebrates. All three of these tissues are vertebrate novelties. While enamel is made by ectodermally derived cells, both bone and dentin are generated from neural crest cells (Chai et al., 2000; Gillis et al., 2017), so GRN cooption between bone and dentin would not involve different germ layers. In the third hypothesis, the first osteoblast coopted a GRN that was used to mineralize cartilage in the ancestor to noncyclostome vertebrates. Intelligent application of modern technologies that characterize GRN structure can be used to test the molecular mechanism through which neural crest evolved the ability to make cartilage and bone.

6.2 THE PLAYERS IN THIS EVOLUTIONARY GAME: CARTILAGE, NOTOCHORD, BONE, AND DENTIN/ENAMEL

One of the challenges in re-building the evolution of skeletogenesis in chordates is that most data on skeletal tissues derive from studies in mammals, which at times might represent a relatively recent vertebrate lineage's adaptations to selective pressures on the skeleton. However, after summarizing these data, we provide some perspective to minimize this potential limitation for a robust understanding of skeletal tissue evolution.

6.2.1 CARTILAGE

Cartilage typically functions as a flexible structural support, and the molecules that comprise cartilage extracellular matrix (ECM) impart these roles. Collagen type 2 (Col2) fibers are the most abundant protein in the ECM of cartilage, providing this tissue with the structural stability to withstand some tension while maintaining flexibility (Aumailley and Gayraud, 1998; Gray and Williams, 1989). These loosely wound Col2 fibers bind Aniline blue of Milligan's Trichrome histological protocol, giving cartilage a light-blue staining pattern (Eames et al., 2007). Other collagens, such as Col9 and Col11, are also abundant in cartilage (Eames et al., 2003). Sulfated proteoglycans (PGs) in the ECM lend cartilage the property of compressive resistance due to the massive amounts of water absorbed by sulfated PGs (Aumailley and Gayraud, 1998; Gray and Williams, 1989). In fact, the swelling of hydrated PGs accounts for the majority of the volume of cartilage ECM (Ham and Cormack, 1987). Sulfated PGs bind to Alcian blue and Safranin O, staining cartilage blue and red in these respective histological protocols (Eames et al., 2007). The most abundant PG expressed in cartilage is Aggrecan (Acan), which is a chondroitin sulfate PG, due to the repeating disaccharide glucuronic acid and N-acetylgalactosamine extending from the Acan core protein (Watanabe et al., 1998). Many other proteins present in cartilage ECM at lower levels, along with hyaluronic acid repeating disaccharide, supplement Col2 and sulfated PGs in providing cartilage with the mechanical properties that determine its function (Aumailley and Gayraud, 1998; Ham and Cormack, 1987).

Establishment and maintenance of cartilage ECM depends upon the action of transcription factors that regulate expression of genes encoding these molecules. Sox9 is the main transcription factor driving cartilage formation. Loss of Sox9 function abrogates cartilage differentiation, while gain of Sox9 function promotes ectopic cartilage (Bi et al., 1999; Eames et al., 2004). Other transcription factors, such as Sox5, Sox6, C/EBPs, and FoxOs, are involved in the expression of cartilage genes, but none of these has as central a role as Sox9 in chondrocyte differentiation (Kurakazu et al., 2019; Okuma et al., 2015; Smits et al., 2001). Sox9 binds to regulatory elements and promotes the expression of many genes that are highly expressed in cartilage, including *Col2a1*, *Col9a1*, *Col11a2*, and *Acan* (Bridgewater et al., 2003; Hu et al., 2012; Ng et al., 1997; Zhang et al., 2003). A Sox9 gene regulatory network (GRN) likely dictates cartilage formation (Cole, 2011; Gomez-Picos and Eames, 2015).

The two previous paragraphs focus on hyaline cartilage, which in addition to fibrocartilage and elastic cartilage, are the three main cartilage types in mammals. All cartilage types share the basic molecular features outlined for hyaline cartilage, but they have significant differences related to their specific functions. Hyaline cartilage occurs predominantly in articulating surfaces of skeletal joints, absorbing compressive forces between the bones (Gray and Williams, 1989; Naumann et al., 2002). Fibrocartilage has much more tensional resistance than hyaline cartilage, so the collagen fibers are increased, while the PGs are decreased (Ham and Cormack, 1987; Naumann et al., 2002). In addition to an increase in fiber quantity, fibrocartilage contains large amounts of Col1, in addition to Col2 (Wachsmuth et al., 2006). Fibrocartilage is found predominantly in the intervertebral discs of the spine and joint menisci (Gray and Williams, 1989). Some parts of the anatomy, such as the external ear, epiglottis, Eustachian tube, and nose, have elastic cartilage, in which abundant elastin fibers impart more flexibility to the Col2-positive tissue (Gray and Williams, 1989; Ham and Cormack, 1987; Naumann et al., 2002). While the mammalian classification system recognizes these three cartilage types, many cartilage tissues exhibit intermediate characteristics among the three types. Therefore—and this becomes very important when expanding an understanding of cartilage types beyond mammals—each skeletal tissue should be considered as merely one instance along a spectrum of possibilities.

Even within the hyaline cartilage type, two versions commonly occur in mammals: immature and mature cartilage. After secretion of the hyaline cartilage matrix described above, some developing chondrocytes (usually in the middle of a developing skeletal element) progress through a series of differentiation steps collectively termed *maturation* (de Crombrugghe et al., 2000; Eames et al., 2003). During cartilage maturation, many chondrocytes undergo hypertrophy and mineralize their matrix, forming mineralized cartilage. During endochondral ossification, some mature chondrocytes also degrade their matrix and die, whereas others can even transdifferentiate into osteoblasts as their ultimate fate (Hammond and Schulte-Merker, 2009; Park et al., 2015; Zhou et al., 2014). By logic, hyaline cartilage that does not mature can be described as *immature* cartilage (Gomez-Picos and Eames, 2015). Many signals, including Hedgehog, Parathyroid hormone, FGF, BMP, and Wnt, regulate cartilage maturation, and mature chondrocytes are characterized by expression of such genes as *Col10a1* and *Indian hedgehog* (Mak et al., 2006; St-Jacques et al., 1999; Vortkamp, 2001; Vortkamp et al., 1996; Yoon and Lyons, 2004). A GRN under control of the transcription factor Runx2 (Cbfa1) likely drives cartilage maturation (Gomez-Picos and Eames, 2015). Loss of Runx2 function abrogates cartilage maturation, and gain of Runx2 function promotes cartilage maturation (Eames et al., 2004; Enomoto et al., 2000; Ueta et al., 2001). Mature and immature hyaline cartilage are not merely transient embryonic tissues; they occur in different zones of adult articular cartilage (Gray and Williams, 1989).

6.2.2 NOTOCHORD

The chordate notochord is a tissue with many similarities to cartilage, but also some interesting differences. The notochord ECM includes Col2, Col9, and Acan (Stemple, 2005). At a cellular level, the notochord is formed by chordoblasts, which secrete the notochord ECM and undergo vacuole formation intracellularly; both of these features impart the mechanical properties of stiffness to the notochord. Sequestration of some PGs in chordoblast vacuoles also might play a major role in notochord mechanical properties (Stemple, 2005). While Sox9 is required for cartilage formation, it appears to be dispensable for notochord formation, but it is required for notochord maintenance (Barrionuevo et al., 2006; Bi et al., 1999). Many regulatory elements drive expression of genes in both notochord and cartilage, indicating that these tissues might rely upon similar GRNs for their differentiation (Bagheri-Fam et al., 2006; Dale and Topczewski, 2011).

6.2.3 BONE

Bone serves as a rigid structure also involved in mineral homeostasis, and molecules in bone ECM impart these features. Similar to cartilage, bone ECM has abundant collagens and PGs, but bone has a much higher collagen-to-PG ratio (Ham and Cormack, 1987). Bone collagen fibers are largely Col1 fibers, which are tightly wound, giving bone a dark-blue stain with Aniline blue in Trichrome histological protocols (Eames et al., 2007). These fibers also provide tensile resistance that is characteristic of bone (Fyhrie and Christiansen, 2015). The rigidity of bone derives from an abundant biomineral, hydroxyapatite, embedded within the ECM (Fyhrie and Christiansen, 2015). Alizarin red, which binds to calcium deposits, is a common histological stain for bone. The massive amount of mineral in bone also contributes to calcium and phosphorus homeostasis in the body (Oldknow et al., 2015).

The formation and maintenance of bone ECM depend upon the expression of genes encoding these molecules that are regulated by transcription factors in a bone GRN. With interesting evolutionary implications (see below), Runx2, mentioned above as the main driver of cartilage maturation, is the main transcription factor involved in bone formation. Loss of Runx2 function abrogates bone differentiation, and gain of Runx2 function promotes ectopic bone formation (Ducy et al., 1997; Eames et al., 2004; Komori et al., 1997; Otto et al., 1997). Other transcription factors, such as Sp7 (Osterix), Msx1, Msx2, Twist1, and Twist2, influence the expression of bone genes, but none of these has as central role as Runx2 in osteoblast differentiation (Bialek et al., 2004; Dodig et al., 1999; Nakashima et al., 2002; Satokata and Maas, 1994). Sp7 also is required for bone formation in mice, but the arrest in osteoblast differentiation in *Sp7* null mice occurs downstream of Runx2, and Runx2 promotes *Sp7* expression (Nakashima et al., 2002). Also, as we discuss below, the ancestral expression Sp7 function might not be required for bone formation. Runx2 binds to regulatory elements and drives high levels of expression of many bone genes, such as *Col1a1*, *Col1a2*, *Mgp*, and *Bglap* (Ducy et al., 1997; Harada et al., 1999; Kern et al., 2001; Sato et al., 1998). A Runx2 GRN might dictate bone formation (Gomez-Picos and Eames, 2015).

6.2.4 DENTIN/ENAMEL

Dentin and enamel comprise the main skeletal tissues of mammalian teeth, wherein their hard ECM functions in mastication and predation. Dentin shares far more features with bone than enamel, including an ECM with tightly-wound Col1 fibers for tensile strength and a high degree of mineralization by hydroxyapatite for rigidity (Ham and Cormack, 1987). Even non-collagenous proteins that are highly enriched in dentin ECM are expressed in bone, including Dentin sialophosphotein, Dentin matrix acidic phosphoprotein 1, and Matrix extracellular phosphoglycoprotein

(Kawasaki et al., 2004). Dentin is formed by mesenchymal cells, similar to bone, while enamel is secreted by overlying epithelial cells. Similar to bone, fibers in dentin bind Aniline blue during Trichrome staining, and the high mineral content of dentin and enamel is reflected by Alizarin red staining. Enamel does not contain large collagen fibers and has more tissue-specific ECM components than dentin with respect to bone (Kawasaki et al., 2004; Moradian-Oldak, 2012). In addition to Amelogenin, for example, enamel contains Ameloblastin and Enamelin.

Critical transcription factors regulating expression of the ECM molecules in dentin and enamel are currently unknown. Several transcription factors affect formation of these tissues, such as Runx2, NF-kappaB, or Pax9, but none of these appear to abrogate dentin or enamel formation completely (Bonczek et al., 2017; D'Souza et al., 1999; Ohazama and Sharpe, 2004). Therefore, a transcription factor that dominates a GRN for odontoblast or ameloblast formation remains to be identified.

6.2.5 PHYLOGENETIC CONSIDERATIONS FOR SKELETAL TISSUE TYPES

The dogma of discrete categories of skeletal tissue that has prevailed from classic studies of mammalian systems is misleading. As more information accumulates on mouse skeletal tissues, examples of intermediate tissues abound. Chondroid bone, for example, which forms in the roof of the mammalian skull, among other places, contains features intermediate between cartilage and bone, similar to, but distinct from, fibrocartilage (Beresford, 1981). While the field of mammalian skeletal biology has begun to appreciate this perspective in the past decade, it has been the accepted view for many decades for skeletal biologists that study nonmammalian species. For example, instead of the three mammalian cartilage types, about ten cartilage types in teleosts were classified almost 30 years ago (Benjamin, 1990). Extant agnathans produce at least two types of cartilage, mucocartilage and (cell-rich) hyaline cartilage, and no clear homolog of mucocartilage exists in other vertebrates (Cattell et al., 2011). Considering the cases of chondroid bone and fibrocartilage, many tissues that are intermediate to "classic mammalian" cartilage and bone were described in teleosts (Benjamin, 1989; Benjamin and Ralphs, 1991; Benjamin et al., 1992). These observations lead to an important perspective on skeletal tissue identity that informs our understanding of how these tissues evolved in the first place: Instead of discrete entities, skeletal tissues across phylogeny often serve as examples along a spectrum of possibilities among the archetypal cartilage, notochord, bone, dentin, and enamel.

6.3 LOCATION, LOCATION, LOCATION

Two major factors impacting discussion of how cartilage, bone, and neural crest evolved are the location in the body where a tissue forms, and how these locations have evolved over time. A major theme in this chapter is that, once the ability to make a tissue is encoded in the genome as a GRN, different populations of cells can coopt expression of this GRN, adding another location in the body where this tissue forms. For example, if cells in the tail evolved the ability to make muscle, then for muscle to appear in the head, head cells must simply coopt expression of the muscle GRN that was employed by the tail cells. How could a different population of cells coopt expression of an entire GRN? It might not be as difficult as it seems, since GRNs typically have one to a few transcription factors that operate near the top of the GRN hierarchy (Davidson and Levine, 2008; Levine and Davidson, 2005). In this case, co-option of a GRN might entail simply co-opting expression of one or two genes, and the downstream gene expression would follow (Halfon, 2017). In the case of cartilage and bone, Sox9 and Runx2 are the prime candidates to sit at the top of their GRN hierarchies, respectively (Gomez-Picos and Eames, 2015).

What evidence exists to support the contention that changing expression of one or two transcription factors can change skeletal tissue location? In threespine stickleback fish, the pelvic fin skeleton of freshwater populations is reduced dramatically, compared to oceanic populations (Peichel et al., 2001). Genetic analyses suggested that noncoding polymorphisms in enhancers near one transcription factor, Pitx1, drove these natural evolutionary differences in skeletal tissues (Thompson et al., 2018). Also, comparison of candidate gene expression across homologous cell types in different regions of the body demonstrates few differences. For example, immature cartilage, mature cartilage, and bone express a core set of genes, including distinct Sox9 and Runx2 patterns, in the limb and in the head (Eames and Helms, 2004). Since skeletal tissues occurred in the head of vertebrates before paired appendages evolved (Janvier, 1996), these data support the idea that limb mesenchyme coopted a GRN dictating skeletal tissue differentiation in head mesenchyme. Other transcription factors, such as those in Hox, Dlx, Pax, and Nkx families, have been proposed as major players in skeletal tissue formation, but this influence is only region-specific, suggesting that these factors are more involved in patterning than differentiation *per se*. For example, Dlx5 and Nkx3.2 are commonly referred to as important transcription factors for bone and cartilage differentiation (Ferrari and Kosher, 2002; Tadic et al., 2002; Zeng et al., 2002). However, neither Dlx5 nor Nkx3.2 loss of function mutations abrogate bone or cartilage throughout the body, which rules these transcription factors out as major players in skeletal cell GRNs. On the other hand, evolution of regulatory elements of SoxE (ancestral to Sox9) might explain the appearance of cellular cartilage in the neural crest–derived vertebrate head (Jandzik et al., 2015).

6.4 SKELETAL TISSUES FROM EXTANT ANIMALS AND FOSSIL RECORD BEFORE THE EMERGENCE OF NCCS

6.4.1 INVERTEBRATE CARTILAGE

Cartilage is known from several lineages of bilaterian metazoans (Figure 6.1). Generally, two cartilage categories with unclear mutual relationships are distinguished: 1) acellular cartilage, and 2) cellular cartilage. Acellular cartilage can be found in the pharynx of two deuterostome groups, hemichordates and cephalochordates (Rychel et al., 2006). It consists of cartilaginous rods (gill bars) with ECM rich in fibrillar collagen, similar to vertebrate Col1 and Col2, and proteoglycans. In contrast to other invertebrate and vertebrate cartilages, the chondrocytes are not embedded in the ECM that they produce, but rather they surround it. Acellular cartilage is presumed to be of endodermal origin in hemichordates, while all three germ layers (endoderm, ectoderm, and mesoderm) might contribute to the formation of acellular cartilage in cephalochordates (Rychel and Swalla, 2007).

Cellular cartilage occurs in six somewhat distantly related metazoan clades (Cole and Hall, 2004; Person and Philpott, 1969). Cellular cartilage can be either cell-rich or matrix-rich depending on relative abundance of chondrocytes versus ECM (Benjamin, 1990). Cellular cartilages of metazoans without NCCs are likely of mesodermal or, less often, putatively endodermal origin. Among protostomes, cellular cartilage occurs in chelicerates (arthropod ecdysozoans), sabellid polychaetes (annelid lophotrochozoans), gastropods, and cephalopods (the latter two being mollusc lophotrochozoans). Among chordate deuterostomes, cellular cartilage can be found in two clades, cephalochordates and vertebrates (vertebrate cartilages are detailed in the next section of this chapter). Besides the acellular cartilage of pharyngeal bars, the cephalochordate amphioxus has been reported to have a cartilaginous skeleton supporting oral cirri (tentacles) that protect the mouth of this sand-burrowing chordate (Jandzik et al., 2015). This cellular cartilage is cell-rich, with chondrocytes organized in a stack-of-coins configuration, and contains collagen (as suggested by in situ hybridization) and proteoglycans (as suggested by Alcian blue staining).

The cellular cartilages of three protostome groups are not uniform, instead showing considerable variation, even within a single taxon. For example, the horseshoe crab (*Limulus polyphemus*), a representative of chelicerate arthropods, has four different types of cartilaginous structures: endosternite, opistosomatic endplates, branchial cartilages, and chilarial cartilages (Cole and Hall, 2004). The endosternite cartilage is dominated by extracellular matrix, with scattered small spherical or irregularly shaped chondrocytes interconnected via plasmatic processes. The ECM contains irregularly organized collagen fibers and is rich in acidic glycosaminoglycans, and it generally resembles vertebrate fibrocartilage. The other cartilaginous structures of horseshoe crabs are cell-rich with vacuolated chondrocytes and minimal ECM that contains elastin, collagen, and mucopolysaccharides, specifically chondroitin-4-sulfate and chondroitin sulfate K (Cole and Hall, 2004; Hall, 2015).

Cellular cartilage of at least 30 species and 18 genera of sabellid polychaetes, or feather duster worms, supports their branching feeding tentacles (Capa et al., 2011). The cartilaginous tissue is similar to non-endosternite cartilages of the horseshoe crab, being composed of large chondrocytes with vacuoles and scant vascularized ECM containing elastin and collagenous fibers and chondroitin sulfate mucopolysaccharides (Hall, 2015). In distal parts of the tentacle, the cartilage could form acellular rods comprised solely of ECM (Capa et al., 2011).

Cephalopod cartilage is probably the best-known invertebrate cartilage. It can be found supporting or protecting various parts of squid, octopus, and cuttlefish bodies, being mostly associated with brain, eye, mantle, and dermis (Cole and Hall, 2004). Cephalopod cartilage resembles "typical" hyaline cartilage with sparsely distributed chondrocytes embedded in abundant ECM. The chondrocytes are connected by processes through canaliculi in matrix, however, which is reminiscent of vertebrate osteocytes. Besides these cell–cell connections, the chondrocytes also form junctions with the ECM (Bairati et al., 1998). In contrast to most other known cartilages, cephalopod cartilage is commonly vascularized (Hall, 2015). Besides elastin, the fibrillar part of cephalopod cartilage ECM contains several types of collagens, some resembling vertebrate Col1, Col2, and Col4, along with an additional type that appears to be cephalopod-specific (Hall, 2015). The mucopolysaccharides in cephalopod cartilage ECM are surprisingly diverse, although their occurrence varies across cephalopod taxa. They include hyaluronan and various chondroitin sulfates, such as chondroitin-4-sulfate and chondroitin sulfate E (Hall, 2015).

Most gastropods have either odontophore or subradular cartilages that support their radulae, a rasping toothed organ (Guralnick and Smith, 1999). Subradular cartilage is typical for some basal forms and might be homologous to the dorsal odontophore cartilage of more derived forms (Golding et al., 2009). The best-studied gastropod cartilage is odontophore cartilage of the sea snail, *Busycon*. It is composed of large vacuolated chondrocytes with very little ECM that is rich in collagen (Hall, 2015), thus morphologically very similar to the branchial cartilage of the horseshoe crab. The ECM does not contain chondroitin sulfate, but is instead composed of a chitin-like polyglucose sulfate. Interestingly, gastropod chondrocytes contain some amount of myoglobin, and cartilages of some other snails, aside from chondrocytes, are also formed by muscle cells (Hall, 2015; Person and Philpott, 1969).

Besides acellular and cellular cartilages, some invertebrates, including brachiopods, polychaetes, and urochordates, have tissues with cartilage-like properties that vary in the structure and organization, or even presence, of chondrocytes. These tissues are usually collectively referred to as *chondroid connective tissue*. Simpler structure, phylogenetic distribution, and variable relative contents of cells, mucopolysaccharides, and fibers indicate that chondroid connective tissue is an ancient tissue type (Cole and Hall, 2004).

The fossil record does not further expand the number of large bilaterian clades with cartilage. Cartilaginous structures, particularly those that mineralize, can be found in fossil remnants of various chordate groups and are usually of strong evolutionary importance. Perhaps most notably, the stem chordate *Haikouella* demonstrates cartilage very similar in appearance to lamprey cartilage (Chen et al., 1999; Mallatt and Chen, 2003). Other fossil examples include cephalochordates of the genus *Cathaymyrus* and stem vertebrates such as *Myllokunmingia*, *Metaspriggina*, and *Haikouichthys* (Hall, 2015; Morris and Caron, 2014). Despite the fact that a mineralized endoskeleton can be found among protostomes, their cartilage is not known to be capable of mineralization *in vivo* (Hall, 2015), which may explain why protostome cartilage has not been well-described in the fossil record to date.

6.4.2 NOTOCHORD

The notochord, or *chorda dorsalis*, is a defining structure of chordates, a deuterostome group comprising three main lineages: cephalochordates, tunicates, and vertebrates (vertebrate notochord is discussed in Section 6.5.2). It forms an axial hydrostatic skeleton against which longitudinal musculature contracts during bilateral, undulating movement typical of early chordate forms and their larvae (Annona et al., 2015). The main portion of the notochord is formed by vacuolated cells, called *chordocytes*, connected by desmosomes and surrounded by a thick acellular perinotochordal sheath (Hall, 2015). Chordocytes are filled with liquid, and their pressure against the stiff sheath facilitates the mechanical function of the notochord. The notochord persists throughout life in cephalochordates and appendicularian tunicates, but forms only transiently in the remaining tunicate lineages (ascidians and thaliaceans; [Hall, 2015]). Ascidians and thaliaceans lose their notochord during regressive metamorphosis. Embryonically, the notochord develops from chordamesoderm. This embryonic tissue is morphologically and molecularly distinct from the remaining mesoderm, and its origin in cephalochordates lies in the dorsal organizer that arises during gastrulation (Stemple, 2005). Aside from a mechanical role in chordates, the notochord also plays an important role as an embryonic signaling center, strongly affecting all three germ layers, most notably in initiating neural tube formation and axial patterning (Corallo et al., 2015; Hall, 2015).

There are some significant differences in notochord structure among chordates. Antero-posteriorly, the notochord runs along almost the entire body in cephalochordates. In tunicates, the notochord does not extend past the hindbrain posteriorly (Annona et al., 2015). Chordocytes of cephalochordates and tunicates are discoidal and stacked like coins. Unique to the cephalochordate notochord, chordocytes contain transverse myofilaments. These large vacuolated chordocytes are surrounded by a layer of thin epithelial cells. In tunicates, the perinotochordal sheath is relatively simple, formed by a basal (also called external) lamina, which is supplemented by two additional collagenous layers in cephalochordates. The notochord of late tunicate larvae, mainly its posterior portion, has a different structure with cells surrounding the lumen filled with liquid (Annona et al., 2015).

The notochord shares several characteristics with cartilage. It contains structural proteins and glycosaminoglycans typical for cartilage, such as Aggrecan, Chondromodulin-1, hyaluronic acid, chondroitin sulfate, and SPARC (Osteonectin), and the perinotochordal sheaths are predominantly formed by collagenous fibers composed of Col2 (Hall, 2015; Stemple, 2005; Vasan, 1987; Welsch et al., 1991). Cells of the forming notochord express cartilage collagen genes and transcription factors *Tbxt (Brachyury)* and *Sox9*, which are also involved in cartilage development.

6.5 SKELETAL TISSUES FROM EXTANT ANIMALS AND FOSSIL RECORD AFTER THE EMERGENCE OF NCCS

6.5.1 CARTILAGE

Both vertebrates and invertebrates have cartilage (Figure. 6.1). Consistent with a GRN expression cooption story, vertebrate head cartilage has most of the same molecular characteristics as invertebrate cartilage, but they are formed by a different embryonic population of cells. Most head and pharyngeal cartilages in vertebrates are neural crest–derived, cellular cartilage (Couly et al., 1993). In fact, neural crest cells have contributed to the formation of many structurally distinct cartilages in different vertebrate clades from cyclostomes to jawed vertebrates. These cartilages display some variations in gene expression patterns, but the basic molecular machinery composed of SoxE and Runt (ancestor to Runx2) transcription factors, as well as collagenous and proteoglycan-rich extracellular matrix, is conserved among vertebrates (Cattell et al., 2011; Hecht et al., 2008; Jandzik et al., 2015; McCauley and

Bronner-Fraser, 2006; Zhang et al., 2006). Indeed, cyclostome cartilage is composed of Col2, sharing this feature with extant gnathostomes.

Cyclostome cartilages are found in the head, fin rays, and dorsoanterior axial skeleton (Ota et al., 2011). Cartilage types described in hagfish and lamprey include mucocartilage, soft cartilage, and hard cartilage (Cattell et al., 2011; Zhang et al., 2009; Zhang et al., 2006). Mucocartilage is composed of elastin-like molecules and fibroblasts surrounded by a perichondrium (Wright et al., 2001; Wright and Youson, 1982). Agnathan soft cartilage is composed of hypertrophic chondrocytes surrounded by a thin extracellular matrix (i.e., cell-rich), whereas hard cartilage contains smaller chondrocytes surrounded by abundant extracellular matrix (i.e., matrix-rich; Zhang et al., 2009]). A potentially lamprey-specific extracellular matrix protein, Lamprin, also can be found in many cartilages (Robson et al., 1993). Although mineralized cartilage has been described in some specimens of extant lamprey of presumed older age and some agnathan fossils of unknown affinity to cyclostomes, such as *Euphanerops longaevus* and *Palaeospondylus gunni*, clear proof of mineralization in cyclostome cartilages has only come from *in vitro* experiments (Bardack and Zangerl, 1968; Hirasawa et al., 2016; Janvier and Arsenault, 2002; Johanson et al., 2010; Langille and Hall, 1993).

A wealth of information has been coming out recently about skeletal tissues in cartilaginous fish, or chondrichthyans. Three main extant lineages of chondrichthyans include chimaera (holocephalans), sharks (selachians), and skates and rays (batoids; selachians and batoids compose elasmobranchs) (Daniel, 1934). Chondrichthyan skeletons are thought to be entirely cartilaginous, mostly composed of matrix-rich hyaline cartilage (Dean and Summers, 2006; Eames et al., 2007; Hall, 2005). While most vertebrates have a limited amount of mineralized cartilage as adults, it serves as a major structure that provides support and acts as a mineral reservoir in cartilaginous fish (Daniel, 1934). Most of the volume of cartilage in the chondrichthyan skeleton is unmineralized, but most cartilages typically contain a surface layer of mineralized blocks, called *tesserae*, surrounded by a fibrous perichondrium (Eames et al., 2007; Seidel et al., 2016). Tesserae are composed of two main components: the cap (more superficial) and body (more deep) zones (Kemp and Westrin, 1979). The molecular components of chondrichthyan cartilage are only becoming known with recent work. The cap zone actually contains a bone-like tissue with abundant Col1 extracellular matrix (Seidel et al., 2017). On the other hand, the body zone appears to be mineralized cartilage with an extracellular matrix rich in Col2 and Col10(Seidel et al., 2017). The mineralized centrum of the vertebrae appears like a mineralized fibrocartilage, whereas there is an odd bone-like tissue in the neural arches of the vertebrae, reminiscent of perichondral bone (Bordat, 1987; Criswell et al., 2017a; Eames et al., 2007; Enault et al., 2015; Peignoux-Deville et al., 1982). While sharks received much focus in previous research, within the past 10 years, a number of groups has recently expanded considerably what is known about chondrichthyan skeletal tissues. Tesseral features of rays are similar to sharks (Dean and Summers, 2006; Seidel et al., 2017; Seidel ct al., 2016). Indeed, the exact same mineralized tissues (tesserae, centra, and neural arches) have been described in both shark and skate species, arguing that these skeletal features were present in the last common ancestor to elasmobranchs (Atake et al., 2019). Almost no published data exist on holocephalan skeletal tissues, but many researchers are currently addressing this shortcoming.

Many lineages of early-derived ray-finned bony fish (i.e., actinopterygians) have living members today, and their skeletal tissues, like those of chondrichthyans, could provide a glimpse into the variety of skeletal tissues that were present in stem vertebrates. However, this is also an area for future research, as there are limited publications on skeletal tissues in these species. The publication of the longnose gar genome has helped push forward interest in this clade (Braasch et al., 2016). Gar matrix–rich hyaline cartilage is abundant throughout the head and expresses genes typical of mammalian cartilage, including *Col2a1*, *Col11a2*, and *Col10a1* (Eames et al., 2012). These extremely limited published data illustrate the unexploited goldmine of molecular studies into the skeletal tissues of gar, sturgeon, paddlefish, bichir, and bowfin fishes.

Teleost fish have a very rich diversity of cartilage types, perhaps due to the teleost-specific genome duplication that resulted in more cartilage differentiation genes, including the Sox family, although no current data support this speculation (Voldoire et al., 2017). Many different cell-rich and matrix-rich cartilages occur in these fish (Benjamin, 1989, 1990; Benjamin and Ralphs, 1991; Benjamin et al., 1992), but very little is known about their molecular composition. Unlike cell-rich cartilages, matrix-rich cartilages are composed of cells occupying less than 50% of total volume. So far, five types of cell-rich cartilage have been described in teleosts, based upon histology and some fiber-type analyses: hyaline-cell rich cartilage, cell-rich hyaline cartilage, fibro/cell cartilage, elastin/cell-rich cartilage, and the cell-rich cartilage Schaffer's Zellknorpel. Hyaline-cell cartilage is composed of small chondrocytes and little extracellular matrix surrounding them. This cartilage subtype is often present in the lips and the head (Benjamin, 1989). (Distribution of all of these cartilage subtypes has only been described in the head, but all of these tissues likely occur throughout the body.) In contrast, cell-rich hyaline cartilage, which is often present in the neurocranium and gill arches, is composed of large lacunae that occupy a significant percentage of the cell volume (Benjamin, 1989, 1990). Fibro/cell cartilage is a nonhyaline type of cartilage composed of large cells surrounded by a matrix rich in collagen, which often forms articular tissues (Benjamin et al., 1992; Kapoor and Khanna, 2004). Elastic/cell-rich cartilage matrix, another type of nonhyaline cartilage, is rich in elastin fibers surrounded by a thick perichondrium, and it is commonly found in the barbels (Benjamin, 1990; Benjamin et al., 1992; Kapoor and Khanna, 2004). The last type of cell-rich cartilage in teleost fish is Schaffer's Zellknorpel, which has a rigid structure and is composed of shrunken cells laying in lacunae (Benjamin, 1989, 1990). In addition to these five cell-rich cartilages, a familiar three matrix-rich cartilages compose the teleost skeleton: matrix-rich hyaline cartilage, fibrocartilage, and elastic cartilage (see Section 6.2.1). Matrix-rich hyaline cartilage is found in the neurocranium and gill arches of many species and is structurally similar to hyaline cartilage in tetrapods, providing the main source of support and growth (Benjamin, 1990). Teleost fibrocartilage lacks a perichondrium, contains irregularly arranged cells, and can be found in the oromandibular region of teleosts (Benjamin and Evans, 1990). The matrix-rich elastic cartilage is composed mostly of elastin fibers. Most fibrocartilages and elastic cartilages, however, are cell-rich in teleosts (Kapoor and Khanna, 2004).

In extant tetrapods, three types of matrix-rich cartilages occur: hyaline cartilage, elastic cartilage, and fibrocartilage, as outlined in Section 6.2.1.

6.5.2 NOTOCHORD

Chordoid (notochord-like) tissue is present in all chordates, and the vertebrate notochord shares most features with cephalochordates and tunicates described above (Section 6.4.2). Similar to tunicates, the notochord is a transient structure in most vertebrates (exceptions include cyclostomes and sturgeons; [Hall, 2005]). In contrast to tunicates, in which the notochord does not extend past the hindbrain posteriorly, the vertebrate notochord does not extend past the hindbrain anteriorly (Annona et al., 2015). In various groups of vertebrates, the notochord may mineralize or be replaced by cartilage and/or bone, or some portions may become parts of intervertebral discs. Similar to cephalochordates, the embryonic origin of the notochord in vertebrates lies in the dorsal organizer that arises during gastrulation (Stemple, 2005). In contrast to the discoidal and stacked-like-coins chordocytes of cephalochordates and tunicates, chordocytes of the vertebrate notochord do not show obvious stacking. Similar to cephalochordates, the vertebrate notochord has multiple epithelial layers surrounding the vacuolated chordocytes (Annona et al., 2015).

A major difference between vertebrate and invertebrate notochords is that the vertebrate notochord can mineralize. A common feature of many vertebrates is the chordacentrum, which is a typically thin mineralization in the fibrous sheet of the notochord (Arratia et al., 2001). Despite the widespread distribution of chordacentra among various vertebrate clades, they are not considered strictly homologous. The centrum of tetrapods, for example, derives from migrating somitic mesoderm, the centrum of teleosts appears to derive from notochordal sheath cells directly, while the centrum of chondrichthyans is also somite-derived (Criswell et al., 2017b). Phylogenetically, the distribution of various forms of the centrum around the notochord is patchy, further weakening an argument for strict homology. Many genes involved in bone formation, such as *runx2* and *sp7*, might regulate chordacentrum formation in teleosts (Renn and Winkler, 2014). However, the fossil record and sparse phylogenetic distribution do not support the idea that the bone GRN was coopted from a centrum GRN.

6.5.3 BONE

Only vertebrates make bone, but its exact anatomical location varies (Figure 6.1). Based upon fossils, bone is thought to have originated on the mesenchymal side of the basement membrane in the pharynx or epidermis, in tooth-like structures termed *odontodes* (Donoghue and Sansom, 2002; Smith and Hall, 1990). Odontodes have been argued to appear first in the fossil record in pharyngeal regions of agnathan conodonts (~515 Mya), while epidermal odontodes appear later as protective head shields in other agnathan clades, who likely had a soft cartilage endoskeleton. Perhaps the first experiment of NC making mineralized tissues, conodonts might represent the earliest form of craniate vertebrates, although more recent hypotheses place conodonts closer to gnathostomes phylogenetically (Donoghue and Rucklin, 2016; Sansom et al., 1992). The feeding apparatus in these fossil species are comparable with those present in vertebrates, and their soft tissues suggest a relationship to hagfish (Donoghue et al., 2000). Odontodes were composed of dentin and bone, thus becoming the first example of bone in the fossil record (Huysseune and Sire, 1998; Reif, 1982; Smith and Hall, 1990). The presence of bone in tooth-like structures of conodonts, however, is debated (Donoghue and Rucklin, 2016). The first unequivocal evidence of bone in the fossil record is in the agnathan clade Pteraspidormorphi (~480 Mya), historically considered a superclass of the ostracoderm group (Donoghue et al., 2006; Janvier, 1996).

Due to the association of odontodes with epithelia and the absence of adjacent cartilage, bone is argued to have formed first via intramembranous ossification, whereas endochondral ossification evolved later and gradually (Janvier, 1996; Mundlos and Olsen, 1997; Smith and Hall, 1990, 1993; Wagner and Aspenberg, 2011). In the endoskeleton, perichondral bone formation occurred first, perhaps using the same genetic programs used to produce the dermal skeleton, whereas cartilage degradation and endochondral bone deposition occurred later (Donoghue and Sansom, 2002; Wagner and Aspenberg, 2011). While pteraspidomorphi were heavily armored by external odontodes, some fossils show both mineralized cartilage and perichondral bone in their endoskeleton, reflecting evolutionary progress leading to endochondral ossification (Janvier, 1996; Zhang et al., 2009).

The location of bone in vertebrates evolved. Critically for understanding the role of neural crest during evolution of bone, extant and fossil Cyclostomata (i.e., lamprey and hagfish) never appeared to have mineralized cartilage or bone. As outlined above, fossilized bone first appeared in exoskeletal head shields of the pteraspidomorphs, presumably formed by neural crest. Current data suggest that perichondral bone later appeared in the braincase of ostracoderms, and then perichondral (and perhaps endochondral) bone appeared in the vertebrae of some placoderms (e.g. petalichthyids and arthodires) (Janvier, 1996; Trinajstic et al., 2015). This would imply that after the evolution of neural crest, cephalic mesoderm and then somitic mesoderm acquired the ability to make bone. Whether extant chondrichthyans generate bone is currently under investigation (Atake et al., 2019; Eames et al., 2007; Enault et al., 2015), but fossil primitive chondrichthyans and especially their acanthodian ancestors had abundant perichondral and perhaps endochondral bone (Coates et al., 1998; Long et al., 2015; Zangerl, 1966). Together, these studies suggest that endochondral bone is not an osteichthyan synapomorphy. Whether all examples of endochondral bone in vertebrates are homologous is complicated by another possible scenario, in which early osteichthyan chondrocytes acquired the ability to transdifferentiate into osteoblasts in response to higher mechanical forces and to provide a new osteoblast reservoir (Cervantes-Diaz et al., 2017).

Histologically, all examples of bone in vertebrates, whether they be dermal, perichondral, or endochondral, appear very similar, if not identical, when animal groups are intercompared. Certainly, some tissues, such as chondroid bone or the mineralizing chondrichthyan tissues, share some, but not all, of these histological features, making their designation as bone difficult. Molecular studies might reveal significant differences (Arendt, 2008; Gomez-Picos and Eames, 2015).

6.5.4   DENTIN/ENAMEL

Dentin and enamel are vertebrate-specific tissues (Figure 6.1). Dentin and enamel/enameloid, together with bone, composed the primitive skeleton in early jawless vertebrates around 500 Mya (Gans and Northcutt, 1983). Similar to bone, it is generally accepted that dentin and enamel/enameloid first appeared in odontodes of Ordovician agnathans as a protective shield (Janvier, 1996; Smith and Hall, 1990). An alternative theory is that the earliest hard skeletal tissue was dentin, forming the most superficial dermal tissue, only later covered by a hypermineralized layer of enamel/enameloid. Conodonts (~515 Mya) potentially represent the earliest form of craniate vertebrates whose skeleton was composed of dentin- and enamel-like tissues (Sansom et al., 1992). Although neural crest-like cells likely existed prior to vertebrates (Hall and Gillis, 2013), it is still unknown whether conodonts possessed neural crest cells. After conodonts, heterostracans, an extinct group of pteraspidomorphs, show traces of dentin tubercles capped with enamel/enameloid in their exoskeleton (Janvier, 1996; Keating et al., 2015). Dentin and enamel/enameloid are present in placoid scales of living chondrichthyans, and in teeth of all other extant vertebrates, including sharks, mammals, teleosts, and reptiles (Kawasaki et al., 2004). Enameloid in osteichthyans appears to serve the same function as mammalian enamel, but enameloid and enamel might actually be convergent tissues. Enameloid appears to be deposited by mesenchymal, not epithelial, cells, and the major proteins in mammalian enamel have not been identified in osteichthyans (Kawasaki et al., 2004). These and other features argue that enameloid in fish and enamel in mammals are a result of convergent evolution (Kawasaki et al., 2004). Although experimental data in the trunk region of vertebrates are limited to one paper, dentin in teeth and dermal denticles are neural crest-derived (Chai et al., 2000; Gillis et al., 2017). Enamel is produced by ectodermal or pharyngeal epithelia, while enameloid is presumably neural crest-derived (Kawasaki et al., 2004).

6.6   EVOLUTION OF GRN UNDERLYING CARTILAGE

6.6.1   CELLULAR CARTILAGE EVOLVED FROM AN ACELLULAR CARTILAGE GRN

The cellular cartilage of vertebrates might have evolved by modifying a GRN that directed acellular cartilage formation (Figure 6.2A). Similar features of agnathan and gnathostome cellular cartilage suggested that a cartilage GRN (presumably driven by SoxE, the ancestor to Sox9) is a vertebrate synapomorphy (Ohtani et al., 2008; Zhang et al., 2006), but this concept has been revised substantially. Acellular cartilage of hemichordates might be homologous to acellular cartilage of cephalochordates, given their general resemblance and close phylogenetic distribution, in which case the tissue was subsequently lost in echinoderms, urochordates, and vertebrates (Hall and Gillis, 2013). Structures similar to acellular cartilaginous rods of hemichordates and cephalochordates were proposed in hypothetical stem vertebrates as antecedents of vertebrate cellular cartilage of the pharyngeal arches (Rychel and Swalla, 2007). According to this hypothesis, NCCs that migrate to the pharynx might have gradually replaced the acellular cartilage by co-opting and modifying

A. Cellular cartilage evolved from acellular cartilage

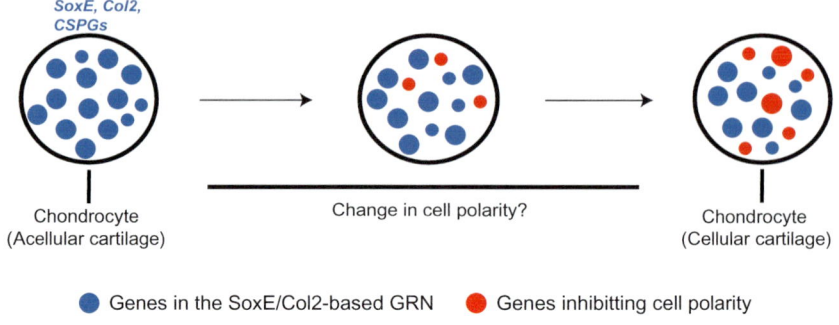

B. Cellular cartilage evolved from chordoid tissue

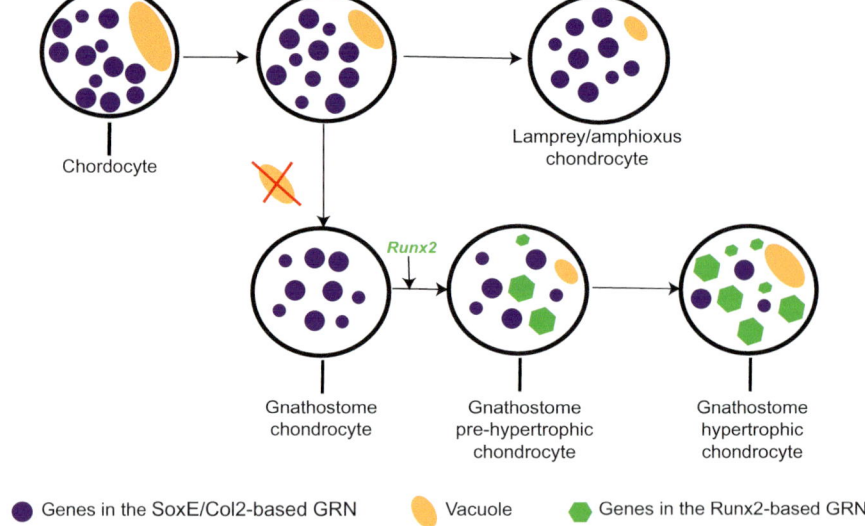

FIGURE 6.2 Evolution of the GRN underlying cartilage. A) Cellular cartilage evolved from acellular cartilage. NCCs acquired expression of the acellular cartilage SoxE/Collagen type II–based cartilage GRN and subsequently modified that GRN to make cellular cartilage. Runx and Barx genes were later coopted by these cells consolidating the cellular cartilage GRN (Cattell et al. 2011). B) Cartilage evolved from chordoid tissue. NCCs acquired expression of the notochord fibrillar collagen based GRN and was then modified to make cellular cartilage. Chordocytes have a single large vacuole and low amounts of glycosaminoglycan-rich ECM. While chondrocytes in amphioxus and lamprey are vacuolated, chondrocytes in vertebrates are avacuolated and have abundant ECM. However, when vertebrate cartilage becomes hypertrophic and degenerates (under the influence of a Runx2 GRN), the ECM frequently becomes vacuolated. Cellular cartilage and notochord might share a common origin since they share the expression of several genes including Sox9, Col2, Acan, Col9, and Cnmd (Jonasson et al. 2012).

the SoxE GRN that presumably operated in the endodermal epithelia of ancestors to hemichordates and cephalochordates. To be clear, this would involve two steps (of uncertain order): 1) acquiring expression of the acellular cartilage SoxE GRN in NCCs; and 2) modifying that GRN to form cellular cartilage. Of course, the acellular cartilage GRN operating in ancestors to amphioxus might have acquired expression first in presumably cephalic mesoderm of larval forms of ancestors to amphioxus, then was modified to form cellular cartilage, and finally, expression of this cellular cartilage GRN was coopted by vertebrate NCCs (Jandzik et al., 2015). However, acellular cartilage might be less similar to vertebrate cartilage than other protostome cellular cartilages. Cellular cartilage appears in many disparate phylogenetic groups and presumably evolved independently at least three times (Hall and Gillis, 2013). The striking similarities in morphology, function, development, and gene expression patterns, though, suggest the existence of underlying homologous mechanisms with deep metazoan origins (i.e., deep homology). For example, the endosternite cartilage of horseshoe crab and the funnel cartilage of cuttlefish share significant parts of the GRN driving hyaline cartilage in vertebrates (Tarazona et al., 2016). Interestingly, the chondrocytes in some portions of the sabellid polychaete tentacles are arranged in a stack-of-coins fashion, reminiscent of cellular cartilage in oral cirri of amphioxus and pharyngeal arches of lamprey and zebrafish (Cattell et al., 2011; Cole and Hall, 2004; Hall, 2015; Jandzik et al., 2015; Kimmel et al., 1998). In summary, acellular cartilage exists in hemichordates, cephalochordates, and agnathans, while cellular cartilage is found in cephalochordates, vertebrates, and various protostomes. Therefore, if vertebrate cellular cartilage evolved from an acellular cartilage GRN, then the independence of this event from the appearance of cellular cartilage in various protostomes would have to be addressed.

6.6.2 Cellular Cartilage Evolved from a Notochord GRN

Some fossils and other features of extant nonvertebrate chordates provide the basis for an evolutionary hypothesis that the GRN for cellular cartilage evolved from a notochord GRN (Figure 6.2B). Since the notochord is usually a diagnostic character of fossil chordates, it is present in all stem chordates in which cellular cartilage has been described, but importantly for this hypothesis, the notochord is apparent in samples with no identifiable cartilage (e.g. *Pikaia*; [Morris and Caron, 2012]). Of living animals, enteropneust hemichordates have a diverticulum of the gut called a *stomochord*, which has some functional and histologic similarities to the chordate notochord (Satoh et al., 2014). These animals only have acellular cartilage, so they might represent an evolutionary snapshot where a notochord GRN was being established prior to cellular cartilage. However, the homology of the stomochord and notochord is questionable (Annona et al., 2015; Lowe et al., 2015; Minarik et al., 2017). The phylogenetically closest group of animals to vertebrates are urochordates, such as sea squirts, some of whose larval forms have a true notochord (Delsuc et al., 2006). However, urochordates lack cartilage, and they are thought to be very derived, which led to them being designated as the sister taxon to vertebrates only relatively recently. The cephalochordate amphioxus has a notochord but also contains both acellular and cellular cartilage, so no living animals represent a transition from notochord only to notochord with cellular cartilage. Despite the lack of an animal phylogenetically positioned to directly support the hypothesis that the cellular cartilage GRN evolved from the notochord GRN, functional, structural, and molecular similarities support the hypothesis that the notochord was a primitive type of cartilage. For example, chondrocytes secrete a highly hydrated ECM, which gives cartilage its structural properties, while chordocytes (notochord cells) retain liquid in their vacuoles (Stemple, 2005). Regarding the vacuolar chordocyte, the mature chondrocyte also undergoes vacuolization during hypertrophy, which, under the evolutionary scenario of chordocyte to chondrocyte, suggests a rather parsimonious recovery of vacuoles, perhaps under Runx2 control (Figure 6.2B). In addition to molecular and developmental similarities between vertebrate and cephalochordate notochords, amphioxus has cellular cartilage that shares properties with vertebrate cellular cartilage. This evolutionary model proposes cooption of the developmental program of originally mesodermal cartilage of an amphioxus-like vertebrate ancestor into the developmental repertoire of the cranial NCCs of vertebrates (Jandzik et al., 2015). Stemple (2015) actually proposed the opposite relationship between the notochord and cartilage, arguing that the notochord coopted cartilage properties. Of course, the "cellular cartilage evolved from the notochord GRN" hypothesis also would have to be reconciled with the appearance of cellular cartilage in various protostomes.

6.7   EVOLUTION OF GRN UNDERLYING BONE

How could bone have evolved within vertebrates, since there is no obvious antecedent tissue in cyclostomes or vertebrate sister groups? Here, we explore three parsimonious explanations, which involve simple cooption by NCCs of a preexisting GRN that directed formation of: 1) mineralized tissues in nonvertebrates; 2) dentin/enamel; or 3) mature cartilage. In principle, option #1 could have preceded options #2 or 3.

6.7.1   Bone Evolved from a Deuterostome Mineralization GRN

The first hypothesis claims that a mineralization GRN operating in mesenchymal cells of nonvertebrate deuterostomes was coopted by NCCs and modified to form bone in the ancestor to noncyclostome agnathan vertebrates (Figure 6.3a). However, many issues complicate this scenario. First, outside of vertebrates within deuterostomes, only echinoderms have extensive mineralization of skeletal tissues, calling into question whether a mineralization GRN could be homologous among deuterostomes. Second, the form of mineral deposited in vertebrate skeletal tissues is mostly hydroxyapatite, which is a calcium phosphate crystal, while most mineral in nonvertebrate skeletal tissues is calcium carbonate–based (Matsushiro and Miyashita, 2004; Wilt et al., 2003). Therefore, at least the portions of the mineralization GRN that drive biochemical synthesis of mineral would be different among echinoderms and vertebrates. Finally, there are few clear orthologs among protein components of the mineralized matrix in echinoderms and vertebrates. Matrix proteins in sea urchin teeth crossreacted with antibodies to a variety of bone and tooth proteins in mammals (Veis et al., 2002). However, most of the proteins associated with spicules (mineralized structures) in sea urchin are in the spicule matrix (SM) group of the mesenchyme-specific cell surface glycoprotein (MSP) gene family, which have no orthologs in vertebrates (Livingston et al., 2006).

Despite these obstacles, proteins in the secretory calcium-binding phosphoprotein (SCPP) family provide an interesting story for the evolution of a mineralization GRN associated with bone formation. SCPP proteins bind to Ca^{++} (when phosphorylated) and facilitate hydroxyapatite crystal nucleation or modulate its growth (Kawasaki et al., 2004). The SCPP proteins Matrix gla protein (MGP) and Bone gamma-carboxyglutamate protein (BGLAP, previously referred to as *Osteocalcin*) are proposed to play important roles in vertebrate mineralization, although loss-of-function models indicate that they function as inhibitors (Ducy et al., 1996; Luo et al., 1997). An SCPP protein, Secreted protein acidic and rich in cysteine (SPARC, previously called *Osteonectin*), is the most abundant non-collagenous protein in bone ECM (Termine et al., 1981). As opposed to MGP and BGLAP, SPARC loss-of-function models have osteopenia (Delany et al., 2000), suggesting perhaps that SPARC is actually a positive regulator of mineralization. Regarding the significance of this hypothesis, SPARC is found in both protostomes and deuterostomes. The family of SCPP proteins is thought to have originated from tandem duplication of SPARC-like 1 (*SPARCL1*) in the ancestor to osteichthyans following their divergence from chondrichthyans (Kawasaki et al., 2004). Interestingly, a group of putative mineralization (spicule)-associated genes in sea urchin also underwent tandem duplications (Livingston et al., 2006). SPARC and a SPARCL were identified in sea urchin, although this SPARCL was not a clear ortholog to vertebrate SPARCL (Livingston et al., 2006). Amphioxus express SPARC/SPARCL in skeletogenic cells, even though they do not mineralize their tissues (Yong and Yu, 2016).

Perhaps a remaining hope for identifying conservation among vertebrate and non-vertebrate mineralization GRNs focusses on regulatory genes upstream of the biochemical and secreted components (Livingston et al., 2006). Genes involved in sea urchin spicule differentiation include *pmar1* (represses *hairy*), *delta, ets, b-catenin/tcf*, and *cart1/alx3/alx4(alx1)* (Wilt et al., 2003). Orthologs of these genes are expressed in bone, and mutations in many of them lead to skeletal patterning/mineralization defects in mouse (Beverdam et al., 2001; Itoh et al., 2012; Li et al., 2018; Mavrogiannis et al., 2001; Shao et al., 2018). Perhaps the acquisition of a vertebrate-specific NC skeletogenic competency factor, Ventx, facilitated co-option of the deuterostome mineralization program under this scenario (Scerbo and Monsoro-Burq, 2020).

6.7.2 Bone Evolved from a Dentin/Enamel GRN

Dentin, enamel, and bone are closely related tissues, so they may share evolutionary histories. For example, an ancestral GRN gave rise to one of these tissues first, and then (perhaps after genome duplication events) this ancestral GRN was modified, giving rise to other mineralized tissues (Figure 6.3B; Fisher and Franz-Odendaal, 2012). According to one scenario, dentin- and enamel-like tissues first originated around the basement membrane of odontodes, and subsequent spread of mineralization deeper into the dermis gave rise to bone (Donoghue et al., 2006). Both dentin and bone are formed by NCCs (presumably even in odontodes of primitive vertebrates), so these

A. Bone evolved from a deuterostome mineralization GRN

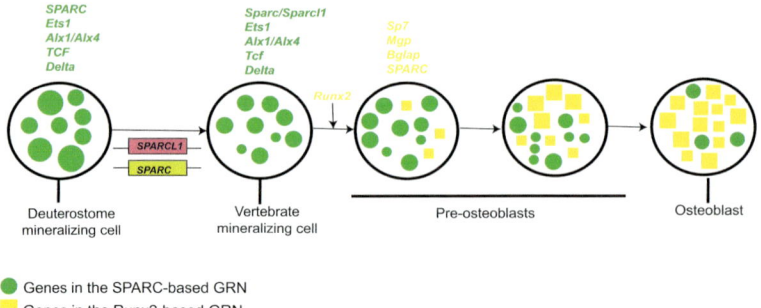

● Genes in the SPARC-based GRN
■ Genes in the Runx2-based GRN

B. Bone evolved from a dentin/enamel GRN

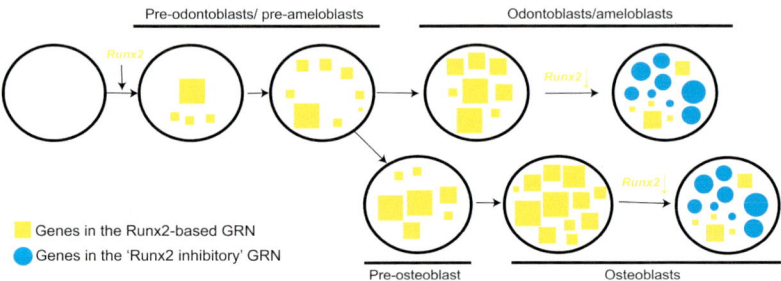

■ Genes in the Runx2-based GRN
● Genes in the 'Runx2 inhibitory' GRN

C. Bone evolved from a mature cartilage GRN

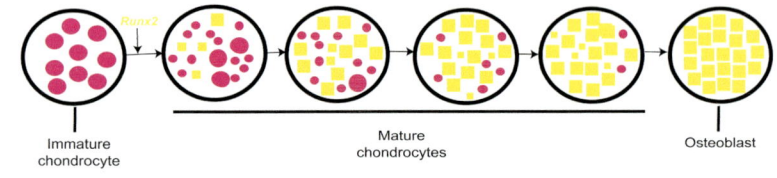

● Genes in the Sox9-based GRN
■ Genes in the Runx2-based GRN

FIGURE 6.3 Evolution of the GRN underlying bone. A) Bone evolved from a deutero-stome mineralization GRN. A mineralization GRN, likely involving *SPARC*, operating in mesenchymal cells of a deuterostome ancestor was coopted and modified in NCCs of an ancestral vertebrate. This ancestral GRN contained many mineralization genes conserved within deuterostomes (i.e., *SPARC*, *Ets1*, *Alx1/Alx4*, *Tcf*, and *Delta*). During vertebrate evo-lution, a duplication event of the ancestral *SPARC* gene generated two paralogs and gave rise to the *SCPP* gene family. Since Runx2 drives the expression of many mineralization genes in vertebrates (i.e., *Sp7*, *Bglap*, and *Mgp*), it is likely that a Runx2-mediated GRN was later established in these NCC populations to give rise to the osteoblast. B) Odontoblasts, ameloblasts, and osteoblasts are closely related cell types that may have derived one from another or from a common precursor. Runx2 is essential for odontoblast, ameloblast, and

FIGURE 6.3 (CONTINUED)

osteoblast differentiation as it regulates the expression of many bone- and tooth-related genes. During early differentiation of dentin/enamel, Runx2 is highly expressed in odonto-blasts/ameloblasts, and then this transcription factor subsequently has to be downregulated at later stages of differentiation. According to this evolutionary scenario for the appearance of bone, another population of neural crest cells coopted the dentin/enamel GRN (regulated by Runx2) to produce bone, the later stages of which also include Runx2 downregulation. C) Immature chondrocytes, developing under a Sox9 GRN, provided a structural and molecular "buffer" for the gradual development of a Runx2 GRN. After the establishment of the Runx2 GRN in mature chondrocytes, the osteoblast appeared when another population of cells coopted this GRN.

cells could have expanded the spatial domain of the dentin GRN expression (deeper in the mesenchyme) within an odontode and modified that GRN to produce bone. Many facts support the hypothesis that a bone GRN evolved by modifications to a dentin GRN. Dentin and bone share many features, including a histologically similar structure and a dense, Col1-enriched matrix that is heavy calcified by hydroxyapatite (Goldberg et al., 2011). Runx2 is essential for both osteoblast and odontoblast differentiation, regulating the expression of many bone and tooth related genes (Chen et al., 2005). In transgenic mice that manipulate Runx2 expression levels specifically in odontoblasts, which normally downregulate Runx2 after early stages of differentiation, osteoblast-like cells embedded in a bone-like matrix were seen instead of normal odontoblasts (Komori, 2010; Li et al., 2011; Miyazaki et al., 2008). By demonstrating the similarity of the GRNs driving odontoblast and osteoblast formation, these studies support the feasibility of the hypothesis that one GRN evolved from the other.

Apart from embryological and histological differences, the bone GRN also might have evolved from an enamel GRN. Enamel is formed by ectoderm-derived ameloblasts, which secrete the enamel-specific matrix proteins Ameloblastin, Enamelin, and Amelogenin (Sire et al., 2007). Therefore, the hypothesis that the bone GRN evolved from the enamel GRN requires that NCCs coopted expression of an enamel GRN operating in ectodermal cells and modified it to form bone. Runx2 is required for the proper differentiation and function of ameloblasts, and it is a key regulator of ameloblast-specific genes (Chu et al., 2018; Gaikwad et al., 2001). Tooth development in Runx2$^{-/-}$ mice is arrested before odontoblast and ameloblast formation in the developing tooth, perhaps due to a role for Runx2 during enamel knot formation (Camilleri and McDonald, 2006; D'Souza et al., 1999). In later stages of enamel mineralization, Runx2 regulates odontogenic ameloblast–associated protein (ODAM) expression and amelotin promoter activity (Lee et al., 2010; Liu et al., 2018). Runx2 downregulation might be involved in late stages of all mineralizing skeletal cells. Due to the number of enamel-specific proteins present in the enamel GRN, however, the modifications required to transform it into the bone GRN presumably were more significant than the modifications required to transform the dentin GRN into one making bone.

Whether bone evolved from dentin or enamel (or *vice versa*) is still debated, but the two hypothetical scenarios on the evolution of the osteoblast (i.e., bone evolved from either dentin or enamel) presented above support the idea that a Runx2 GRN appeared *de novo* to mineralize early vertebrate tissues, independent of cartilage. These models are consistent with saltational evolution, in which large-scale genomic changes facilitate the evolution of novelty over a short period of time (Gould, 2002).

6.7.3 BONE EVOLVED FROM A MATURE CARTILAGE GRN

Recently, we proposed an alternative scenario for the appearance of bone that takes advantage of molecular genetic and developmental biology principles to offer a more parsimonious explanation than the saltational models outlined above (Figure 6.3C; Gomez-Picos and Eames, 2015). The fossil record generally supports the idea that bone evolved before mature/mineralized cartilage, but some exceptions suggest that mineralized cartilage might have preceded bone (Donoghue et al., 2006; Janvier, 1996; Smith and Hall, 1990). For example, the fossil *Palaeospondylus gunni* (~385 Mya) has an entire endoskeleton composed of hypertrophic, mineralized cartilage, but there is no evidence of bone in its skeleton (Johanson et al., 2010). In addition, it is possible that trace amounts of perichondral bone were lost during traditional fossil preparation techniques, in which case, only the more robust odontodes persisted as the first example of bone in vertebrate ancestors.

We hypothesized that the bone GRN evolved from a cartilage maturation GRN, and this assertion complements the previous saltational models by providing a gradual evolutionary model (Gomez-Picos and Eames, 2015). The many stages of cartilage maturation occurring today during endochondral ossification likely evolved gradually over time. Hypertrophy and mineralization appeared first, followed by matrix degradation, and finally vasculogenesis, fat formation, and endochondral bone deposition occurred (Hall, 1975; Smith and Hall, 1990). In our evolutionary scenario, a Runx2-dependent GRN slowly evolved these various additional maturation stages to the Sox9 GRN that drives immature cartilage. Later, ectopic expression of this Runx2 GRN outside of cartilage resulted in the evolution of the bone GRN. For example, expansion of the Runx2 GRN expression domain slightly outside of the Sox9 GRN expression domain of cartilage might produce perichondral bone (Gomez-Picos and Eames, 2015). In support of this hypothesis, some osteoblasts actually transdifferentiate from chondrocytes, showing the overlapping nature of the underlying GRNs (Hammond and Schulte-Merker, 2009; Park et al., 2015; Zhou et al., 2014). In addition to providing strong molecular genetic support for the hypothesis that bone evolved from a mature cartilage GRN, these recent findings also fall in line with the old adage that ontogeny recapitulates phylogeny. Those osteoblasts that transdifferentiate from mature chondrocytes during development might reflect those exact events during evolution. Alternatively, expression of the cartilage maturation Runx2 GRN in dermal tissues might have produced odontode-associated bone.

Again, the philosophical advantages of this evolutionary hypothesis for the origin of bone is its gradualistic nature. Instead of requiring the *de novo,* saltational appearance of a Runx2 GRN driving bone formation, cartilage provided a "buffer" tissue in which a novel Runx2 GRN could evolve that drives production of a heavily mineralized ECM. Once this Runx2 GRN was established, the misexpression of Runx2 in another tissue (neural crest-derived?) would have resulted in the first instance of bone in agnathan vertebrates. Perhaps the many intermediate forms of cartilage and bone highlighted earlier, such as chondroid bone and fibrocartilage, are remnants of the gradual elaboration of the Runx2 GRN as it modified immature cartilage. Interestingly, we note that osteoblasts of earlier-derived vertebrates, such as fish and frog, appear to express more "cartilage" genes than osteoblasts of later-derived vertebrates, such as chick and mouse (Aldea et al., 2013; Eames et al., 2012; Eames and Helms, 2004; Nguyen and Eames, 2020).

6.8 CONCLUSIONS

Traditional studies on non-model organisms, along with the advent of the modern molecular era, have revised considerably previous hypotheses on the origins of cartilage and bone, as well as the relationship of those events to the appearance of vertebrates. Studies of hemichordates and amphioxus have clearly demonstrated that an ability to form cartilage preceded vertebrates in the chordate ancestor. Bone and dentin are vertebrate-specific tissues, and their formation is tied intimately to the evolved features of neural crest cells. Molecular identification of the GRNs underlying the formation of skeletal tissues offers tantalizing clues about the molecular mechanisms underlying the evolution of cartilage and bone. The common theme for the evolutionary scenarios highlighted here is cooption of GRN expression. Once a GRN for differentiating a particular cell type was established, additional cell populations, sometimes from different germ layers, could coopt expression of one or two genes sitting at the top of the GRN hierarchy. As a result, new cell types could evolve, or new modifications to that cell type could evolve. We are very excited that comparative transcriptomics, combined with functional genetics, can resolve among the evolutionary scenarios highlighted here, ultimately providing a clear story of how cartilage and bone evolved.

REFERENCES

Aldea, D., Hanna, P., Munoz, D., Espinoza, J., Torrejon, M., Sachs, L., Buisine, N., Oulion, S., Escriva, H., Marcellini, S., 2013. Evolution of the vertebrate bone matrix: An expression analysis of the network forming collagen paralogues in amphibian osteoblasts. *J Exp Zool B Mol Dev Evol* 320(6), 375–384.

Annona, G., Holland, N.D., D'Aniello, S., 2015. Evolution of the notochord. *EvoDevo* 6, 30.

Arendt, D., 2008. The evolution of cell types in animals: Emerging principles from molecular studies. *Nat Rev Genet* 9(11), 868–882.

Arratia, G., Schultze, H.P., Casciotta, J., 2001. Vertebral column and associated elements in dipnoans and comparison with other fishes: Development and homology. *J Morphol* 250(2), 101–172.

Atake, O.J., Cooper, D.M.L., Eames, B.F., 2019. Bone-like features in skate suggest a novel elasmobranch synapomorphy and deep homology of trabecular mineralization patterns. *Acta Biomater* 84, 424–436.

Aumailley, M., Gayraud, B., 1998. Structure and biological activity of the extracellular matrix. *J Mol Med (Berl)* 76(3–4), 253–265.

Bagheri-Fam, S., Barrionuevo, F., Dohrmann, U., Gunther, T., Schule, R., Kemler, R., Mallo, M., Kanzler, B., Scherer, G., 2006. Long-range upstream and downstream enhancers control distinct subsets of the complex spatiotemporal Sox9 expression pattern. *Dev Biol* 291(2), 382–397.

Bairati, A., Comazzi, M., Gioria, M., Rigo, C., 1998. The ultrastructure of chondrocytes in the cartilage of *Sepia officinalis* and *Octopus vulgaris* (Mollusca, Cephalopoda). *Tissue Cell* 30(3), 340–351.

Bardack, D., Zangerl, R., 1968. First fossil lamprey: A record from the Pennsylvanian of Illinois. *Science* 162(3859), 1265–1267.

Barrionuevo, F., Taketo, M.M., Scherer, G., Kispert, A., 2006. Sox9 is required for notochord maintenance in mice. *Dev Biol* 295(1), 128–140.

Benjamin, M., 1989. Hyaline-cell cartilage (chondroid) in the heads of teleosts. *Anat Embryol (Berl)* 179(3), 285–303.

Benjamin, M., 1990. The cranial cartilages of teleosts and their classification. *J Anat* 169, 153–172.

Benjamin, M., Evans, E.J., 1990. Fibrocartilage. *J Anat* 171, 1–15.

Benjamin, M., Ralphs, J.R., 1991. Extracellular matrix of connective tissues in the heads of teleosts. *J Anat* 179, 137–148.

Benjamin, M., Ralphs, J.R., Eberewariye, O.S., 1992. Cartilage and related tissues in the trunk and fins of teleosts. *J Anat* 181(1), 113–118.

Beresford, W.A., 1981. *Chondroid Bone, Secondary Cartilage, and Metaplasia.* Urban & Schwarzenberg, Baltimore, MD.

Beverdam, A., Brouwer, A., Reijnen, M., Korving, J., Meijlink, F., 2001. Severe nasal clefting and abnormal embryonic apoptosis in Alx3/Alx4 double mutant mice. *Development* 128(20), 3975–3986.

Bi, W., Deng, J.M., Zhang, Z., Behringer, R.R., de Crombrugghe, B., 1999. Sox9 is required for cartilage formation. *Nat Genet* 22(1), 85–89.

Bialek, P., Kern, B., Yang, X., Schrock, M., Sosic, D., Hong, N., Wu, H., Yu, K., Ornitz, D.M., Olson, E.N., Justice, M.J., Karsenty, G., 2004. A twist code determines the onset of osteoblast differentiation. *Dev Cell* 6(3), 423–435.

Bonczek, O., Balcar, V.J., Sery, O., 2017. PAX9 gene mutations and tooth agenesis: A review. *Clin Genet* 92(5), 467–476.

Bordat, C., 1987. Ultrastructural study of the vertebrae of the selachian *Scyliorhinus canicula. Can J Zool* 65(6), 1435–1444.

Braasch, I., Gehrke, A.R., Smith, J.J., Kawasaki, K., Manousaki, T., Pasquier, J., Amores, A., Desvignes, T., Batzel, P., Catchen, J., Berlin, A.M., Campbell, M.S., Barrell, D., Martin, K.J., Mulley, J.F., Ravi, V., Lee, A.P., Nakamura, T., Chalopin, D., Fan, S., Wcisel, D., Canestro, C., Sydes, J., Beaudry, F.E., Sun, Y., Hertel, J., Beam, M.J., Fasold, M., Ishiyama, M., Johnson, J., Kehr, S., Lara, M., Letaw, J.H., Litman, G.W., Litman, R.T., Mikami, M., Ota, T., Saha, N.R., Williams, L., Stadler, P.F., Wang, H., Taylor, J.S., Fontenot, Q., Ferrara, A., Searle, S.M., Aken, B., Yandell, M., Schneider, I., Yoder, J.A., Volff, J.N., Meyer, A., Amemiya, C.T., Venkatesh, B., Holland, P.W., Guiguen, Y., Bobe, J., Shubin, N.H., Di Palma, F., Alfoldi, J., Lindblad-Toh, K., Postlethwait, J.H., 2016. The spotted gar genome illuminates vertebrate evolution and facilitates human-teleost comparisons. *Nat Genet* 48(4), 427–437.

Bridgewater, L.C., Walker, M.D., Miller, G.C., Ellison, T.A., Holsinger, L.D., Potter, J.L., Jackson, T.L., Chen, R.K., Winkel, V.L., Zhang, Z., McKinney, S., de Crombrugghe, B., 2003. Adjacent DNA sequences modulate Sox9 transcriptional activation at paired Sox sites in three chondrocyte-specific enhancer elements. *Nucleic Acids Res* 31(5), 1541–1553.

Camilleri, S., McDonald, F., 2006. Runx2 and dental development. *Eur J Oral Sci* 114(5), 361–373.

Capa, M., Nogueira, J.M., Rossi, M.C., 2011. Comparative internal structure of dorsal lips and radiolar appendages in Sabellidae (Polychaeta) and phylogenetic implications. *J Morphol* 272(3), 302–319.

Cattell, M., Lai, S., Cerny, R., Medeiros, D.M., 2011. A new mechanistic scenario for the origin and evolution of vertebrate cartilage. *PLOS ONE* 6(7), e22474.

Cervantes-Diaz, F., Contreras, P., Marcellini, S., 2017. Evolutionary origin of endochondral ossification: The transdifferentiation hypothesis. *Dev Genes Evol* 227(2), 121–127.

Chai, Y., Jiang, X., Ito, Y., Bringas, P., Jr., Han, J., Rowitch, D.H., Soriano, P., McMahon, A.P., Sucov, H.M., 2000. Fate of the mammalian cranial neural crest during tooth and mandibular morphogenesis. *Development* 127(8), 1671–1679.

Chen, J.-Y., Huang, D.-Y., Li, C.-W., 1999. An early Cambrian craniate-like chordate. *Nature* 402(6761), 518–522.

Chen, S., Santos, L., Wu, Y., Vuong, R., Gay, I., Schulze, J., Chuang, H.H., MacDougall, M., 2005. Altered gene expression in human cleidocranial dysplasia dental pulp cells. *Arch Oral Biol* 50(2), 227–236.

Chu, Q., Gao, Y., Gao, X., Dong, Z., Song, W., Xu, Z., Xiang, L., Wang, Y., Zhang, L., Li, M., Gao, Y., 2018. Ablation of Runx2 in ameloblasts suppresses enamel maturation in tooth development. *Sci Rep* 8(1), 9594.

Coates, M.I., Finarelli, J.A., Sansom, I.J., Andreev, P.S., Criswell, K.E., Tietjen, K., Rivers, M.L., La Riviere, P.J., 2018. An early chondrichthyan and the evolutionary assembly of a shark body plan. *Proc Biol Sci* 285(1870).

Coates, M.I., Sequeira, S.E.K., Sansom, I.J., Smith, M.M., 1998. Spines and tissues of ancient sharks. *Nature* 396(6713), 729–730.

Cole, A.G., 2011. A review of diversity in the evolution and development of cartilage: The search for the origin of the chondrocyte. *Eur Cells Mater* 21, 122–129.

Cole, A.G., Hall, B.K., 2004. The nature and significance of invertebrate cartilages revisited: Distribution and histology of cartilage and cartilage-like tissues within the Metazoa. *Zoology (Jena)* 107(4), 261–273.

Corallo, D., Trapani, V., Bonaldo, P., 2015. The notochord: Structure and functions. *Cell Mol Life Sci* 72(16), 2989–3008.

Couly, G.F., Coltey, P.M., Douarin, N.M.L., 1992. The developmental fate of the cephalic mesoderm in quail-chick chimeras. *Development* 114(1), 1–15.

Couly, G.F., Coltey, P.M., Le Douarin, N.M., 1993. The triple origin of skull in higher vertebrates: A study in quail-chick chimeras. *Development* 117(2), 409–429.

Criswell, K.E., Coates, M.I., Gillis, J.A., 2017a. Embryonic development of the axial column in the little skate, *Leucoraja erinacea*. *J Morphol* 278(3), 300–320.

Criswell, K.E., Coates, M.I., Gillis, J.A., 2017b. Embryonic origin of the gnathostome vertebral skeleton. *Proc Biol Sci* 284(1867).

D'Souza, R.N., Aberg, T., Gaikwad, J., Cavender, A., Owen, M., Karsenty, G., Thesleff, I., 1999. Cbfa1 is required for epithelial-mesenchymal interactions regulating tooth development in mice. *Development* 126(13), 2911–2920.

Dale, R.M., Topczewski, J., 2011. Identification of an evolutionarily conserved regulatory element of the zebrafish col2a1a gene. *Dev Biol* 357(2), 518–531.

Daniel, J.F., 1934. *The Elasmobranch Fishes.* University of California Press, Berkeley, CA.

Davidson, E.H., Levine, M.S., 2008. Properties of developmental gene regulatory networks. *Proc Natl Acad Sci U S A* 105(51), 20063–20066.

de Crombrugghe, B., Lefebvre, V., Behringer, R.R., Bi, W., Murakami, S., Huang, W., 2000. Transcriptional mechanisms of chondrocyte differentiation. *Matrix Biol J Int Soc Matrix Biol* 19(5), 389–394.

Dean, M.N., Summers, A.P., 2006. Mineralized cartilage in the skeleton of chondrichthyan fishes. *Zoology (Jena)* 109(2), 164–168.

Delany, A.M., Amling, M., Priemel, M., Howe, C., Baron, R., Canalis, E., 2000. Osteopenia and decreased bone formation in osteonectin-deficient mice. *J Clin Invest* 105(7), 915–923.

Delsuc, F., Brinkmann, H., Chourrout, D., Philippe, H., 2006. Tunicates and not cephalochordates are the closest living relatives of vertebrates. *Nature* 439(7079), 965–968.

Dodig, M., Tadic, T., Kronenberg, M.S., Dacic, S., Liu, Y.H., Maxson, R., Rowe, D.W., Lichtler, A.C., 1999. Ectopic Msx2 overexpression inhibits and Msx2 antisense stimulates calvarial osteoblast differentiation. *Dev Biol* 209(2), 298–307.

Donoghue, P.C., Forey, P.L., Aldridge, R.J., 2000. Conodont affinity and chordate phylogeny. *Biol Rev Camb Philos Soc* 75(2), 191–251.

Donoghue, P.C., Rucklin, M., 2016. The ins and outs of the evolutionary origin of teeth. *Evol Dev* 18(1), 19–30.

Donoghue, P.C., Sansom, I.J., 2002. Origin and early evolution of vertebrate skeletonization. *Microsc Res Tech* 59(5), 352–372.

Donoghue, P.C., Sansom, I.J., Downs, J.P., 2006. Early evolution of vertebrate skeletal tissues and cellular interactions, and the canalization of skeletal development. *J Exp Zool B Mol Dev Evol* 306(3), 278–294.

Ducy, P., Desbois, C., Boyce, B., Pinero, G., Story, B., Dunstan, C., Smith, E., Bonadio, J., Goldstein, S., Gundberg, C., Bradley, A., Karsenty, G., 1996. Increased bone formation in osteocalcin-deficient mice. *Nature* 382(6590), 448–452.

Ducy, P., Zhang, R., Geoffroy, V., Ridall, A.L., Karsenty, G., 1997. Osf2/Cbfa1: A transcriptional activator of osteoblast differentiation. *Cell* 89(5), 747–754.

Dunn, C.W., Giribet, G., Edgecombe, G.D., Hejnol, A., 2014. Animal phylogeny and its evolutionary implications. *Annual Review of Ecology, Evolution, and Systematics* 45(1), 371–395.

Eames, B.F., Allen, N., Young, J., Kaplan, A., Helms, J.A., Schneider, R.A., 2007. Skeletogenesis in the swell shark *Cephaloscyllium ventriosum*. *J Anat* 210(5), 542–554.

Eames, B.F., Amores, A., Yan, Y.L., Postlethwait, J.H., 2012. Evolution of the osteoblast: Skeletogenesis in gar and zebrafish. *BMC Evol Biol* 12, 27.

Eames, B.F., de la Fuente, L., Helms, J.A., 2003. Molecular ontogeny of the skeleton. *Birth Defects Res C Embryo Today* 69(2), 93–101.

Eames, B.F., Helms, J.A., 2004. Conserved molecular program regulating cranial and appendicular skeletogenesis. *Dev Dyn Off Publ Am Assoc Anatomists* 231(1), 4–13.

Eames, B.F., Sharpe, P.T., Helms, J.A., 2004. Hierarchy revealed in the specification of three skeletal fates by Sox9 and Runx2. *Dev Biol* 274(1), 188–200.

Enault, S., Munoz, D.N., Silva, W.T., Borday-Birraux, V., Bonade, M., Oulion, S., Venteo, S., Marcellini, S., Debiais-Thibaud, M., 2015. Molecular footprinting of skeletal tissues in the catshark *Scyliorhinus canicula* and the clawed frog *Xenopus tropicalis* identifies conserved and derived features of vertebrate calcification. *Front Genet* 6, 283.

Enomoto, H., Enomoto-Iwamoto, M., Iwamoto, M., Nomura, S., Himeno, M., Kitamura, Y., Kishimoto, T., Komori, T., 2000. Cbfa1 is a positive regulatory factor in chondrocyte maturation. *J Biol Chem* 275(12), 8695–8702.

Ferrari, D., Kosher, R.A., 2002. Dlx5 is a positive regulator of chondrocyte differentiation during endochondral ossification. *Dev Biol* 252(2), 257–270.

Fisher, S., Franz-Odendaal, T., 2012. Evolution of the bone gene regulatory network. *Curr Opin Genet Dev* 22(4), 390–397.

Fyhrie, D.P., Christiansen, B.A., 2015. Bone material properties and skeletal fragility. *Calcif Tissue Int* 97(3), 213–228.

Gaikwad, J.S., Cavender, A., D'Souza, R.N., 2001. Identification of tooth-specific downstream targets of Runx2. *Gene* 279(1), 91–97.

Gans, C., Northcutt, R., 1983. Neural crest and the origin of vertebrates: A new head. *Science* 220(4594), 268–274.

Gillis, J.A., Alsema, E.C., Criswell, K.E., 2017. Trunk neural crest origin of dermal denticles in a cartilaginous fish. *Proc Natl Acad Sci U S A* 114(50), 13200–13205.

Goldberg, M., Kulkarni, A.B., Young, M., Boskey, A., 2011. Dentin: Structure, composition and mineralization. *Front Biosci (Elite Ed)* 3, 711–735.

Golding, R.E., Ponder, W.F., Byrne, M., 2009. Three-dimensional reconstruction of the odontophoral cartilages of Caenogastropoda (Mollusca: Gastropoda) using micro-CT: Morphology and phylogenetic significance. *J Morphol* 270(5), 558–587.

Gomez-Picos, P., Eames, B.F., 2015. On the evolutionary relationship between chondrocytes and osteoblasts. *Front Genet* 6, 297.

Goudemand, N., Orchard, M.J., Urdy, S., Bucher, H., Tafforeau, P., 2011. Synchrotron-aided reconstruction of the conodont feeding apparatus and implications for the mouth of the first vertebrates. *Proc Natl Acad Sci U S A* 108(21), 8720–8724.

Gould, S.J., 2002. *The Structure of Evolutionary Theory*. Belknap Press of Harvard University Press, Cambridge, MA.

Gray, H., Williams, P.L., 1989. *Gray's Anatomy*. C. Livingstone, Edinburgh/New York, NY.

Guralnick, R., Smith, K., 1999. Historical and biomechanical analysis of integration and dissociation in molluscan feeding, with special emphasis on the true limpets (Patellogastropoda: Gastropoda). *J Morphol* 241(2), 175–195.

Halfon, M.S., 2017. Perspectives on gene regulatory network evolution. *Trends Genet* 33(7), 436–447.

Hall, B.K., 1975. Evolutionary consequences of skeletal differentiation. *Am Zool* 15(2), 329–350.

Hall, B.K., 2005. *Bones and Cartilage: Developmental and Evolutionary Skeletal Biology.* Elsevier Academic Press, San Diego, CA/London, pp. xxviii, 760 p.

Hall, B.K., 2015. *Bones and Cartilage: Developmental and Evolutionary Skeletal Biology*, 2nd ed. Elsevier/AP, Academic Press is an imprint of Elsevier, Amsterdam.

Hall, B.K., Gillis, J.A., 2013. Incremental evolution of the neural crest, neural crest cells and neural crest-derived skeletal tissues. *J Anat* 222(1), 19–31.

Ham, A.W., Cormack, D.H., 1987. *Ham's Histology*, 9th ed. Lippincott, Philadelphia.

Hammond, C.L., Schulte-Merker, S., 2009. Two populations of endochondral osteoblasts with differential sensitivity to Hedgehog signalling. *Development* 136(23), 3991–4000.

Harada, H., Tagashira, S., Fujiwara, M., Ogawa, S., Katsumata, T., Yamaguchi, A., Komori, T., Nakatsuka, M., 1999. Cbfa1 isoforms exert functional differences in osteoblast differentiation. *J Biol Chem* 274(11), 6972–6978.

Hecht, J., Stricker, S., Wiecha, U., Stiege, A., Panopoulou, G., Podsiadlowski, L., Poustka, A.J., Dieterich, C., Ehrich, S., Suvorova, J., Mundlos, S., Seitz, V., 2008. Evolution of a core gene network for skeletogenesis in chordates. *PLOS Genet* 4(3), e1000025.

Hirasawa, T., Oisi, Y., Kuratani, S., 2016. Palaeospondylus as a primitive hagfish. *Zool Lett* 2(1), 20.

His, W., 1868. *Untersuchungen über die erste Anlage des Wirbelthierleibes: die erste Entwickelung des Hühnchens im Ei.* FCW Vogel.

Horigome, N., Myojin, M., Ueki, T., Hirano, S., Aizawa, S., Kuratani, S., 1999. Development of cephalic neural crest cells in embryos of *Lampetra japonica*, with special reference to the evolution of the jaw. *Dev Biol* 207(2), 287–308.

Hu, G., Codina, M., Fisher, S., 2012. Multiple enhancers associated with ACAN suggest highly redundant transcriptional regulation in cartilage. *Matrix Biol J Int Soc Matrix Biol* 31(6), 328–337.

Huysseune, A., Sire, J.Y., 1998. Evolution of patterns and processes in teeth and tooth-related tissues in non-mammalian vertebrates. *Eur J Oral Sci* 106 Suppl 1, 437–481.

Irisarri, I., Baurain, D., Brinkmann, H., Delsuc, F., Sire, J.Y., Kupfer, A., Petersen, J., Jarek, M., Meyer, A., Vences, M., Philippe, H., 2017. Phylotranscriptomic consolidation of the jawed vertebrate timetree. *Nat Ecol Evol* 1(9), 1370–1378.

Itoh, T., Ando, M., Tsukamasa, Y., Akao, Y., 2012. Expression of BMP-2 and Ets1 in BMP-2-stimulated mouse pre-osteoblast differentiation is regulated by microRNA-370. *FEBS Lett* 586(12), 1693–1701.

Jandzik, D., Garnett, A.T., Square, T.A., Cattell, M.V., Yu, J.K., Medeiros, D.M., 2015. Evolution of the new vertebrate head by co-option of an ancient chordate skeletal tissue. *Nature* 518(7540), 534–537.

Janvier, P., 1996. *Early Vertebrates*. Oxford University Press, Oxford.

Janvier, P., Arsenault, M., 2002. Palaeobiology: Calcification of early vertebrate cartilage. *Nature* 417(6889), 609.

Johanson, Z., Kearsley, A., den Blaauwen, J., Newman, M., Smith, M.M., 2010. No bones about it: An enigmatic Devonian fossil reveals a new skeletal framework--A potential role of loss of gene regulation. *Semin Cell Dev Biol* 21(4), 414–423.

Jonasson, K.A., Russell, A.P., Vickaryous, M.K., 2012. Histology and histochemistry of the gekkotan notochord and their bearing on the development of notochordal cartilage. *J Morph* 273(6), 596–603.

Kapoor, B.G., Khanna, B., 2004. *Ichthyology Handbook*. Springer-Verlag, Berlin Heidelberg.

Kawasaki, K., Suzuki, T., Weiss, K.M., 2004. Genetic basis for the evolution of vertebrate mineralized tissue. *Proc Natl Acad Sci U S A* 101(31), 11356–11361.

Keating, J.N., Marquart, C.L., Donoghue, P.C.J., 2015. Histology of the heterostracan dermal skeleton: Insight into the origin of the vertebrate mineralised skeleton. *J Morphol* 276(6), 657–680.

Kemp, N.E., Westrin, S.K., 1979. Ultrastructure of calcified cartilage in the endoskeletal tesserae of sharks. *J Morphol* 160(1), 75–109.

Kern, B., Shen, J., Starbuck, M., Karsenty, G., 2001. Cbfa1 contributes to the osteoblast-specific expression of type I collagen genes. *J Biol Chem* 276(10), 7101–7107.

Kimmel, C.B., Miller, C.T., Kruze, G., Ullmann, B., BreMiller, R.A., Larison, K.D., Snyder, H.C., 1998. The shaping of pharyngeal cartilages during early development of the zebrafish. *Dev Biol* 203(2), 245–263.

Komori, T., 2010. Regulation of bone development and extracellular matrix protein genes by RUNX2. *Cell Tissue Res* 339(1), 189–195.

Komori, T., Yagi, H., Nomura, S., Yamaguchi, A., Sasaki, K., Deguchi, K., Shimizu, Y., Bronson, R.T., Gao, Y.H., Inada, M., Sato, M., Okamoto, R., Kitamura, Y., Yoshiki, S., Kishimoto, T., 1997. Targeted disruption of Cbfa1 results in a complete lack of bone formation owing to maturational arrest of osteoblasts. *Cell* 89(5), 755–764.

Kurakazu, I., Akasaki, Y., Hayashida, M., Tsushima, H., Goto, N., Sueishi, T., Toya, M., Kuwahara, M., Okazaki, K., Duffy, T., Lotz, M.K., Nakashima, Y., 2019. FOXO1 transcription factor regulates chondrogenic differentiation through transforming growth factor β1 signaling. *J Biol Chem* 294(46), 17555–17569.

Langille, R.M., Hall, B.K., 1993. Pattern formation and the neural crest. In: *The Skull*, Hanken, J., Hall, B.K. (Eds.). The University of Chicago Press, Chicago, IL, pp. 77–111.

Lee, H.K., Lee, D.S., Ryoo, H.M., Park, J.T., Park, S.J., Bae, H.S., Cho, M.I., Park, J.C., 2010. The odontogenic ameloblast-associated protein (ODAM) cooperates with RUNX2 and modulates enamel mineralization via regulation of MMP-20. *J Cell Biochem* 111(3), 755–767.

Leprevost, A., Azaïs, T., Trichet, M., Sire, J.Y., 2017. Vertebral development and ossification in the Siberian sturgeon (*Acipenser baerii*), with new insights on bone histology and ultrastructure of vertebral elements and scutes. *Anat Rec* (Hoboken) 300(3), 437–449.

Levine, M., Davidson, E.H., 2005. Gene regulatory networks for development. *Proc Natl Acad Sci U S A* 102(14), 4936–4942.

Li, S., Kong, H., Yao, N., Yu, Q., Wang, P., Lin, Y., Wang, J., Kuang, R., Zhao, X., Xu, J., Zhu, Q., Ni, L., 2011. The role of runt-related transcription factor 2 (Runx2) in the late stage of odontoblast differentiation and dentin formation. *Biochem Biophys Res Commun* 410(3), 698–704.

Li, Z., Xu, Z., Duan, C., Liu, W., Sun, J., Han, B., 2018. Role of TCF/LEF transcription factors in bone development and osteogenesis. *Int J Med Sci* 15(12), 1415–1422.

Liu, X., Wang, Y., Zhang, L., Xu, Z., Chu, Q., Xu, C., Sun, Y., Gao, Y., 2018. Combination of Runx2 and Cbfbeta upregulates Amelotin gene expression in ameloblasts by directly interacting with cisenhancers during amelogenesis. *Mol Med Rep* 17(4), 6068–6076.

Livingston, B.T., Killian, C.E., Wilt, F., Cameron, A., Landrum, M.J., Ermolaeva, O., Sapojnikov, V., Maglott, D.R., Buchanan, A.M., Ettensohn, C.A., 2006. A genome-wide analysis of biomineralization-related proteins in the sea urchin *Strongylocentrotus purpuratus*. *Dev Biol* 300(1), 335–348.

Long, J.A., Burrow, C.J., Ginter, M., Maisey, J.G., Trinajstic, K.M., Coates, M.I., Young, G.C., Senden, T.J., 2015. First shark from the Late Devonian (Frasnian) Gogo formation, Western Australia sheds new light on the development of tessellated calcified cartilage. *PLOS ONE* 10(5), e0126066.

Lowe, C.J., Clarke, D.N., Medeiros, D.M., Rokhsar, D.S., Gerhart, J., 2015. The deuterostome context of chordate origins. *Nature* 520(7548), 456–465.

Luo, G., Ducy, P., McKee, M.D., Pinero, G.J., Loyer, E., Behringer, R.R., Karsenty, G., 1997. Spontaneous calcification of arteries and cartilage in mice lacking matrix gla protein. *Nature* 386(6620), 78–81.

Mak, K.K., Chen, M.H., Day, T.F., Chuang, P.T., Yang, Y., 2006. Wnt/{beta}-catenin signaling interacts differentially with Ihh signaling in controlling endochondral bone and synovial joint formation. *Development* 133(18), 3695–3707.

Mallatt, J., Chen, J.Y., 2003. Fossil sister group of craniates: Predicted and found. *J Morphol* 258(1), 1–31.

Marletaz, F., Peijnenburg, K., Goto, T., Satoh, N., Rokhsar, D.S., 2019. A new spiralian phylogeny places the enigmatic arrow worms among Gnathiferans. *Curr Biol* 29, 312–318. e3.

Matsushiro, A., Miyashita, T., 2004. Evolution of hard-tissue mineralization: Comparison of the inner skeletal system and the outer shell system. *J Bone Miner Metab* 22(3), 163–169.

Mavrogiannis, L.A., Antonopoulou, I., Baxova, A., Kutilek, S., Kim, C.A., Sugayama, S.M., Salamanca, A., Wall, S.A., Morriss-Kay, G.M., Wilkie, A.O., 2001. Haploinsufficiency of the human homeobox gene ALX4 causes skull ossification defects. *Nat Genet* 27(1), 17–18.

McCauley, D.W., Bronner-Fraser, M., 2003. Neural crest contributions to the lamprey head. *Development* 130(11), 2317–2327.

McCauley, D.W., Bronner-Fraser, M., 2006. Importance of SoxE in neural crest development and the evolution of the pharynx. *Nature* 441(7094), 750–752.

Minarik, M., Stundl, J., Fabian, P., Jandzik, D., Metscher, B.D., Psenicka, M., Gela, D., Osorio-Perez, A., Arias-Rodriguez, L., Horacek, I., Cerny, R., 2017. Pre-oral gut contributes to facial structures in non-teleost fishes. *Nature* 547(7662), 209–212.

Miyazaki, T., Kanatani, N., Rokutanda, S., Yoshida, C., Toyosawa, S., Nakamura, R., Takada, S., Komori, T., 2008. Inhibition of the terminal differentiation of odontoblasts and their transdifferentiation into osteoblasts in Runx2 transgenic mice. *Arch Histol Cytol* 71(2), 131–146.

Moradian-Oldak, J., 2012. Protein-mediated enamel mineralization. *Front Biosci (Landmark Ed)* 17, 1996–2023.

Morris, S.C., Caron, J.B., 2012. Pikaia gracilens Walcott, a stem-group chordate from the Middle Cambrian of British Columbia. *Biol Rev Camb Philos Soc* 87(2), 480–512.

Morris, S.C., Caron, J.B., 2014. A primitive fish from the Cambrian of North America. *Nature* 512(7515), 419–422.

Mundlos, S., Olsen, B.R., 1997. Heritable diseases of the skeleton. Part I: Molecular insights into skeletal development-transcription factors and signaling pathways. *FASEB J* 11(2), 125–132.

Nakashima, K., Zhou, X., Kunkel, G., Zhang, Z., Deng, J.M., Behringer, R.R., de Crombrugghe, B., 2002. The novel zinc finger-containing transcription factor osterix is required for osteoblast differentiation and bone formation. *Cell* 108(1), 17–29.

Naumann, A., Dennis, J.E., Awadallah, A., Carrino, D.A., Mansour, J.M., Kastenbauer, E., Caplan, A.I., 2002. Immunochemical and mechanical characterization of cartilage subtypes in rabbit. *J Histochem Cytochem* 50(8), 1049–1058.

Ng, L.J., Wheatley, S., Muscat, G.E., Conway-Campbell, J., Bowles, J., Wright, E., Bell, D.M., Tam, P.P., Cheah, K.S., Koopman, P., 1997. SOX9 binds DNA, activates transcription, and coexpresses with type II collagen during chondrogenesis in the mouse. *Dev Biol* 183(1), 108–121.

Nguyen, J.K.B. and Eames, B.F., 2020. Evolutionary repression of chondrogenic genes in the vertebrate osteoblast. The FEBS Journal.

Nikitina, N., Bronner-Fraser, M., Sauka-Spengler, T., 2009. DiI cell labeling in lamprey embryos. *Cold Spring Harb Protoc* 2009, pdb.prot5124.

Noden, D.M., 1988. Interactions and fates of avian craniofacial mesenchyme. *Development* 103 Suppl, 121–140.

Ohazama, A., Sharpe, P.T., 2004. TNF signalling in tooth development. *Curr Opin Genet Dev* 14(5), 513–519.

Ohtani, K., Yao, T., Kobayashi, M., Kusakabe, R., Kuratani, S., Wada, H., 2008. Expression of Sox and fibrillar collagen genes in lamprey larval chondrogenesis with implications for the evolution of vertebrate cartilage. *J Exp Zool B Mol Dev Evol* 310(7), 596–607.

Okuma, T., Hirata, M., Yano, F., Mori, D., Kawaguchi, H., Chung, U.I., Tanaka, S., Saito, T., 2015. Regulation of mouse chondrocyte differentiation by CCAAT/enhancer-binding proteins. *Biomed Res* 36(1), 21–29.

Oldknow, K.J., MacRae, V.E., Farquharson, C., 2015. Endocrine role of bone: Recent and emerging perspectives beyond osteocalcin. *J Endocrinol* 225(1), R1–19.

Ota, K.G., Fujimoto, S., Oisi, Y., Kuratani, S., 2011. Identification of vertebra-like elements and their possible differentiation from sclerotomes in the hagfish. *Nat Commun* 2, 373.

Ota, K.G., Kuraku, S., Kuratani, S., 2007. Hagfish embryology with reference to the evolution of the neural crest. *Nature* 446(7136), 672–675.

Otto, F., Thornell, A.P., Crompton, T., Denzel, A., Gilmour, K.C., Rosewell, I.R., Stamp, G.W., Beddington, R.S., Mundlos, S., Olsen, B.R., Selby, P.B., Owen, M.J., 1997. Cbfa1, a candidate gene for cleidocranial dysplasia syndrome, is essential for osteoblast differentiation and bone development [see comments]. *Cell* 89(5), 765–771.

Park, J., Gebhardt, M., Golovchenko, S., Perez-Branguli, F., Hattori, T., Hartmann, C., Zhou, X., deCrombrugghe, B., Stock, M., Schneider, H., von der Mark, K., 2015. Dual pathways to endochondral osteoblasts: A novel chondrocyte-derived osteoprogenitor cell identified in hypertrophic cartilage. *Biol Open* 4(5), 608–621.

Peichel, C.L., Nereng, K.S., Ohgi, K.A., Cole, B.L., Colosimo, P.F., Buerkle, C.A., Schluter, D., Kingsley, D.M., 2001. The genetic architecture of divergence between threespine stickleback species. *Nature* 414(6866), 901–905.

Peignoux-Deville, J., Lallier, F., Vidal, B., 1982. Evidence for the presence of osseous tissue in dogfish vertebrae. *Cell Tissue Res* 222(3), 605–614.

Person, P., Philpott, D.E., 1969. The nature and significance of invertebrate cartilages. *Biol Rev Camb Philos Soc* 44(1), 1–16.

Reif, W.-E., 1982. Evolution of dermal skeleton and dentition in vertebrates. In: *Evolutionary Biology*: 15, Hecht, M.K., Wallace, B., Prance, G.T. (Eds.). Springer US, Boston, MA, pp. 287–368.

Renn, J., Winkler, C., 2014. Osterix/Sp7 regulates biomineralization of otoliths and bone in medaka (*Oryzias latipes*). *Matrix Biol J Int Soc Matrix Biol* 34, 193–204.

Robson, P., Wright, G.M., Sitarz, E., Maiti, A., Rawat, M., Youson, J.H., Keeley, F.W., 1993. Characterization of lamprin, an unusual matrix protein from lamprey cartilage. Implications for evolution, structure, and assembly of elastin and other fibrillar proteins. *J Biol Chem* 268(2), 1440–1447.

Rychel, A.L., Smith, S.E., Shimamoto, H.T., Swalla, B.J., 2006. Evolution and development of the chordates: Collagen and pharyngeal cartilage. *Mol Biol Evol* 23(3), 541–549.

Rychel, A.L., Swalla, B.J., 2007. Development and evolution of chordate cartilage. *J Exp Zool B Mol Dev Evol* 308(3), 325–335.

Sansom, I.J., Smith, M.P., Armstrong, H.A., Smith, M.M., 1992. Presence of the earliest vertebrate hard tissue in conodonts. *Science* 256(5061), 1308–1311.

Sato, M., Morii, E., Komori, T., Kawahata, H., Sugimoto, M., Terai, K., Shimizu, H., Yasui, T., Ogihara, H., Yasui, N., Ochi, T., Kitamura, Y., Ito, Y., Nomura, S., 1998. Transcriptional regulation of osteopontin gene in vivo by PEBP2alphaA/CBFA1 and ETS1 in the skeletal tissues. *Oncogene* 17(12), 1517–1525.

Satoh, N., Tagawa, K., Lowe, C.J., Yu, J.K., Kawashima, T., Takahashi, H., Ogasawara, M., Kirschner, M., Hisata, K., Su, Y.H., Gerhart, J., 2014. On a possible evolutionary link of the stomochord of hemichordates to pharyngeal organs of chordates. *Genesis* 52(12), 925–934.

Satokata, I., Maas, R., 1994. *Msx1* deficient mice exhibit cleft palate and abnormalities of craniofacial and tooth development. *Nat Genet* 6(4), 348–356.

Scerbo, P., Monsoro-Burq, A.H., 2020. The vertebrate-specific VENTX/NANOG gene empowers neural crest with ectomesenchyme potential. *Science Advances* 6(18): eaaz146.

Seidel, R., Blumer, M., Zaslansky, P., Knotel, D., Huber, D.R., Weaver, J.C., Fratzl, P., Omelon, S., Bertinetti, L., Dean, M.N., 2017. Ultrastructural, material and crystallographic description of endophytic masses – A possible damage response in shark and ray tessellated calcified cartilage. *J Struct Biol* 198(1), 5–18.

Seidel, R., Lyons, K., Blumer, M., Zaslansky, P., Fratzl, P., Weaver, J.C., Dean, M.N., 2016. Ultrastructural and developmental features of the tessellated endoskeleton of elasmobranchs (sharks and rays). *J Anat* 229(5), 681–702.

Shao, J., Zhou, Y., Xiao, Y., 2018. The regulatory roles of Notch in osteocyte differentiation via the crosstalk with canonical Wnt pathways during the transition of osteoblasts to osteocytes. *Bone* 108, 165–178.

Sire, J.Y., Davit-Beal, T., Delgado, S., Gu, X., 2007. The origin and evolution of enamel mineralization genes. *Cells Tissues Organs* 186(1), 25–48.

Sire, J.Y., Donoghue, P.C., Vickaryous, M.K., 2009. Origin and evolution of the integumentary skeleton in non-tetrapod vertebrates. *J Anat* 214(4), 409–440.

Smith, M.M., Hall, B.K., 1990. Development and evolutionary origins of vertebrate skeletogenic and odontogenic tissues. *Biol Rev Camb Philos Soc* 65(3), 277–373.

Smith, M.M., Hall, B.K., 1993. A developmental model for evolution of the vertebrate exoskeleton and teeth. In: *Evolutionary Biology*, Hecht, M.K. (Ed.). Plenum Press, New York, NY.

Smits, P., Li, P., Mandel, J., Zhang, Z., Deng, J.M., Behringer, R.R., de Croumbrugghe, B., Lefebvre, V., 2001. The transcription factors L-Sox5 and Sox6 are essential for cartilage formation. *Dev Cell* 1(2), 277–290.

St-Jacques, B., Hammerschmidt, M., McMahon, A.P., 1999. Indian hedgehog signaling regulates proliferation and differentiation of chondrocytes and is essential for bone formation. *Genes Dev* 13(16), 2072–2086.

Stemple, D.L., 2005. Structure and function of the notochord: An essential organ for chordate development. *Development* 132(11), 2503–2512.

Tadic, T., Dodig, M., Erceg, I., Marijanovic, I., Mina, M., Kalajzic, Z., Velonis, D., Kronenberg, M.S., Kosher, R.A., Ferrari, D., Lichtler, A.C., 2002. Overexpression of Dlx5 in chicken calvarial cells accelerates osteoblastic differentiation. *J Bone Miner Res* 17(6), 1008–1014.

Tarazona, O.A., Slota, L.A., Lopez, D.H., Zhang, G., Cohn, M.J., 2016. The genetic program for cartilage development has deep homology within Bilateria. *Nature* 533(7601), 86–89.

Termine, J.D., Belcourt, A.B., Conn, K.M., Kleinman, H.K., 1981. Mineral and collagen-binding proteins of fetal calf bone. *J Biol Chem* 256(20), 10403–10408.

Thompson, A.C., Capellini, T.D., Guenther, C.A., Chan, Y.F., Infante, C.R., Menke, D.B., Kingsley, D.M., 2018. A novel enhancer near the Pitx1 gene influences development and evolution of pelvic appendages in vertebrates. *eLife* 7.

Trinajstic, K., Boisvert, C., Long, J., Maksimenko, A., Johanson, Z., 2015. Pelvic and reproductive structures in placoderms (stem gnathostomes). *Biol Rev Camb Philos Soc* 90(2), 467–501.

Ueta, C., Iwamoto, M., Kanatani, N., Yoshida, C., Liu, Y., Enomoto-Iwamoto, M., Ohmori, T., Enomoto, H., Nakata, K., Takada, K., Kurisu, K., Komori, T., 2001. Skeletal malformations caused by overexpression of Cbfa1 or its dominant negative form in chondrocytes. *J Cell Biol* 153(1), 87–100.

Vasan, N.S., 1987. Somite chondrogenesis: The role of the microenvironment. *Cell Differ* 21(3), 147–159.

Veis, A., Barss, J., Dahl, T., Rahima, M., Stock, S., 2002. Mineral-related proteins of sea urchin teeth: *Lytechinus variegatus*. *Microsc Res Tech* 59(5), 342–351.

Voldoire, E., Brunet, F., Naville, M., Volff, J.N., Galiana, D., 2017. Expansion by whole genome duplication and evolution of the sox gene family in teleost fish. *PLOS ONE* 12(7), e0180936.

Vortkamp, A., 2001. Interaction of growth factors regulating chondrocyte differentiation in the developing embryo. *Osteoarthr Cartil* 9 Suppl A, S109–117.

Vortkamp, A., Lee, K., Lanske, B., Segre, G.V., Kronenberg, H.M., Tabin, C.J., 1996. Regulation of rate of cartilage differentiation by Indian hedgehog and PTH-related protein [see comments]. *Science* 273(5275), 613–622.

Wachsmuth, L., Soder, S., Fan, Z., Finger, F., Aigner, T., 2006. Immunolocalization of matrix proteins in different human cartilage subtypes. *Histol Histopathol* 21(5), 477–485.

Wagner, D.O., Aspenberg, P., 2011. Where did bone come from? *Acta Orthop* 82(4), 393–398.

Watanabe, H., Yamada, Y., Kimata, K., 1998. Roles of aggrecan, a large chondroitin sulfate proteoglycan, in cartilage structure and function. *J Biochem* 124(4), 687–693.

Welsch, U., Erlinger, R., Potter, I.C., 1991. Proteoglycans in the notochord sheath of lampreys. *Acta Histochem* 91(1), 59–65.

Whelan, N.V., Kocot, K.M., Moroz, T.P., Mukherjee, K., Williams, P., Paulay, G., Moroz, L.L., Halanych, K.M., 2017. Ctenophore relationships and their placement as the sister group to all other animals. *Nat Ecol Evol* 1(11), 1737–1746.

Wilt, F.H., Killian, C.E., Livingston, B.T., 2003. Development of calcareous skeletal elements in invertebrates. *Differentiation* 71(4–5), 237–250.

Wright, G.M., Keeley, F.W., Robson, P., 2001. The unusual cartilaginous tissues of jawless craniates, cephalochordates and invertebrates. *Cell Tissue Res* 304(2), 165–174.

Wright, G.M., Youson, J.H., 1982. Ultrastructure of mucocartilage in the larval anadromous sea lamprey, *Petromyzon marinus* L. *Am J Anat* 165(1), 39–51.

Yong, L.W., Yu, J.K., 2016. Tracing the evolutionary origin of vertebrate skeletal tissues: Insights from cephalochordate amphioxus. *Curr Opin Genet Dev* 39, 55–62.

Yoon, B.S., Lyons, K.M., 2004. Multiple functions of BMPs in chondrogenesis. *J Cell Biochem* 93(1), 93–103.

Zangerl, R., 1966. A new shark in the family Edestidae, *Ornithoprion hertwigi* from the Pennsylvania Mecca and Logan Quarry Shales of Indiana. *Fieldiana Geol* 16, 1–43.

Zeng, L., Kempf, H., Murtaugh, L.C., Sato, M.E., Lassar, A.B., 2002. Shh establishes an Nkx3.2/Sox9 autoregulatory loop that is maintained by BMP signals to induce somitic chondrogenesis. *Genes Dev* 16(15), 1990–2005.

Zhang, G., Eames, B.F., Cohn, M.J., 2009. Chapter 2. Evolution of vertebrate cartilage development. *Curr Top Dev Biol* 86, 15–42.

Zhang, G., Miyamoto, M.M., Cohn, M.J., 2006. Lamprey type II collagen and Sox9 reveal an ancient origin of the vertebrate collagenous skeleton. *Proc Natl Acad Sci U S A* 103(9), 3180–3185.

Zhang, P., Jimenez, S.A., Stokes, D.G., 2003. Regulation of human COL9A1 gene expression. Activation of the proximal promoter region by SOX9. *J Biol Chem* 278(1), 117–123.

Zhou, X., von der Mark, K., Henry, S., Norton, W., Adams, H., de Crombrugghe, B., 2014. Chondrocytes transdifferentiate into osteoblasts in endochondral bone during development, postnatal growth and fracture healing in mice. *PLOS Genet* 10(12), e1004820.

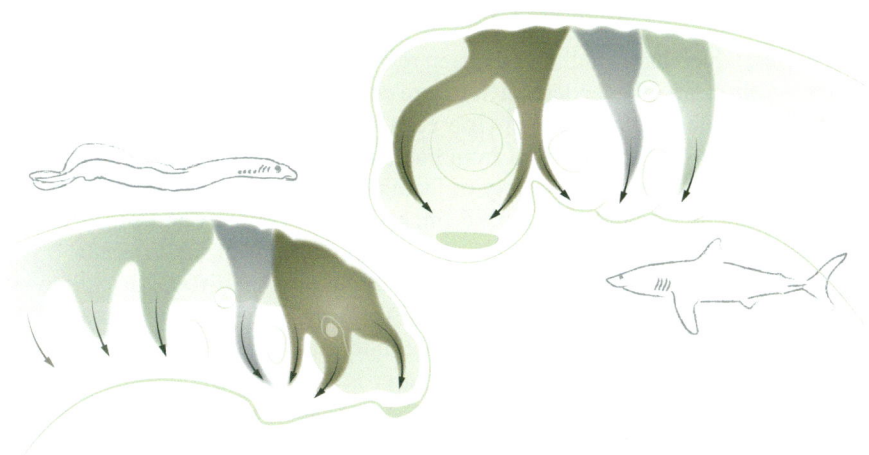

7 Neural Crest and Craniofacial Evolution of Early Vertebrates

Shigeru Kuratani

CONTENTS

7.1   INTRODUCTION—VERTEBRATES AND THE NEURAL CREST

Vertebrates, which have traditionally been regarded as a subphylum of chordates but more recently as an independent phylum, are defined by the possession of a vertebral column and endoskeletal cranium (Irie et al., 2018). Developmentally, vertebrates have specific cell lineages derived from the neural crest (NC) and placodes that are thought to have resulted in a variety of features specific to vertebrates—especially structures uniquely found in the head, including the cranium and sensory organs (Gans and Northcutt, 1983; Northcutt and Gans, 1983; Noden, 1986, 1988).

Vertebrate embryos have an extensive mesenchyme that differentiates into musculoskeletal elements as structural bases for the large size of these animals in both the embryonic and adult stages. The vertebrate embryonic mesenchyme has two sources: the mesoderm and the NC. The latter is an ectoderm-derived, multipotent cell lineage that gives rise not only to skeletogenic cells but also to various other cell types, including pigment cells, peripheral neurons, and supporting cells (Le Douarin, 1982; Le Douarin and Kalcheim, 1999). During development, delaminated cephalic NC (CNC) cells migrate extensively in the embryonic head to occupy specific sites, forming craniofacial primordia–containing NC-derived ectomesenchyme (reviewed by Abramyan and Richman, 2018). A large portion of the cranium develops in these primordia (Noden, 1986, 1988; Le Douarin, 1982). This chapter summarizes the evolution of the vertebrate cranium from the perspective of the evolutionary development of CNC cells.

7.2   VERTEBRATE EVOLUTION AND THE NC

Living vertebrates consist of two monophyletic groups, cyclostomes and gnathostomes (Mallatt and Sullivan, 1998; Kuraku et al., 1999; Takezaki et al., 2003; Kuraku, 2008; Heimberg et al., 2010). Cyclostomes are living jawless vertebrates and are often also called *agnathans*. However, the term *agnathans* refers primarily to a jawless "grade" seen in a paraphyletic group that contains all the jawless vertebrates, including cyclostomes and advanced fossil forms called *ostracoderms*. The latter animals appeared as earlier-derived groups of gnathostomes after the splitting-off of the cyclostomes. Thus, the earlier-derived gnathostomes were jawless (Janvier, 1996), and the biting jaw was obtained only secondarily in the lineage of gnathostomes. In craniofacial evolution, therefore, the jaw is a later-derived feature of crown or living gnathostomes. This does not necessarily mean that the cyclostomes can be viewed as representing an ancestral, plesiomorphic state. Rather, as suggested by Sewertzoff (1931) and Jollie (1971, 1977, 1981), cyclostomes and gnathostomes are more likely to have experienced two different paths of evolution, through which their craniofacial patterning mechanisms exhibit both shared and differentiated patterns of development. Simultaneously, in principle, the craniofacial patterns of neither of the two lineages can be derived directly from those of the other. Moreover, because ancestral developmental constraints can often be lost or modified in evolution, morphological homology can be lost through the acquisition of evolutionary novelties (Müller and Wagner, 1991; Wagner and Müller, 2002). Therefore, direct counterparts or precursors of the upper and lower jaws are not always evident, or even cannot be expected, in cyclostomes (if homologs of the upper and lower jaws were found, cyclostomes would no longer be agnathans); only an undifferentiated and plesiomorphic primordium such as the mandibular arch can be identified in common. Thus, the evolution of craniofacial patterning in evolutionary developmental biology (evo-devo) studies can be formulated through comparative analyses aimed at understanding when, where, and how the development of NC cells is regulated in the embryo.

7.3 PAN-VERTEBRATE PATTERN OF CREST-CELL MIGRATORY STREAMS

As has been extensively studied, the early distribution of CNC cells is highly conserved among jawed vertebrates (reviewed by Kuratani et al., 2018, in press, and references therein). The majority of CNC cells migrate along a dorsolateral pathway found beneath the surface ectoderm to populate the ventral part of the head, including the pharyngeal arches (CNC cell populations are anteroposteriorly segmented into the pharyngeal arches; see Figure 7.1A1, B1). Therefore, the distribution of the CNC is more lateral and ventral than that of the majority of trunk NC cells, which are found more medially in the paraxial mesoderm (segmented into somites) along the ventrolateral pathway (Figure 7.1A1, B1). The migration and distribution of NC cells together prefigure the distinct morphology of the spinal and branchiomeric nerves (Johnston, 1966; Noden, 1978; Loring and Erickson, 1987). The interface between the head and trunk, therefore, may be reflected in the distribution of the two types of NC cell populations, forming an S-shaped boundary along the course of the hypoglossal and vagus nerves (Kuratani, 1997; reviewed by Kuratani et al., 2018; Figure 7.1A1, B1). This domain, in amniotes, serves as an embryonic basis for acquisition of the neck domain (Matsuoka et al., 2005; Kuratani et al., 2018; Figure 7.1B1).

CNC cells are categorized into three major cell populations called, from anterior to posterior, the *trigeminal*, *hyoid*, and *circumpharyngeal* NC cells (Kuratani and Kirby, 1991; Kuratani, 1997; reviewed by Fish, 2017; Figure 7.1B). The trigeminal NC cells are so called because their distribution domains coincide with the innervation domains of trigeminal nerve branches (reviewed by Higashiyama and Kuratani, 2014; also see Oisi et al., 2013a). The basic distribution pattern of the trigeminal NC cells is similar between cyclostome and jawed vertebrate embryos (Figure 7.1A2, B2; axonal growth patterns are often prefigured by preexisting crest-cell streams Johnston, 1966; Noden, 1978; Kuratani, 1997); the trigeminal NC cell population can be divided anteroposteriorly into one stream that leads to the mandibular arch and another leading to the more rostral region, generally called the *premandibular* region (Figure 7.1.1A2, B2).

The premandibular NC cell population is further subdivided into preoptic and postoptic streams (Figure 7.1.1A2, B2; Fig 7.2A3, B3). This triple stream pattern of the trigeminal CNC cells is widespread; it is seen not only in the early pharyngula of various jawed vertebrate species, but also in cyclostome embryos (reviewed by Kuratani et al., 2001; Kuratani, 2012, 2018; and by Fish, 2017). Therefore, this pattern can be recognized as a panvertebrate migratory pattern of the trigeminal NC cells (Figure 7.1). Simultaneously, the division of the premandibular and mandibular

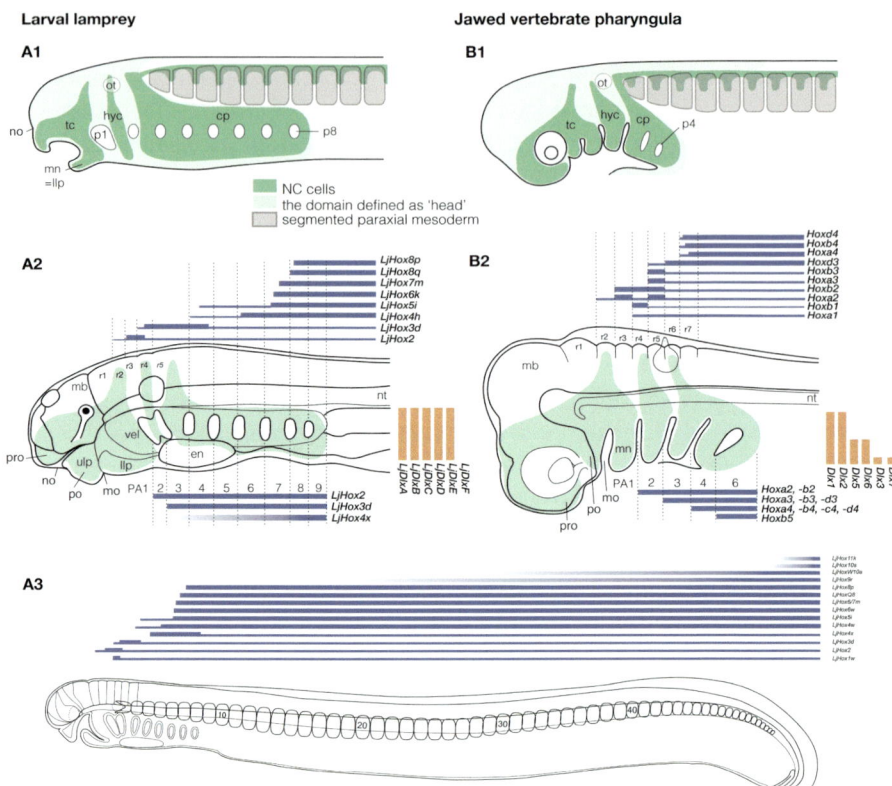

FIGURE 7.1 Comparison of lamprey and jawed vertebrate developmental patterns. **A1** to **A3**. Lamprey embryos. **B1** and **B2**. Generalized jawed vertebrate embryos. Highly schematized. **A1** and **B1**. Major distribution domain of neural crest (NC) cells are green, and the head region defined by the cephalic neural crest (CNC) cell distribution is light green. Trunk NC cell populations are segmentally separated by the presence of somites, defining the trunk region of the embryos. The interface between the head and trunk forms an S-shaped boundary. CNC cells form three ectomesenchymal populations, namely the trigeminal NC cells (tc), hyoid NC cells (hyc), and circumpharyngeal NC cells (cp). Redrawn and modified from Kuratani et al. (2018): page S63. **A2** and **B2**. Comparison of homeobox gene expression domains. In both the lamprey and jawed vertebrate embryonic head, *Hox* gene transcripts show nested anteroposterior patterns of distribution (Hox code; light blue bars) along the neuraxis as well as in the pharyngeal arches (PAs); the mandibular arch (PA1) expresses no *Hox* genes (Hox code default state). The Hox codes in the pharyngeal arch ectomesenchyme are similar in the rostral three arches (PA1 to 3) in both embryos. Trigeminal NC cells populate not only the mandibular arch, but also the premandibular region, where the NC cells are also Hox negative. Premandibular CNC cells are further categorized into preoptic NC cells (pro) and postoptic NC cells (po). Although the basic patterns of the trigeminal NC cells are comparable between the lamprey and jawed vertebrates, the relative positions of the NC cell populations, mouth (mo), and external nostril are slightly different (for details, see Shigetani et al., 2002; also see Figure 7.2). Mandibular arch NC cells (mn) form mesenchyme in the velum (vel) and lower lip (llp) in the lamprey, and in the upper and lower jaws in jawed vertebrates. Dlx genes (orange) are also expressed in CNC cells in the pharyngeal arches. Dlx

FIGURE 7.1 (CONTINUED)
gene transcripts show a dorsoventrally nested pattern in the PAs of jawed vertebrates, but not in lamprey PAs. This scheme is based on the work of Kuraku et al. (2010), but other authors suggest the presence of a dorsoventrally symmetrical nested pattern of expression (Cerny et al., 2010). Redrawn and modified from Kuratani (2012): page 76. **A3**. Hox code in the neural tube of the lamprey, showing the anteroposteriorly nested pattern of expression. Based on the work of Takio et al. (2007). Other abbreviations: en, endostyle; mb, midbrain; no, nostril; nt, notochord; ot, otocyst; p1 to p8, pharyngeal pouches; r1 to r7, rhombomeres; ulp, upper lip.

arch streams prefigures the distribution of trigeminal sensory nerve branches associated with the ophthalmic and maxillomandibular nerves, respectively; the maxillomandibular nerve primarily supplies the mandibular arch derivatives, and the ophthalmic nerve primarily supplies the premandibular domain, in both cyclostomes and gnathostomes (see Kuratani et al., 1997; Oisi et al., 2013a; Higashiyama and Kuratani, 2014). For its curious innervation pattern in both the mandibular arch derivative and premandibular domain, the maxillary nerve—the second branch of the trigeminal nerve—has been suggested to belong primarily to the premandibular nerve, which secondarily incorporated the upper jaw sensory fibers in jawed vertebrates (Higashiyama and Kuratani, 2013).

7.4 JAWED VERTEBRATE–SPECIFIC PATTERN

The late development of the trigeminal NC cells in jawed vertebrates has been well documented; namely, the craniofacial primordia arise in concert with the above-noted CNC cell streams (Johnston, 1966; reviewed by Abramyan and Richman, 2018). The preoptic stream of cells populates either side of the nasal pit to form the mesenchyme of the lateral and medial nasal prominences; the postoptic stream carries NC cells of the trabecular anlage, and the mandibular arch stream fills the mandibular arch, which will give rise to both the maxillary and mandibular processes. The medial nasal prominence is also called the frontonasal process and contains the anlage for the premaxilla and the intertrabecula (see below; reviewed by Wada et al., 2011; Figure 7.2B3). In addition, there is another premandibular CNC cell population—the postoptic NC cells found rostral to the mandibular arch—which gives rise to the paired trabeculae in later development (Figure 7.2B3; Shigetani et al., 2000, 2002; Kuratani et al., 2013). The boundary between this cell population and the mandibular arch NC cell population is hard to recognize, but *Dlx* gene expression is found only in the latter (Figure 7.2B1; Kuratani et al., 2013).

Development of the trabecular complex is central to understanding the prechordal part of the neurocranium (i.e., the prechordal cranium) in jawed vertebrates (Couly et al., 1993; reviewed by Kuratani and Ahlberg, 2018). It has long been known that the jawed vertebrate neurocranium is composed of two portions: the rostral, CNC-derived part and the posterior, mesodermally derived part (Hörstadius and Sellman, 1946; reviewed by Hall and Hörstadius, 1988; Couly et al., 1993; McBratney-Owen et al., 2008). The mesodermal portion, which is derived from the parachordal cartilages, is also known as the *chordal cranium*, because its anteroposterior axis is coextensive with the notochord. Just as the somitic part is induced by notochord-derived

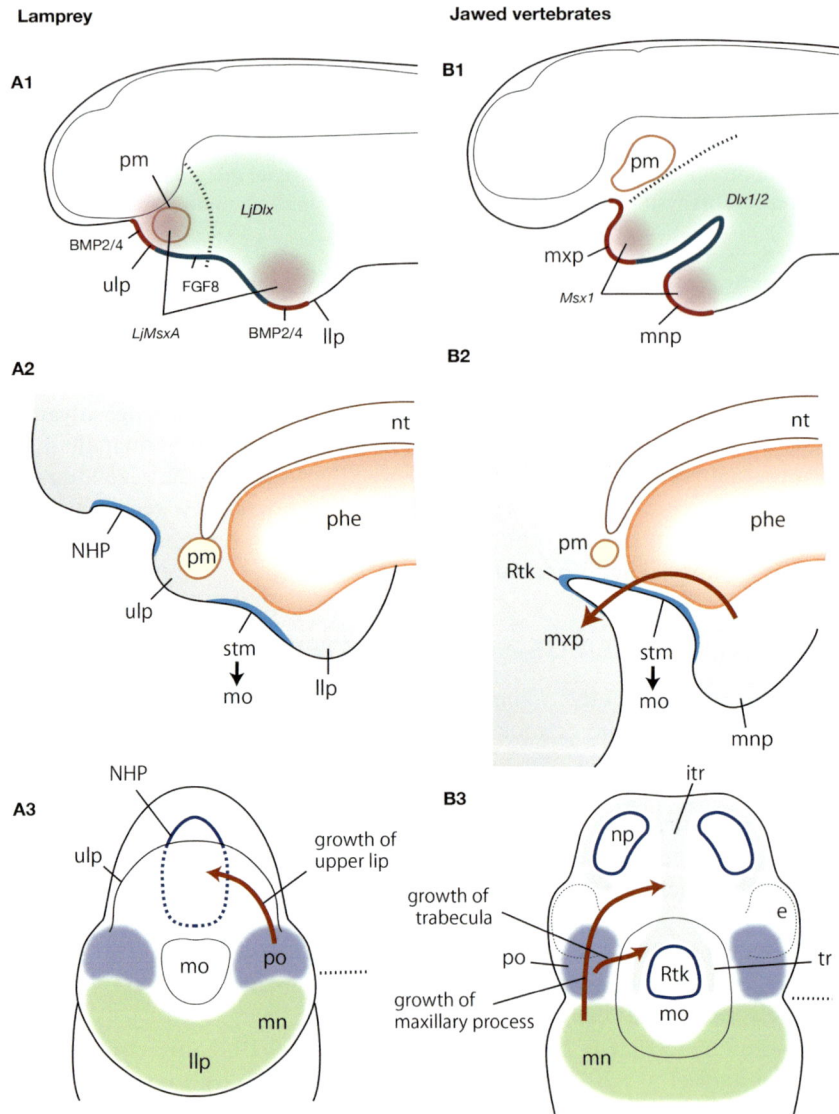

FIGURE 7.2 Comparison of craniofacial developmental patterns. **A1** to **A3**. Lamprey larvae. **B1** to **B3**. Generalized jawed vertebrate embryo. **A1** and **B1**. Proximodistal specification of oral apparatus ectomesenchyme through ectodermal–ectomesenchymal interaction. Left lateral views. Ectoderm covering the distal portions of the oral apparatus (distal tips of lips in the lamprey, those of jaws in jawed vertebrates) releases a growth factor, BMP2/4, to induce the expression of a downstream target gene, *Msx*, in the underlying mesenchyme. In the proximal part of the apparatus, the blue-colored ectoderm expresses FGF8-encoding genes, and the proximal ectomesenchyme is induced to upregulate *Dlx* genes. However, these homologous genes are not expressed in morphologically homologous tissues, because the boundaries of premandibular–mandibular ectomesenchyme (broken lines) are not found at equivalent

FIGURE 7.2 (CONTINUED)
levels in lamprey and jawed vertebrate embryos. Redrawn and modified from Kuratani (2012): page 78. **A2** and **B2**. Formation of the mouth and placodes. In the lamprey, the stomodaeum (stm) arises posterior to the nasohypophyseal plate (NHP; comparable to nasal placodes plus Rathke's pouch [Rtk], or adenohypophyseal placode, in jawed vertebrates), and the anterior margin of the mouth (mo) is defined by the upper lip (ulp), which grows between the NHP and stomodaeum. In jawed vertebrates, the mouth is defined secondarily by the maxillary process (mxp), which grows secondarily from the mandibular arch (arrow). Note that in jawed vertebrates, the adenohypophyseal placode (Rtk) arises as part of the oral ectoderm, which in the lamprey is separate from the NHP. Redrawn and modified from Kuratani (2005): page 20. **A3** and **B3**. Ectomesenchymal growth in craniofacial primordia. Note that the upper lip in the lamprey is comparable not to the upper jaw (maxillary process) in jawed vertebrates but to the trabecula derived from the postoptic ectomesenchyme. Upper lip and upper jaw are, therefore, morphologically not homologous to each other. Redrawn and modified from Kuratani (2012): page 82. Other abbreviations: e, eye; itr, intertrabecula (derived from preoptic neural crest cells); llp, lower lip; mn, mandibular arch ectomesenchyme; mnp, mandibular process; np, nasal placode (derivatives); nt, notochord; phe, pharyngeal endoderm; pm, premandibular mesoderm; tr, trabecula.

factor to differentiate into skeletogenic sclerotome, the mesodermal neurocranium (or the rostral part of the parachordals) arises in the head mesoderm in the vicinity of the notochord (Couly et al., 1993; reviewed by Kuratani, 2018). Thus, the rostralmost part of the chordal cranium is identified in the premandibular mesoderm that abuts the rostral tip of the notochord, differentiating into the orbital cartilage and later into the pila antotica or dorsum sellae in the fully formed amniote chondrocranium (Couly et al., 1993). Other mesodermal elements, such as the ala hypochiasmatica in mammals and the supratrabecula in the reptilian chondrocranium, are also thought to originate from the same domain. As will be shown below, the dual origin of the neurocranium is a later-derived and shared feature of the jawed vertebrates.

The major portion of the CNC-derived prechordal cranium consists of the trabecular complex, from which the nasal septum, interorbital septum, and rostral part of the cranial base (rostral to the hypophyseal foramen) will develop (De Beer, 1937 and references therein; reviewed by Wada et al., 2011). Developmentally, this complex consists of an anterior median component, the intertrabecular; and posterior paired elements, the trabeculae. Specifically, Wada et al. (2011) have shown that the CNC cells along the preoptic stream contribute to the formation of the intertrabecula, whereas the paired trabeculae are derived from the cells in the postoptic stream (Figure 7.2B3). Developmental patterning of these two types of primordia appears to be regulated by independent genetic control (Eberhart et al., 2008; reviewed by Wada et al., 2011).

Patterning of the jaw also characterizes the jawed vertebrate cranium. Again, the jaw is a composite structure in most living gnathostomes, consisting of dorsoventrally specified mandibular arch derivatives and a premaxillary part of the upper jaw derived from the frontonasal mesenchyme (Shigetani et al., 2000; Cerny et al., 2004; Lee et al., 2004; also see Soukup et al., 2013 and Minarik et al., 2017 for evolution of the vertebrate oral region). The exceptions are elasmobranchs and sturgeons, the upper jaw of which is derived only from the mandibular arch–derived palatoquadrate.

As part of the developmental program for specification of the pharyngeal arch ectomesenchyme, the gnathostome jaw is patterned through positional information provided by the Cartesian-grid-like expression of homeobox genes—the so-called Dlx-Hox code (Figure 7.1A2, B2; Rijli et al., 1993; Depew et al., 2002; Gillis et al., 2013; Takechi et al., 2013). The mandibular arch–derived portion of the jaw is regionally specified by the Dlx code, namely the dorsoventrally nested pattern of *Dlx* genes that is established downstream of endothelin signaling in the rostral endoderm of embryos (Couly et al., 2002; Depew et al., 2002). Thus, deletion of the ventrally expressed *Dlx5* and *-6* genes results in the transformation of the lower jaw into the morphological identity of the upper jaw (Depew et al., 2002). In contrast, antero-posterior specification of the pharyngeal arches is mediated by nested expression of *Hox* genes (i.e., by the Hox code), and patterning of the jaw depends on the absence of *Hox* gene transcripts (the Hox-code default state). Disruption of the *Hoxa-2* gene, which functions in the hyoid arch and posteriorly, results in transformation of the second arch into the identity of the first (Rijli et al., 1993; also see Couly et al., 1998). Many other genes are also known to be involved in the positional and mor-phological specification of mandibular arch derivatives. For example, the ectoder-mally expressed *Fgf8* and *Bmp2/4* are involved in specification of the proximodistal patterning of the mandibular-arch part of the jaw by upregulating their downstream target genes, *Dlx1* and *Msx2*, respectively, in the distal ectomesenchyme. *Bapx1* is expressed specifically at the junction of the upper and lower jaw (i.e., at the primary jaw joint); its expression is retained in the incudomalleolar joint of the mammalian middle ear (Miller et al., 2003; Tucker et al., 2004; Kitazawa et al., 2015).

7.5   NEURAL CREST IN THE LAMPREY

Of the two major groups of vertebrates, cyclostomes are further classified into two sister groups, lampreys and hagfish, which are the only jawless vertebrates living today. The hagfish was once excluded from the vertebrates for its lack of vertebrate-defining traits including vertebral elements, lens of the eye, lateral lines, and so on. However, recent molecular phylogenetic analyses strongly support the monophyly of the cyclostomes, and it has now become accepted that the hagfish have secondarily lost their vertebral column for most of the body axis, although vestigial vertebral elements are present in the tail (reviewed by Janvier, 1996 and by Ota and Kuratani, 2006; Ota et al., 2011, 2013, 2014). Because previous studies of NC cells were made mainly on jawed vertebrates, and because of the curious phylogenetic position of cyclostomes, their NC cells have drawn the interest of researchers in the field of evo-devo for the past two decades.

CNC cells have been studied extensively in the lamprey. These cells delaminate and emigrate from the epithelial NC developing dorsal to the neural tube (Horigome et al., 1999; Shigetani et al., 2002; McCauley and Bronner-Fraser, 2003), showing gene expression comparable to that known in gnathostome CNC cells (Horigome et al., 1999; Myojin et al., 2001; Shigetani et al., 2002; Meulemans et al., 2003; Takio et al., 2004, 2007; McCauley and Bronner-Fraser, 2004, 2006; Rahimi et al., 2009; Cerny et al., 2010; Yao et al., 2011; Lakiza et al., 2011; Takechi et al., 2013; Lee et al., 2016; York et al., 2017, 2018; reviewed by McCauley et al., 2015). Especially in regard to early specification of the premigratory NC, comprehensive gene expression analyses have been performed in the lamprey; they have revealed that the basic gene regulatory networks for basic NC specification appear to have been established long before the split of the cyclostome and gnathostome lineages (Sauka-Spengler et al., 2007; Sauka-Spengler and Bronner-Fraser, 2008; Betancur et al., 2010). In scanning electron micrographs, emigrating NC cells in the lamprey appear as flattened cells that are clearly distinguishable from the neurepithelial cells, which in the head originate along the entire neuraxis; the emigrating NC cells secondarily adhere only to even-numbered rhombomeres. This pattern is shared by gnathostome embryos (Horigome et al., 1999; Kuratani et al., 1999). The predicted migratory routes and distributions of NC cells are also common to lamprey and jawed vertebrate embryos; CNC cells and trunk NC cells are juxtaposed at the postotic level, forming an S-shaped head–trunk interface (Figure 7.1A1, A2; Kuratani et al., 1997; Shigetani et al., 2002; McCauley and Bronner-Fraser, 2006).

The above findings imply strongly that not only the evolutionary origin of the NC but also the basic developmental program of migratory NC cells dates back more than 500 mya, before the split of cyclostomes and gnathostomes. According to recent studies, the NC and placode may have been the same cell lineage in a tunicate-like ancestor and would have differentiated into distinct two lineages secondarily (Horie et al., 2018). Also, it appears likely that, in the latest common ancestor of all vertebrates living today, there must have already been a distinction of the cephalic and trunk NC cells; the CNC cells would have been specified as the source of both the cranial peripheral nervous system and the cephalic ectomesenchyme, which differentiates into various mesenchymal derivatives in the head, including the endoskeletal cranium consisting of the neurocranium and viscerocranium. Thus, apparent lack of skeletogenic potential in trunk NC cells would also have originated from the common ancestor of cyclostomes and gnathostomes.

Craniofacial development in the lamprey begins with three distinct primordia, namely the mandibular process, postoptic process, and preoptic process (Figure 7.3, bottom; Figure 7.4A; homologs of the maxillary process and the medial and lateral nasal prominences are missing from the lamprey embryo). These processes are filled by ectomesenchyme derived from the mandibular arch NC cells, postoptic NC cells, and preoptic crest cells, respectively (Figures 7.1 and 7.2). In the lamprey, unlike in jawed vertebrate embryos, because the nasal and adenohypophyseal placodes never divide into three placodes but stay fused as a single median placode called the *naso-hypophyseal plate* (NHP), the preoptic ectomesenchyme is not divided into three portions but stays as a single NC cell mass rostral to the plate (Figure 7.3, bottom; Figure 7.4A). This difference results in the conspicuously different craniogenesis in lamprey development thereafter (Figure 7.2).

First, the mandibular arch does not divide clearly into dorsal and ventral halves but instead into dorsal and ventral moieties, namely the velum and lower lip of ammocoete larvae. These structures are not homologous with the upper and lower jaws. *Bapx1*, the jaw-joint-specifier gene, is not expressed at the junction of these structures (Cerny et al., 2010). The rest of the pharyngeal arches develop visceral skeletons that are more or less dorsoventrally symmetrical (Martin et al., 2009; Oisi et al., 2013b). The expression patterns of lamprey *Dlx* genes, which are not clearly dorsoventrally nested, are consistent with this morphology (Figure 7.1). In addition, trigeminal nerve innervation pattern does not support the jaw-like nature of the ammocoete velum or lower lip, either (Murakami and Kuratani, 2008; Higashiyama and Kuratani, 2014).

The postoptic process of the lamprey embryo differentiates into the upper lip (Kuratani et al., 2001; Shigetani et al., 2002). Although it resembles the jawed vertebrate's upper jaw in terms of apparent function as well as *Bmp2/4* and *Msx* gene expression patterns, this ectomesenchyme belongs to the premandibular domain (Figure 7.2A1).

A curious question regarding the lamprey cranium is the developmental origin of the structure often called the *trabecula*. This skeletal element consists of a pair of rods, each connected rostrally with its counterpart in the middle (like an inverted U-shape) beneath the rostral brain, and its overall morphology resembles that of the jawed vertebrate trabecula (Parker, 1883a, b; De Beer, 1937; Johnels, 1948). However, previous developmental observations and labeling experiments have shown that a major part of this skeletal element is derived from head mesoderm at the level of the mandibular arch (Johnels, 1948; Kuratani et al., 2004; but also see Newth, 1951, 1956; Langille and Hall, 1988), and that it is more likely a homolog of parachordals in jawed vertebrates, secondarily extended rostrally. In terms of the original positions of the craniofacial primordia, the true homolog of the jawed vertebrate trabecula is found in the upper lip mesenchyme of ammocoete larvae (Figures 7.2A3, 7.2B3, 7.4A). Preoptic CNC cells could be the developmental origin of the commissure region of the lamprey "trabecula," but there is as yet no direct evidence in support of this possibility.

As noted above, the lamprey and jawed vertebrates have very diversified patterns of basic cranial composition. Specifically, the cranial base of the lamprey is made mostly of mesodermal mesenchyme, unlike in the jawed vertebrates, in which the neurocranium is half of NC origin and half of mesoderm. Another issue is the composition of the oral apparatus. In both the lamprey and jawed vertebrates, the mandibular arch and premandibular region are involved in the patterning of the oral apparatus, and its proximodistal specification is mediated by *Bmp* and *Fgf8* genes in both animal groups (Shigetani et al., 2002). However, the postoptic NC cells are not involved in jaw patterning; the dorsal and ventral processes of the oral apparatus (upper and lower jaws, upper and lower lips) are derived from different embryonic components in lampreys and jawed vertebrates (Figure 7.2A1, B1). The heterotopy theory explains the origin of the vertebrate jaw by a topographic shift in epithelial–ectomesenchymal interactions, not by simple modification of the ancestral mandibular arch into a dorsoventrally articulated biting structure (Figure 7.2A1, B1; Shigetani et al., 2002; see below for details). Importantly, this theory explains the origin of the trabecula in jawed vertebrates as well: once the postoptic NC cells had differentiated into the dorsal part of the oral apparatus in some ostracoderms and earlier-derived placoderms, this part was released from its original function and had the opportunity to form a secondary cranial base to adapt to the presence of an enlarged forebrain (Dupret et al., 2014, 2017). Acquisition of a jaw may have then enabled further enlargement of the forebrain (Figure 7.3, top).

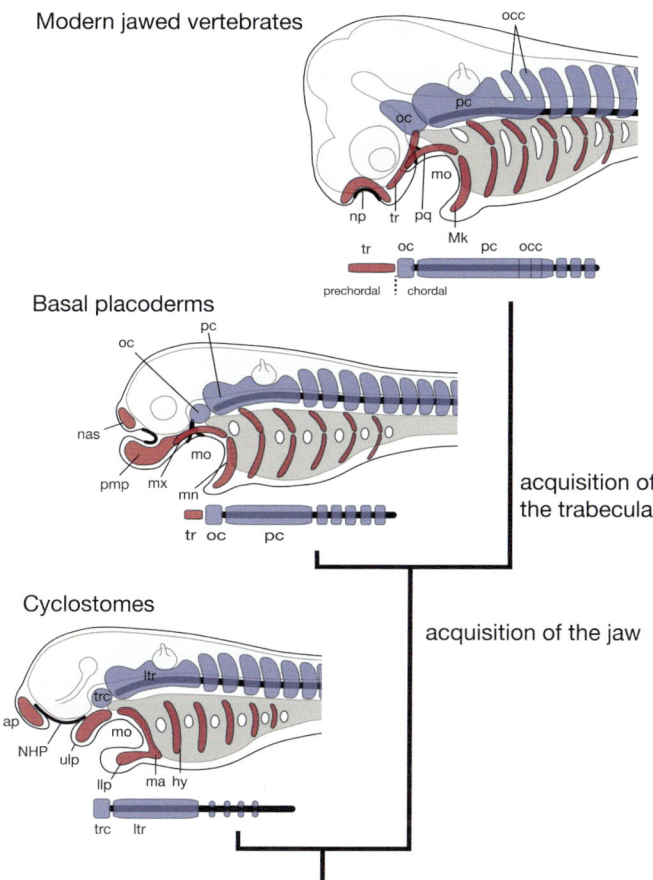

FIGURE 7.3 Evolution of the neurocranium. The evolutionary polarity of the vertebrate neurocranium is suggested in this scheme to take the form of sequential addition of pre-mandibular cephalic neural crest (CNC)-derived components (trabeculae) rostral to the noto-chord. Similar to the modern cyclostomes (bottom), the common ancestor of all vertebrates is assumed to have possessed mesodermal neurocranium and a neural crest (NC)-derived oroviscerocranium. Creation of the prechordal cranium became possible only through the release of postoptic NC cells that once functioned as part of the oral apparatus, after the acquisition of the mandibular-arch-derived jaw. Even after the acquisition of the jaw, as seen in primitive placoderms (middle) the trabecula was absent, and its precursor (postoptic NC cells) resembles the oral hood of the lamprey and some ostracoderms. An increase in the size of the prechordal cranium allowed enlargement of the forebrain, which reached its highest level in modern jawed vertebrates (top). Redrawn and modified from Kuratani and Ahlberg (2018): page 8. Abbreviations: hy, hyoid arch; ltr, lamprey trabecula; ma, mandibular arch; Mk, Meckel's cartilage; mn, mandibular process; mo, mouth; mx, maxillary process; nas, nasal capsule; NHP, nasohypophyseal plate; np, nasal placode; oc, orbital cartilage; occ, occipital; pc, parachordal cartilage; pmp, premandibular process; pq, palatoquadrate; tr, tra-becula (of jawed vertebrates); trc, trabecular commissure (in the lamprey).

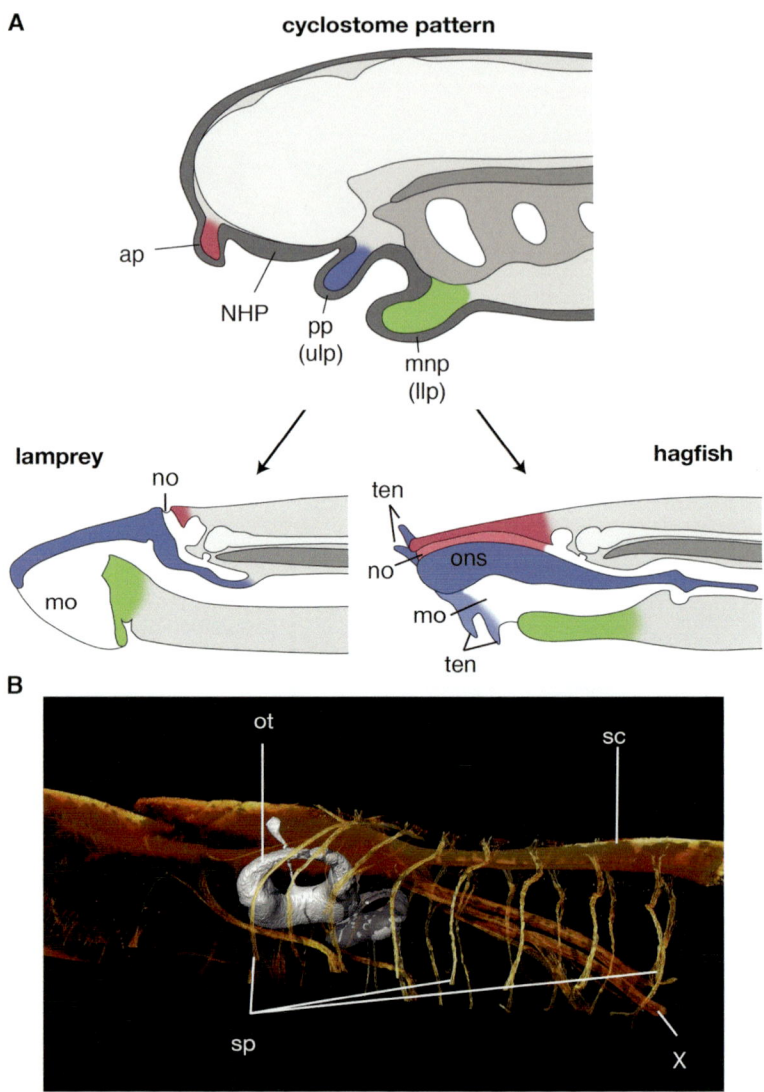

FIGURE 7.4 **A** Comparison of craniofacial morphogenesis between lamprey and hagfish. In the late pharyngula stages of lamprey and hagfish development, the morphological patterns of the embryonic head become very similar; this pattern is not shared by jawed vertebrate embryos (top). It can be regarded as cyclostome-specific developmental pattern. At this stage, the craniofacial primordia consist of anterior (ap) and posterior (pp) processes, and a mandibular (arch) process (mnp). Note that the position of the nasohypophyseal plate (NHP) ends up as the nasal opening in the adult (no). (Adult stages are shown at the bottom in A.) The anterior process is filled with preoptic neural crest (NC) cells and the posterior process by postoptic crest cells. On the basis of this common pattern, craniofacial skeletal elements and various cranial structures can be homologized between the lamprey and hagfish. Specifically, the posterior process differentiates into the upper lip (ulp) in the lamprey larva, whereas in

FIGURE 7.4 (CONTINUED)
the hagfish it becomes the hagfish-specific oronasal septum (ons) as well as tentacles (ten). The root of the posterior process is secondarily absorbed in the hagfish to make the nasal cavity and pharynx confluent, indicating that the previous classification of Hyperoartia (lampreys) and Hyperotreti (hagfish) was based on an unexpectedly trivial morphological difference. For detailed comparison of chondrocranial elements and their cell lineage origins, see Oisi et al. (2013b) and Kuratani et al. (2016). **B.** Modified anatomical pattern of the head and trunk in the hagfish. Reconstruction of Dean's stage 53 *Eptatretus burgeri* embryo to show the embryonic patterns of peripheral nerves (left lateral view). Rostral spinal nerves (sp) are covering the otic vesicle (ot) laterally. The spinal nerves are also seen lateral to the vagus nerve (X). This unusual anatomical pattern implies embryonic violation of the head–trunk interface, which becomes more conspicuous in later development (see Oisi et al., 2015 for details). Reconstruction performed by Shinnosuke Higuchi. Other abbreviations: llp, lower lip; mo, mouth; sc, spinal cord.

7.6 NEURAL CREST IN THE HAGFISH

Development of the NC has been most enigmatic in the hagfish since Conel (1942) stated that it arises as an epithelial infolding of ectoderm between the neural tube and the surface ectoderm (reviewed by Hall, 1999). However, it is now known that this observation was due to an artifact (tissue distortion) originating from inappropriate fixation of the embryo: the embryos of *Bdellostoma stouti* were fixed *in toto*, with egg shells attached, in alcohol (Dean, 1899). As a result of histological observations and in situ hybridization of *Sox* gene expression performed in 2007, it has become clear that the hagfish embryo also develops delaminating NC cells that migrate between somites in the trunk to form intersomitic dorsal root ganglia; the location of these ganglia is specific to this animal lineage (Ota et al., 2007).

Less is known about the distribution pattern of CNC cells in the hagfish than in the lamprey. However, in situ hybridization data to detect the expression of *Dlx* genes, as well as histological observations in late-pharyngula-stage embryos of *Eptatretus burgeri*, suggests that the CNC-derived ectomesenchyme of this animal also occupies the ventral part of the head to form the craniofacial primordia (Oisi et al., 2013a; Fujimoto et al., 2013). Furthermore, the development of cranial nerve roots on even-numbered rhombomeres—segmental bulges of the hindbrain—implies selective adhesion of CNC cells in the hindbrain (Oisi et al., 2013a).

One peculiarity about hagfish NC cells is that the vertebrate-defining head–trunk interface is violated early in development, after the establishment of a phylotypic morphological pattern found at pharyngular stage (Figure 7.4A ; Oisi et al., 2015). By Dean's stage 53 of development, the primordia of the cranial and spinal nerves are well dissociated from each other: the spinal nerve anlages are found only postotically (Figure 7.4B), as in other vertebrate species. Later in development, however, spinal nerve components shift rostrally along the neuraxis, invading the preotic region. As a result, the rostral spinal nerves (apparently corresponding to the occipitospinal nerves in jawed vertebrates) are distributed in superficial layers in the head, giving the impression of amphioxus-like peripheral nerve morphology (Oisi et al., 2015). This is unlike the case of the hypoglossal nerve in jawed vertebrates, which represents the secondarily modified and bundled ventral roots of rostral spinal nerves. Such a hagfish-specific phenomenon is apparently associated with the development of hagfish-specific morphology of the hypobranchial musculature in relation to a hagfish-specific caudal shift of the posterior pharynx. Outgroup comparisons, including jawed vertebrates and lampreys, suggest that this pattern is most likely a hagfish-specific later-derived feature introduced secondarily in the hagfish lineage after the split from the other cyclostome (lampreys and unknown fossils) lineages (Oisi et al., 2015). This peculiar morphogenetic process suggests that the typical vertebrate-like NC cell distribution pattern is disturbed in hagfish embryos. Nevertheless, up to the mid-pharyngula stages, hagfish embryos show quite a conserved pattern of development, including the head–trunk interface, likely part of the developmental program already established by the latest ancestor of all vertebrates.

7.7   EMBRYONIC PATTERN SPECIFIC TO CYCLOSTOMES

Historically, the crania of the lamprey and hagfish were thought to be very different from each other, and homology of cranial skeletal elements was very hard to establish between these two animals (Parker, 1883a, b; De Beer, 1937; Neumayer, 1938; Holmgren, 1946; reviewed by Kuratani et al., 2016; but also see Yalden, 1985). Because of differences in the development and distribution of the ectodermal placodes, the distribution pattern of cephalic ectomesenchyme in the cyclostomes differs from that in jawed vertebrates; moreover, it is likely that one-to-one homologies of skeletal elements cannot be established between the crania of cyclostomes and jawed vertebrates. The nasal and adenohypophyseal placodes form a single median plate on the ventral aspect of the early pharyngular head in cyclostomes (the hagfish adenohypophysis also differentiates from the ectoderm; Oisi et al., 2013a), thus forming a single external nostril leading to both the olfactory epithelium and the hypophysis. The premandibular ectomesenchyme forms only two craniofacial primordia anteroposteriorly to the single placode—the NHP—that characterizes cyclostome embryos. Oisi et al. (2013a) found that Tahara's stage 26 lamprey (Tahara, 1988), and Dean's stage 45 hagfish (Dean, 1899) embryo have very similar, mutually comparable configurations, enabling us to homologize the craniofacial structures (see also Heintz, 1963). Curiously, this tripartite craniofacial pattern of cyclostomes can also explain the morphological composition of the cranium of a long enigmatic fossil species, *Palaeospondylus*—a mysterious fish-like animal from Devonian Scotland—raising the possibility that this fossil represents an earlier-derived hagfish lineage (Hirasawa et al., 2016).

The process rostral to the NHP in cyclostomes is called the *anteronasal* (or simply *anterior*) process; it differentiates into the nasal capsules. The posterior (or *posthypophyseal*) process differentiates into the lateral wall of the upper lip (a major portion of the oral funnel) in the lamprey, whereas in the hagfish, it becomes the oronasal septum. This septum initially arises from the primary palate, but it becomes detached secondarily by absorption of the root portion of the growing process, thus making the oronasal cavity and pharynx confluent with each other. Thus, the classical terminology of *Hyperoartia* and *Hyperotreti* in reference to lampreys and hagfishes, respectively (Duméril, 1806), relies merely on the late phase of morphogenesis that secondarily results in craniofacial variations among cyclostomes (Oisi et al., 2013a).

In both the hagfish and the lamprey, the mandibular arch transforms into the velum and the lower lip of the oral apparatus (Oisi et al., 2013b). Thus, the ecto-mesenchyme in this arch is implied to differentiate into the connective tissues and cartilages of these structures. Importantly, the upper jaw in jawed vertebrates and the posterior process derivatives (upper lip/oronasal septum) are not homologous with each other, but they arise from non-comparable primordia. The upper jaw in jawed vertebrates is primarily the dorsal half of the mandibular arch, to which the frontonasal prominence–derived rostral part is added (except in sharks and stur-geons), whereas the upper lip/oronasal septum in cyclostomes originates from a pre-mandibular ectomesenchyme that in jawed vertebrates differentiates into the paired trabeculae; that is, there are no true trabeculae in the cyclostomes.

From the developmental patterns, it is not always easy to tell which of the jawed-vertebrate (diplorhiny) and cyclostome (monorhiny) craniofacial patterns is more apo-morphic than the other; we do not have an appropriate outgroup for use in speculating on the polarity of the character. However, earlier-derived gnathostome fossils—the so-called ostracoderms—often have a single nostril placed close to the eye, resembling the cyclostome condition. This suggests that the cyclostome pattern reflects an ances-tral (plesiomorphic) craniofacial patterning program before the cyclostome–gnathos-tome split, if not accurately, as compared with crown gnathostomes.

KEY READING

Janvier, P. 1996. *Early Vertebrates*. Oxford University Press.
Oisi, Y., Ota, K.G., Kuraku, S., Fujimoto, S., Kuratani, S. 2013a. Craniofacial development of hagfishes and the evolution of vertebrates. *Nature* 493, 175–180.

7.8   NOTES ON THE ORIGINS OF THE JAW

One of the most intriguing issues associated with cyclostome evo-devo studies is the origin of the gnathostome jaw. A number of theories have been proposed in the past, including our "heterotopy theory" (Mallatt, 1996; Shigetani et al., 2002, 2005; Cerny et al., 2004; reviewed by Janvier, 1996; Mallatt, 2008; Kuratani, 2012 ; and by Fish, 2017). In this chapter, I intend to focus only on a few problems related to the concept of "coupling", which has often been neglected (Hall, 1998).

First, although it is true that the major part of the jaw is derived from the man-dibular arch of the embryo—the Hox-negative ectomesenchyme of which is dorso-ventrally polarized by vertebrate-specific endothelin signaling and regionalized by subsequent developmental interactions—other structures are also involved in, or at least associated with, jaw formation and are tightly linked to differentiation of the mandibular arch. Except in lineages such as elasmobranchs and sturgeons, the upper jaw of crown gnathostomes has a premandibular element incorporated into its rostral part. Therefore, the frontonasal prominence–derived premaxillary part is develop-mentally a premandibular element, and this part was present in the earliest jawed vertebrates before the rise of sharks and rays (see Shigetani et al., 2000; Cerny et al., 2004; Lee et al., 2004; Higashiyama and Kuratani, 2014; see also Dupret et al., 2017). This premandibular jaw is not homologous with the upper lip of the lamprey larva, because the latter is topographically more comparable to the paired trabeculae of gnathostomes. Needless to say, acquisition of the double nostril would have predated the acquisition of the premaxillary portion, since the latter is derived from the ectomesenchyme located in the internasal region (medial nasal prominence in the gnathostome embryos). Therefore, the occurrence of a gnathostome-type upper jaw is possible only in embryos with crown gnathostome–type placodal patterning.

Second, because lamprey trabeculae are likely homologous to mesodermal parachordals in gnathostomes, it appears that the jawed vertebrate trabecula would have been obtained only after the establishment of the palatoquadrate and lower jaw cartilage (Johnels, 1948; Kuratani et al., 2004). This scenario is consistent with the morphology of the earlier-derived placoderm cranium (Figure 7.3, middle), in which the neurocranium resembles that of the ostracoderm, with no involvement of the true trabecula, whereas the jaw is already present (Dupret et al., 2014, 2017). In the latter animal, it appears that the postoptic NC cells still formed a cyclostome upper lip–like structure in the dorsal oral region. One fossil record suggests that the beginnings of the true trabecula were already present in advanced ostracoderms, which were starting to possess two nasal sacs (Gai et al., 2011). It is also curious that the positioning of the oral apparatus appears to differ between the cyclostomes and gnathostomes. Heterotopy theory predicts that the gene regulatory module specifying the proximodistal axis of the oral apparatus would have been conserved across all vertebrates, but the mesenchymal domains for the apparatus could have shifted during evolution (Shigetani et al., 2002, 2005). Evolutionary shifts in the relative positions of ectomesenchyme and surface ectoderm containing placodes would then have played a crucial role in the acquisition of novel structures.

KEY READING

Mallatt, J. 2008. Origin of the vertebrate jaw: neoclassical ideas versus newer, development-based ideas. *Zool Sci* 25, 990–998.

Kuratani, S. 2012. Evolution of the vertebrate jaw from developmental perspectives. *Evol Dev* 14, 76–92.

Fish, J.L. 2016. Evolvability of the vertebrate craniofacial skeleton. In *Seminars in Cell & Developmental Biology*. Academic Press.

7.9   CONCLUDING REMARKS

Cyclostomes and jawed vertebrates have shared patterns of regulatory gene expression in the NC and ectomesenchyme. Colinear expression of *Hox* genes appears to have been established in the CNC cells of the early ancestor of all vertebrates (Takio et al., 2004; Pascual-Anaya et al., 2018). The origin of the ectomesenchymal upregulation of *Dlx* genes was also early, but the dorsoventrally nested pattern of their expression is gnathostome specific.[*] Nevertheless, knowledge of the differences in gene regulatory mechanisms alone will not elucidate vertebrate craniofacial evolution, because morphologically the development of these animal lineages follows

---

[*] An alternative opinion is expressed in Square T., Jandzik D., Romasek M., Cerny R., Medeiros D.M. (2017). The origin and diversification of the developmental mechanisms that pattern the vertebrate head skeleton. *Developmental Biology* 427: 219–229.

noncomparable embryonic patterns. Therefore, the craniofacial pattern of one animal lineage cannot be derived directly from that of another; cyclostomes develop according to a cyclostome-typic pattern, which is not the same as the modern jawed-vertebrate pattern.

The major difference in craniofacial patterns between these animals can be ascribed to differences in the midembryonic distribution of the ectodermal placodes in the head. In jawed vertebrates, the preoptic NC cells are categorized into three NC cell populations populating a pair of lateral nasal prominences and a single medial nasal prominence, whereas in cyclostomes, the equivalent NC cell population populates a single cell mass, the anterior process. It is also noteworthy that, in crown gnathostomes, the adenohypophyseal placode becomes separate from the nasal placode to form part of the oral ectoderm, whereas in cyclostomes it stays as the posterior half of the NHP for a long period of development, leading to different topography of the craniofacial structures in terms of the positions of the mouth, oral apparatus (lips and jaws), and adenohypophyseal opening in these two animal groups. This difference results in shifting of the topography of NC cell distribution as well as shifts in epithelial–ectomesenchymal interactions, thus modifying the developmental constraints that have maintained morphological homologies; these are the consequences of heterotopic shift. In this sense, acquisition of the jaw was truly an evolutionary novelty, established not simply by modification of the same ancestral patterns (i.e., idioadaptation), but also by changes in the basic morphological pattern itself through the overriding of ancestral constraints (i.e., aromorphosis) (reviewed by Sewertzoff, 1931).

Precisely when in evolution this shift in placodal patterning took place is an intriguing question. Because some of the earlier-derived gnathostome lineages (ostracoderms) maintained a craniofacial pattern reminiscent of that of cyclostomes, the shift may have been introduced secondarily in the gnathostomes. However, it is important is to realize that this change must have taken place early in the developmental process, because the initial distribution of the placodes would have been identical between cyclostomes and jawed vertebrates (for the early pattern of hagfish placodes, see Oisi et al., 2013a; for lampreys, see McCauley and Bronner-Fraser, 2002; Modrell et al., 2014; Lara-Ramírez et al., 2015; and Mukendi et al., 2016; note that even in jawed vertebrates the adenohypophyseal placode is mapped in juxtaposition with the paired nasal placodes), but the difference in placodal patterns becomes clear before the NC cell influx (see Horigome et al., 1999). Thus, the difference in placodal distribution becomes obvious before the establishment of the vertebrate phylotype, supporting the validity of Sewertzoff's "archallaxis" (a Greek term meaning changes in the early phase of development). Determination of the mechanistic background to modification of the body plan by reprogramming of the developmental program would be an intriguing subject in evo-devo studies.

Curiously, the hypothetical common ancestor of vertebrates would have possessed a developmentally simple cranium, in which the head mesoderm formed the neurocranium and the NC cells the viscerocranium (Figure 7.3). Acquisition of the jaw allowed the postoptic NC cells to form the trabecula, or the prechordal cranium, making the composition of the cranium very complicated. Simultaneously, this shift apparently permitted the growth of the forebrain as seen in modern vertebrates. In cranial evolution, we see the role of developmental changes opening up new possibilities for shapes and functions.

ACKNOWLEDGMENTS

We'd like to thank the neural crest cells, for guiding our studies. Also, grant support for this chapter for BFE and PGP was Natural Sciences and Engineering Research Council (NSERC) RGPIN 435655-201 and RGPIN 2014-05563, while DJ was funded by the Scientific Grant Agency of the Slovak Republic VEGA grant No.1/0415/17.

REFERENCES

Abramyan, J., Richman, J.M., 2018. Craniofacial development: Discoveries made in the chicken embryo. *Int. J. Dev. Biol.* 62(1–2), 97–107.

Betancur, P., Bronner-Fraser, M., Sauka-Spengler, T., 2010. Assembling neural crest regulatory circuits into a gene regulatory network. *Ann. Rev. Cell. Dev. Biol.* 26, 581–603.

Cerny, R., Lwigale, P., Ericsson, R., Meulemans, D., Epperlein, H.H., Bronner-Fraser, M., 2004. Developmental origins and evolution of jaws: New interpretation of "maxillary" and "mandibular". *Dev. Biol.* 276(1), 225–236.

Cerny, R., Cattell, M., Sauka-Spengler, T., Bronner-Fraser, M., Yu, F., Medeiros, D.M., 2010. Evidence for the prepattern/cooption model of vertebrate jaw evolution. *Proc. Natl. Acad. Sci. U.S.A.* 107(40), 17262–17267.

Conel, J.L., 1942. The origin of the neural crest. *J. Comp. Neurol.* 76(2), 191–215.

Couly, G., Grapin-Botton, A., Coltey, P., Ruhin, B., Le Douarin, N.M., 1998. Determination of the identity of the derivatives of the cephalic neural crest: Incompatibility between Hox gene expression and lower jaw development. *Development* 125(17), 3445–3459.

Couly, G., Creuzet, S., Bennaceur, S., Vincent, C., Le Douarin, N.M., 2002. Interactions between Hox-negative cephalic neural crest cells and the foregut endoderm in patterning the facial skeleton in the vertebrate head. *Development* 129(4), 1061–1073.

Couly, G.F., Coltey, P.M., Le Douarin, N.M., 1993. The triple origin of skull in higher vertebrates: A study in quail-chick chimeras. *Development* 117(2), 409–429.

Dean, B., 1899. *On the Embryology of Bdellostoma stouti. A Genera Account of Myxinoid Development from the Egg and Segmentation to Hatching.* Festschrift zum 70ten Geburtstag Carl Von Kupffer, Jena, pp. 220–276.

De Beer, G.R., 1937. *The Development of the Vertebrate Skull.* Oxford, UK: Oxford University Press.

Depew, M.J., Lufkin, T., Rubenstein, J.L., 2002. Specification of jaw subdivisions by Dlx genes. *Science* 298(5592), 371–373.

Duméril, C., 1806. *Zoologie analytique, ou Méthode naturelle de classification des animaux: Rendue plus facile à l'aide de tableaux synoptiques.* Paris, France: Allais.

Dupret, V., Sanchez, S., Goujet, D., Tafforeau, P., Ahlberg, P.E., 2014. A primitive placoderm sheds light on the origin of the jawed vertebrate face. *Nature* 507(7493), 500–503.

Dupret, V., Sanchez, S., Goujet, D., Ahlberg, P.E., 2017. The internal cranial anatomy of *Romundina stellina* Ørvig, 1975 (Vertebrata, Placodermi, Acanthothoraci) and the origin of jawed vertebrates—Anatomical atlas of a primitive gnathostome. *PLOS ONE* 12(2), e0171241.

Eberhart, J.K., He, X., Swartz, M.E., Yan, Y.L., Song, H., Boling, T.C., Kunerth, Walker, A.K., M.B. Kimmel, C.B., Postlethwait, J.H., 2008. MicroRNA Mirn140 modulates pdgf signaling during palatogenesis. *Nat. Genet.* 40(3), 290.

Fish, J.L., 2017. Evolvability of the vertebrate craniofacial skeleton. In: *Seminars in Cell & Developmental Biology.* Academic Press.

Fujimoto, S., Oisi, T., Kuraku, S., Ota, K.G., Kuratani, S., 2013. Non-parsimonious evolution of the Dlx genes in the hagfish. *BMC Evol. Biol.* 13, 15.

Gai, Z., Donoghue, P.C.J., Zhu, M., Janvier, P., Stampanoni, M., 2011. Fossil jawless fish from China foreshadows early jawed vertebrate anatomy. *Nature* 476(7360), 324–327.

Gans, C., Northcutt, R.G., 1983. Neural crest and the origin of vertebrates: A new head. *Science* 220(4594), 268–274.

Gillis, J.A., Modrell, M.S., Baker, C.V., 2013. Developmental evidence for serial homology of the vertebrate jaw and gill arch skeleton. *Nat. Commun.* 4, 1436.

Hall, B.K., 1998. *Evolutionary Developmental Biology*, 2nd ed. Chapman & Hall.

Hall, B. K., 1999. *The Neural Crest in Development and Evolution*. New York, NY: Springer Science & Business Media.

Hall, B.K., Hörstadius, S., 1988. *The Neural Crest*. Oxford, UK: Oxford University Press.

Heimberg, A.M., Cowper-Sal-lari, R., Semon, M., Donoghue, P.C., Peterson, K.J., 2010. MicroRNAs reveal the interrelationships of hagfish, lampreys, and gnathostomes and the nature of the ancestral vertebrate. *Proc. Natl. Acad. Sci. U.S.A.* 107(45), 19379–19383.

Heintz, A., 1963. Phylogenetic aspect of myxinoids. In: A. Brodal & R. Fänge (eds), *The Biology of Myxine*. Universitetsforlaget, Oslo, pp. 9–21.

Higashiyama, H., Kuratani, S., 2014. On the maxillary nerve. *J. Morphol.* 275(1), 17–38.

Hirasawa, T., Oisi, Y., Kuratani, S., 2016. *Palaeospondylus* as a primitive hagfish. *Zool. Lett.* 2(1), 20.

Holmgren, N., 1946. On two embryos of *Myxine glutinosa*. *Act. Zool.* 27(1), 1–90.

Horie, R., Hazbun, A., Chen, K., Cao, C., Levine, M., Horie, T., 2018. Shared evolutionary origin of vertebrate neural crest and cranial placodes. *Nature* 560(7717), 228.

Horigome, N., Myojin, M., Hirano, S., Ueki, T., Aizawa, S., Kuratani, S., 1999. Development of cephalic neural crest cells in embryos of *Lampetra japonica*, with special reference to the evolution of the jaw. *Dev. Biol.* 207(2), 287–308.

Hörstadius, S., Sellman, S., 1946. Experimentelle Untersuchungen uber die Determination Des Knorpeligen Kopfskelettes bei Urodelen. *Nova Acta R. Soc. Scient Upsal* 4, 1–170.

Irie, N., Kuratani, S., Satoh, N., 2018. The phylum vertebrata: A way of zoological recognition. *Zool. Lett.* 4, 32.

Janvier, P., 1996. *Early Vertebrates*. Oxford, UK: Oxford University Press.

Johnels, A.G., 1948. On the development and morphology of the skeleton of the head of *Petromyzon*. *Act. Zool.* 29(1), 139–279.

Johnston, M.C., 1966. A radioautographic study of the migration and fate of cranial neural crests cells in the chick embryo. *Anat. Rec.* 156(2), 143–156.

Jollie, M., 1971. A theory concerning the early evolution of the visceral arches. *Act. Zool.* 52(1), 85–96.

Jollie, M., 1981. Segment theory and the homologizing of cranial bones. *Am. Nat.* 118(6), 785–802.

Jollie, M.T., 1977. Segmentation of the vertebrate head. *Am. Zool.* 17(2), 323–333.

Kitazawa, T., Takechi, M., Hirasawa, T., Hirai, T., Narboux-Nême, N., Kume, H., Oikawa, S., Maeda, K., Miyagawa-Tomita, S., Kurihara, Y., Hitomi, J., Levi, G., Kuratani, S., Kurihara, H., 2015. Developmental genetic bases behind the independent origin of the tympanic membrane in mammals and diapsids. *Nat. Commun.* 6, 6853.

Kuraku, S., 2008. Insights into cyclostome phylogenomics: Pre-2R or post-2R. *Zool. Sci.* 25(10), 960–968.

Kuraku, S., Hoshiyama, D., Katoh, K., Suga, H., Miyata, T., 1999. Monophyly of lampreys and hagfishes supported by nuclear DNA-coded genes. *J. Mol. Evol.* 49(6), 729–735.

Kuraku, S., Takio, Y., Sugahara, F., Takechi, M., Kuratani, S., 2010. Evolution of oropharyngeal patterning mechanisms involving Dlx and endothelins in vertebrates. *Dev. Biol.* 341(1), 315–323.

Kuratani, S., 1997. Spatial distribution of postotic crest cells defines the head/trunk interface of the vertebrate body: Embryological interpretation of peripheral nerve morphology and evolution of the vertebrate head. *Anat. Embryol.* 195(1), 1–13.

Kuratani, S., 2005. Cephalic neural crest cells and the evolution of the craniofacial structures in vertebrates: Morphological and embryological significance of the premandibular-mandibular boundary. *Zoology* 108(1), 13–26.

Kuratani, S., 2012. Evolution of the vertebrate jaw from developmental perspectives. *Evol. Dev.* 14(1), 76–92.

Kuratani, S., 2018. The neural crest and origin of the neurocranium in vertebrates. *Genesis* 56(6–7), e23213.

Kuratani, S., Ahlberg, P.E., 2018. Evolution of the vertebrate neurocranium: Problems of the premandibular domain and the origin of the trabecula. *Zool. Lett.* 4, 1.

Kuratani, S., Ueki, T., Aizawa, S., Hirano, S., 1997. Peripheral development of the cranial nerves in a cyclostome, *Lampetra japonica*: Morphological distribution of nerve branches and the vertebrate body plan. *J. Comp. Neurol.* 384(4), 483–500.

Kuratani, S., Horigome, N., Hirano, S., 1999. Developmental morphology of the head mesoderm and re-evaluation of segmental theories of the vertebrate head: Evidence from embryos of an agnathan vertebrate, *Lampetra japonica. Dev. Biol.* 210(2), 381–400.

Kuratani, S., Nobusada, Y., Horigome, N., Shigetani, Y., 2001. Embryology of the lamprey and evolution of the vertebrate jaw: Insights from molecular and developmental perspectives. *Phil. Trans. Roy. Soc.* 356(1414), 1615–1632.

Kuratani, S., Murakami, Y., Nobusada, Y., Kusakabe, R., Hirano, S., 2004. Developmental fate of the mandibular mesoderm in the lamprey, *Lethenteron japonicum*: Comparative morphology and development of the gnathostome jaw with special reference to the nature of trabecula cranii. *J. Exp. Zool. (Mol. Dev. Evol.)* 302B, 458–468.

Kuratani, S., Adachi, N., Wada, N., Oisi, Y., Sugahara, F., 2013. Developmental and evolutionary significance of the mandibular arch and prechordal/premandibular cranium in vertebrates: Revising the heterotopy scenario of gnathostome jaw evolution. *J. Anat.* 222(1), 41–55.

Kuratani, S., Oisi, Y., Ota, K.G., 2016. Evolution of the vertebrate cranium: Viewed from the hagfish developmental studies. *Zool. Sci.* 33(3), 229–238.

Kuratani, S., Kusakabe, R., Hirasawa, T., 2018. The neural crest and evolution of the head/trunk interface in vertebrates. *Dev. Biol.* 444 Suppl 1, 60–66.

Kuratani, S.C., Kirby, M.L., 1991. Initial migration and distribution of the cardiac neural crest in the avian embryo: An introduction to the concept of the circumpharyngeal crest. *Am. J. Anat.* 191(3), 215–227.

Lakiza, O., Miller, S., Bunce, A., Lee, E.M.J., McCauley, D.W., 2011. SoxE gene duplication and development of the lamprey branchial skeleton: Insights into development and evolution of the neural crest. *Dev. Biol.* 359(1), 149–161.

Langille, R.M., Hall, B.K., 1988. Role of the neural crest in development of the trabeculae and branchial arches in embryonic sea lamprey, *Petromyzon marinus* (L). *Development* 102, 301–310.

Lara-Ramírez, R., Patthey, C., Shimeld, S.M., 2015. Characterization of two neurogenin genes from the brook lamprey *Lampetra planeri* and their expression in the lamprey nervous system. *Dev. Dyn.* 244(9), 1096–1108.

Le Douarin, N.M., 1982. *The Neural Crest.* Cambridge, UK: Cambridge University Press.

Le Douarin, N.M., Kalcheim, C., 1999. *The Neural Crest*, 2nd ed., *Developmental and Cell Biology Series.* Cambridge, UK: Cambridge University Press.

Lee, E.M., Yuan, T., Ballim, R.D., Nguyen, K., Kelsh, R.N., Medeiros, D.M., McCauley, D.W., 2016. Functional constraints on SoxE proteins in neural crest development: The importance of differential expression for evolution of protein activity. *Dev. Biol.* 418(1), 166–178.

Lee, S.H., Bédard, O., Buchtová, M., Fu, K., Richman, J.M., 2004. A new origin for the maxillary jaw. *Dev. Biol.* 276(1), 207–224.

Loring, J.F., Erickson, C.A., 1987. Neural crest cell migratory pathways in the trunk of the chick embryo. *Dev. Biol.* 121(1), 220–236.

Mallatt, J., 1996. Ventilation and the origin of jawed vertebrates: A new mouth. *Zool. J. Linn. Soc.* 117(4), 329–404.

Mallatt, J., 2008. The origin of the vertebrate jaw: Neoclassical ideas versus newer, development-based ideas. *Zool. Sci.* 25(10), 990–998.

Mallatt, J., Sullivan, J., 1998. 28S and 18S rDNA sequences support the monophyly of lampreys and hagfishes. *Mol. Biol. Evol.* 15(12), 1706–1718.

Martin, W.M., Bumm, L.A., McCauley, D.W., 2009. Development of the viscerocranial skeleton during embryogenesis of the sea lamprey, *Petromyzon marinus*. *Dev. Dyn.* 238(12), 3126–3138.

Matsuoka, T., Ahlberg, P.E., Kessaris, N., Iannarelli, P., Dennehy, U., Richardson, W.D., McMahon, A.P., Koentges, G., 2005. Neural crest origins of the neck and shoulder. *Nature* 436(7049), 347–355.

McBratney-Owen, B., Iseki, S., Bamforth, S.D., Olsen, B.R., Morriss-Kay, G.M., 2008. Development and tissue origins of the mammalian cranial base. *Dev. Biol.* 322(1), 121–132.

McCauley, D.W., Bronner-Fraser, M., 2002. Conservation of Pax gene expression in ectodermal placodes of the lamprey. *Gene* 287(1–2), 129–139.

McCauley, D.W., Bronner-Fraser, M., 2003. Neural crest contributions to the lamprey head. *Development* 130(11), 2317–2327.

McCauley, D.W., Bronner-Fraser, M., 2004. Conservation and divergence of BMP2/4 genes in the lamprey: Expression and phylogenetic analysis suggest a single ancestral vertebrate gene. *Evol. Dev.* 6(6), 411–422.

McCauley, D.W., Bronner-Fraser, M., 2006. Importance of SoxE in neural crest development and the evolution of the pharynx. *Nature* 441, 750–752.

McCauley, D.W., Docker, M.F., Whyard, S., Li, W., 2015. Lampreys as diverse model organisms in the genomics era. *BioScience* 65(11), 1046–1056.

Meulemans, D., McCauley, D., Bronner-Fraser, M., 2003. Id expression in amphioxus and lamprey highlights the role of gene cooption during neural crest evolution. *Dev. Biol.* 264(2), 430–442.

Miller, C.T., Yelon, D., Stainier, D.Y., Kimmel, C.B., 2003. Two endothelin 1 effectors, *hand2* and *bapx1*, pattern ventral pharyngeal cartilage and the jaw joint. *Development* 130(7), 1353–1365.

Minarik, M., Stundl, J., Fabian, P., Jandzik, D., Metscher, B.D., Psenicka, M., Gela, D., Osorio-Pérez, A., Arias-Rodriguez, I., Horácek, I., Cerny, R., 2017. Pre-oral gut contributes to facial structures in non-teleost fishes. *Nature* 547(7662), 209–212.

Modrell, M.S., Hockman, D., Uy, B., Buckley, D., Sauka-Spengler, T., Bronner, M.E., Baker, C.V., 2014. A fate-map for cranial sensory ganglia in the sea lamprey. *Dev. Biol.* 385(2), 405–416.

Mukendi, C., Dean, N., Lala, R., Smith, J.J., Bronner, M.E., Nikitina, N.V., 2016. Evolution of the vertebrate claudin gene family: Insights from a basal vertebrate, the sea lamprey. *Int. J. Dev. Biol.* 60(1–3), 39–51.

Müller, G.B., Wagner, G.P., 1991. Novelty in evolution: Restructuring the concept. *Annu. Rev. Ecol. Syst.* 22(1), 229–256.

Myojin, M., Ueki, T., Sugahara, F., Murakami, Y., Shigetani, Y., Aizawa, S., Hirano, S., Kuratani, S., 2001. Isolation of Dlx and Emx gene cognates in an agnathan species, *Lampetra japonica*, and their expression patterns during embryonic and larval development: Conserved and diversified regulatory patterns of homeobox genes in vertebrate head evolution. *J. Exp. Zool. (Mol. Dev. Evol.)* 291(1), 68–84.

Neumayer, L., 1938. Die Entwicklung des Kopfskelettes von *Bdellostoma*. St. L. *Arch. Ital. Anat. Embryol.* 40 Suppl, 1–222.

Newth, D.R., 1951. On the neural crest of the lamprey embryo. *J. Embryol. Exp. Morphol.* 4, 358–375.

Newth, D.R., 1956. Experiments on the neural crest of the lamprey embryo. *J. Exp. Biol.* 28, 247–260.

Noden, D.M., 1978. The control of avian cephalic neural crest cytodifferentiation. II. Neural tissues. *Dev. Biol.* 67(2), 313–329.

Noden, D.M., 1986. Origins and patterning of craniofacial mesenchymal tissues. *J. Craniofac. Genet. Dev. Biol. Suppl.* 2, 15–31.

Noden, D.M., 1988. Interactions and fates of avian craniofacial mesenchyme. *Development* 103 Suppl, 121–140.

Northcutt, R.G., Gans, C., 1983. The genesis of neural crest and epidermal placodes: A reinterpretation of vertebrate origins. *Quart. Rev. Biol.* 58(1), 1–28.

Oisi, Y., Ota, K.G., Kuraku, S., Fujimoto, S., Kuratani, S., 2013a. Craniofacial development of hagfishes and the evolution of vertebrates. *Nature* 493(7431), 175–180.

Oisi, Y., Ota, K.G., Fujimoto, S., Kuratani, S., 2013b. Development of the chondrocranium in hagfishes, with special reference to the early evolution of vertebrates. *Zool. Sci.* 30(11), 944–961.

Oisi, Y., Fujimoto, S., Ota, K.G., Kuratani, S., 2015. On the peculiar morphology and development of the hypoglossal, glossopharyngeal and vagus nerves and hypobranchial muscles in the hagfish. *Zool. Lett.* 1, 6.

Ota, K.G., Kuratani, S., 2006. The history of scientific endeavours towards understanding hagfish embryology. *Zool. Sci.* 23(5), 403–418.

Ota, K.G., Kuraku, S., Kuratani, S., 2007. Hagfish embryology with reference to the evolution of the neural crest. *Nature* 446(7136), 672–675.

Ota, K.G., Fujimoto, S., Oisi, Y., Kuratani, S., 2011. Identification of vertebra-like elements and their possible differentiation from sclerotomes in the hagfish. *Nat. Commun.* 2, 373.

Ota, K.G., Fujimoto, S., Oisi, Y., Kuratani, S., 2013. Late development of the hagfish vertebral elements. *J. Exp. Zool. (Mol. Dev. Evol.)* 320, 129–139.

Ota, K.G., Oisi, Y., Fujimoto, S., Kuratani, S., 2014. The origin of developmental mechanisms underlying vertebral elements: Implications from hagfish evo-devo. *Zoology* 117(1), 77–80.

Parker, K.W., 1883a. On the skeleton of the marsipobranch fishes. Part I. The myxinoids (*Myxine*, and *Bdellostoma*). *Phil. Trans. R. Soc. Lond.* 174, 373–409.

Parker, K.W., 1883b. On the skeleton of the marsipobranch Fishes. Part II. *Petromyzon. Phil. Trans. R. Soc. Lond.* 174, 411–457.

Pascual-Anaya, J., Sato, I., Paps, J., Yandong, R., Sugahara, F., Higuchi, S., Takagi, W., Ruiz-Villalba, A., Ota, K.G., Wang, W., Kuratani, S., 2018. Hagfish and lamprey Hox genes reveal conservation of temporal colinearity in vertebrates. *Nat. Ecol. Evol.* 2(5), 859–866.

Rahimi, R.A., Allmond, J.J., Wagner, H., McCauley, D.W., Langeland, J.A., 2009. Lamprey snail highlights conserved and novel patterning roles in vertebrate embryos. *Dev. Genes Evol.* 219(1), 31–36.

Rijli, F.M., Mark, M., Lakkaraju, S., Dierich, A., Dollé, P., Chambon, P., 1993. A Homeotic transformation is generated in the rostral branchial region of the head by disruption of *Hoxa-2*, which acts as a selector gene. *Cell* 75(7), 1333–1349.

Sauka-Spengler, T., Bronner-Fraser, M., 2008. Evolution of the neural crest viewed from a gene regulatory perspective. *Genesis* 46(11), 673–682.

Sauka-Spengler, T., Meulemans, D., Jones, M., Bronner-Fraser, M., 2007. Ancient evolutionary origin of the neural crest gene regulatory network. *Dev. Cell* 13(3), 405–420.

Sewertzoff, A.N., 1931. *Morphologische Gesetzmässigkeiten der Evolution.* Gustav Fischer, Jena.

Shigetani, Y., Nobusada, Y., Kuratani, S., 2000. Ectodermally-derived FGF8 defines the maxillomandibular region in the early chick embryo: Epithelial–mesenchymal interactions in the specification of the craniofacial ectomesenchyme. *Dev. Biol.* 228(1), 73–85.

Shigetani, Y., Sugahara, F., Kawakami, Y., Murakami, Y., Hirano, S., Kuratani, S., 2002. Heterotopic shift of epithelial-mesenchymal interactions in vertebrate jaw evolution. *Science* 296(5571), 1316–1319.

Shigetani, Y., Sugahara, F., Kuratani, S., 2005. A new evolutionary scenario of the vertebrate jaw. *BioEssays* 27(3), 331–333.

Soukup, V., Horacek, I., Cerny, R., 2013. Development and evolution of the vertebrate primary mouth. *J. Anat.* 222(1), 79–99.

Tahara, Y., 1988. Normal stages of development in the lamprey, *Lampetra reissneri* (Dybowski). *Zool. Sci.* 5, 109–118.

Takechi, M., Adachi, N., Hirai, T., Kuratani, S., Kuraku, K., 2013. The Dlx genes as clues for vertebrate genomics and craniofacial evolution. *Sem. Cell. Dev. Biol.* 24, 110–118.

Takezaki, N., Figueroa, F., Zaleska-Rutczynska, Z., Klein, J., 2003. Molecular phylogeny of early vertebrates: Monophyly of the agnathans as revealed by sequences of 35 genes. *Mol. Biol. Evol.* 20(2), 287–292.

Takio, Y., Pasqualetti, M., Kuraku, S., Hirano, S., Rijli, F.M., Kuratani, S., 2004. Lamprey Hox genes and the evolution of jaws. *Nature* 429, 6989.

Takio, Y., Kuraku, S., Kusakabe, R., Murakami, Y., Pasqualetti, M., Rijli, F.M., Narita, Y., Kuratani, S., Kusakabe, R., 2007. Hox gene expression patterns in *Lethenteron japonicum* embryos insights into the evolution of the vertebrate Hox code. *Dev. Biol.* 308(2), 606–620.

Tucker, A.S., Watson, R.P., Lettice, L.A., Yamada, G., Hill, R.E., 2004. *Bapx1* regulates patterning in the middle ear: Altered regulatory role in the transition from the proximal jaw during vertebrate evolution. *Development* 131(6), 1235–1245.

Wada, N., Nohno, T., Kuratani, S., 2011. Dual origins of the prechordal cranium in the chicken embryo. *Dev. Biol.* 356(2), 529–540.

Wagner, G.P., Müller, G.B., 2002. Evolutionary innovations overcome ancestral constraints: A re-examination of character evolution in male sepsid flies (Diptera: Sepsidae). *Evol. Dev.* 4(1), 1–6.

Yalden, D.W., 1985. Feeding mechanisms as evidence for cyclostome monophyly. *Zool. J. Linn. Soc.* 84(3), 291–300.

Yao, T., Ohtani, K., Kuratani, S., Wada, H., 2011. Development of lamprey mucocartilage and its dorsal–ventral patterning by endothelin signaling, with insight into vertebrate jaw evolution. *J. Exp. Zool. (Mol. Dev. Evol.)* 316B, 339–346.

York, J.R., Yuan, T., Zehnder, K., McCauley, D.W., 2017. Lamprey neural crest migration is Snail-dependent and occurs without a differential shift in cadherin expression. *Dev. Biol.* 428(1), 176–187.

York, J.R., Yuan, T., Lakiza, O., McCauley, D.W., 2018. An ancestral role for Semaphorin3F-Neuropilin signaling in patterning neural crest within the new vertebrate head. *Development* 145(14), dev164780.

8 Neural Crest in Fossil Vertebrates

What, If Anything, Can We Know?

Per Erik Ahlberg and Tatjana Haitina

CONTENTS

8.1　INTRODUCTION

The desire to investigate the role of the neural crest in the development, morphogenesis, and evolution of fossil vertebrates comes up hard against a seemingly insuperable obstacle: the cell population we are interested in cannot be observed in fossils. Everything we know about the neural crest derives from observations and experiments on living organisms. The early migration of the cranial neural crest streams can in some cases be observed directly because the cell masses are large enough to be visible in SEM preparations of embryos (Steffek et al. 1979, Erickson & Weston 1983). DiI injection in living embryos allows the migration to be traced much further, to the point where the neural crest cells become incorporated into the developing anatomical structures of the animal (Serbedzija et al. 1992, Gillis et al. 2017). Ablation experiments can achieve similar results "in negative" by removing specific portions of pre-migratory crest, such as individual rhombomeres, and mapping the resulting anatomical effects (Saldivar et al. 1997, Rinon et al. 2007). Finally, the recognition of certain neural markers (for example *Wnt1*) allows the creation of permanently labelled transgenic cell lineages that permit the final destination of the migrating crest to be mapped with single-cell precision (e.g. Matsuoka et al. 2005, Yoshida et al. 2008, reviewed by Debbache et al. 2018). But none of these techniques can be used on fossils. The vertebrate fossil record presents us with a cavalcade of static morphologies, at best providing us with a partial record of the postembryonic ontogeny of certain features such as the ossification sequence (Schoch 2004; Hawthorn et al. 2007) or dental development (Chen et al. 2016, 2017), but usually not even that. The hard-tissue histology is often preserved with perfect three-dimensional fidelity (Figure 8.1), allowing enclosed cell types such as odontoblasts and osteocytes to be identified by the shapes of the cavities that once housed them (Donoghue et al. 2006; Sanchez et al. 2012), and it is becoming increasingly clear that soft-tissue preservation of complex biomolecules including identifiable remnants of proteins is possible over time spans of hundreds of millions of years (Lindgren et al. 2012, 2018). Nevertheless, molecular identification of neural crest cells in fossils will probably never be possible. Comparison of crest-derived and mesoderm-derived versions of particular skeletal tissues in areas where the two occur side by side, such as for example in the skull bones and shoulder girdles of mammals (Matsuoka et al. 2005), reveal no diagnostic histological differences. In short, there is no technique available at present or in the foreseeable future that would allow us to directly interrogate fossil tissue and determine whether it is of neural crest origin. We appear to have reached an impasse. Can we know anything at all about neural crest in fossil vertebrates?

8.2　HISTOLOGY AND ANATOMY

In fact, it is possible to reach beyond this formidable obstacle and arrive at quite detailed and informative conclusions about the role of neural crest in vertebrate evolution. Two different routes are available, one relying on direct evidence from tissue histology and the other on anatomical landmarks.

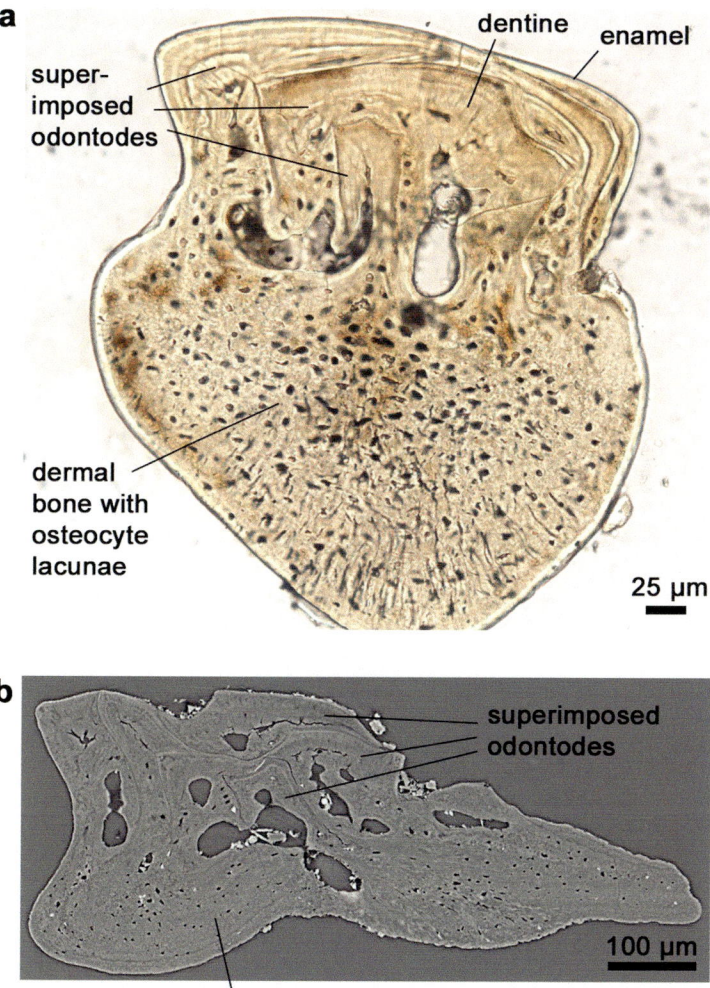

FIGURE 8.1 **a**, Transverse thin section of scale of the Silurian stem osteichthyan *Andreolepis*, viewed by transmitted light, showing well-preserved bone, dentine, and enamel. **b**, longitudinal section of scale of *Andreolepis*, taken from a propagation phase contrast synchrotron microtomography data set, generated at the European Synchrotron Radiation Facility (ESRF) in Grenoble, France. (Images generously provided by Qingming Qu.)

Because fossil hard tissues frequently preserve detailed information about cell types and matrix deposition patterns, they allow in principle for direct determination of neural crest origin, if the corresponding tissue in extant vertebrates can be shown to be specifically and exclusively produced by neural crest cells. The only hard tissue for which such a claim can be made at present is dentine, which appears to be produced by neural crest–derived odontoblasts both on the head and the trunk

(Chai et al. 2000; Gillis et al. 2017). Nondentinous hard tissues such as dermal bone, endoskeletal bone, and calcified cartilage do not carry an unambiguous cell lineage identity. For these tissues, and any anatomical structures built from them, it is necessary to use a different approach based on the identification of anatomical landmarks in a phylogenetic framework. In essence, we identify such tissues and structures as neural crest–derived, not on the basis of their histology, but on their position in the organism. If carefully done, this approach yields methodologically robust results. It is possible to say that the inferred neural crest maps are implicit in the data, to the extent that any alternative would require some form of *ad hoc* special pleading, such as an unsupported claim that the pattern of crest migration in the fossil organism differed substantially from that of all living vertebrates. However, even though the inferences are robust, they are not strictly speaking testable, because the fossil organisms cannot be studied experimentally. This might seem to limit their utility, but set against these limitations is the unique value provided by fossils: direct glimpses of morphologies and histologies from the deep past.

8.3 THE PHYLOGENETIC FRAMEWORK

The robustness of a phylogenetically based landmark analysis is of course entirely dependent on the reliability of the phylogeny. Fortunately, over the past decade, phylogenetic analyses based on molecular and morphological data have converged on a stable large-scale phylogeny of vertebrates, incorporating both living and fossil groups (Figure 8.2) (Shimeld & Donoghue 2012, and references therein). Extant vertebrates split into two clades, the jawless Cyclostomata and the jawed Gnathostomata. The Cyclostomata comprise two subclades, Myxini (hagfishes) and Petromyzontida (lampreys), while the Gnathostomata comprise the Chondrichthyes (cartilaginous fishes) and Osteichthyes (bony fishes and tetrapods). Gnathostome monophyly, and the sister-group relationship of osteichthyans and chondrichthyans, have never been controversial. Molecular data overwhelmingly support cyclostome monophyly, but the morphological evidence is more ambiguous, and many nonmolecular analyses have in the past recovered a paraphyletic Cyclostomata with lampreys as the sister group of gnathostomes (e.g. Janvier 1996; Donoghue et al. 2006). However, recent investigations of hagfish anatomy and development have shown that they are not as different from lampreys as had been thought (e.g. Ota et al. 2007, 2011; Kuratani & Ota 2008; Oisi et al. 2013a,b; Kuratani et al. 2016). For the purposes of this chapter, we will accept the phylogeny in Figure 8.2 as true, although there is some dispute about the position of the Anaspida, which some authors place in the cyclostome stem group (Miyashita et al. 2019).

The key concept of the comparative landmark approach is the *extant phylogenetic bracket* (EPB) (Witmer 1995). This states that a characteristic which is conserved between two living groups of organisms, and can thus be inferred to have been inherited from their last common ancestor, can also be attributed to any fossil taxa that fall within this bracket. To take a simple example (Figure 8.3a), osteichthyans and chondrichthyans—like all other vertebrates—share the possession of haemoglobin, which was presumably inherited from their last common ancestor; the fossil fish *Akmonistion* (which lived during the Carboniferous period, approximately 325–230

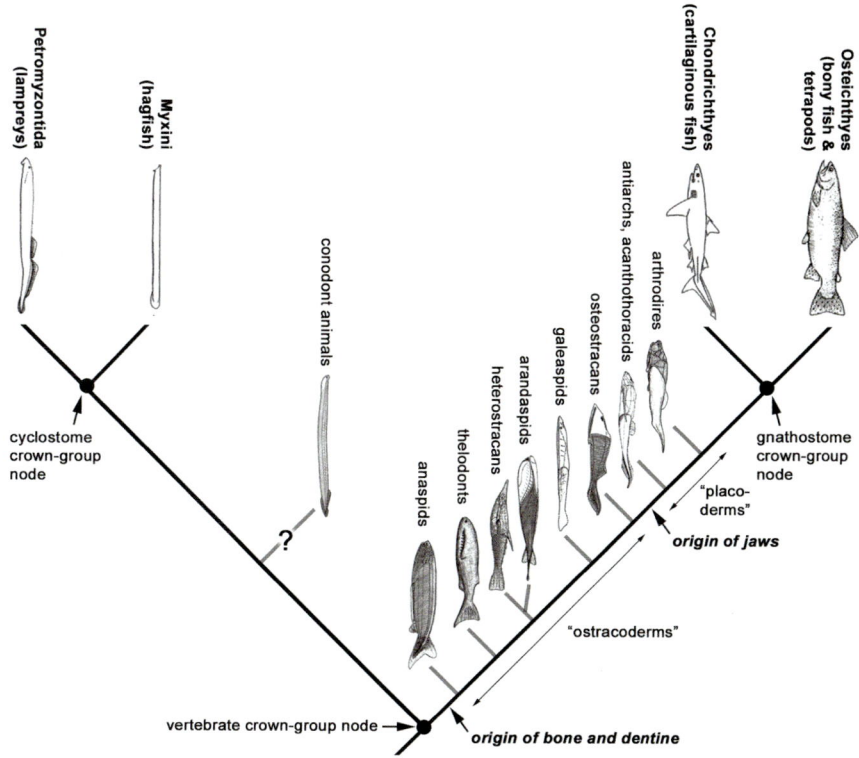

FIGURE 8.2 Phylogeny of the vertebrate crown group, including a selection of fossil stem gnathostomes and stem cyclostomes. Branches carrying fossil groups are shown in gray. (Thumbnail images of animals from Miyashita (2016).)

million years ago) falls within the osteichthyan-chondrichthyan bracket, being an early member of the chondrichthyan lineage (Coates & Sequeira 2001); *Akmonistion* can therefore be inferred to have possessed hemoglobin, even though this biomolecule is not preserved in the fossils.

Before proceeding to consider the specific EPBs that are relevant for reconstructing neural crest maps of extinct vertebrates, we need to introduce three terminological concepts that facilitate discussion of the placement and significance of fossils on a phylogeny: the crown group, the total group, and the stem group (Figure 8.3b). A *crown group* comprises all the living members of a clade, plus their last common ancestor, and any fossil taxa that fall within this clade. A *total group* comprises all taxa, living or fossil, which are more closely related to the relevant crown group than to any other living group. The gnathostome total group thus includes not only all crown gnathostomes but also all fossil taxa that are more closely related to gnathostomes than to cyclostomes. A total group is always larger than the crown group it contains, because it also includes a basal branch segment linking the crown group to the last common ancestor with the nearest living relative of the crown group; this basal branch segment is the *stem group*. In the present example, the gnathostome

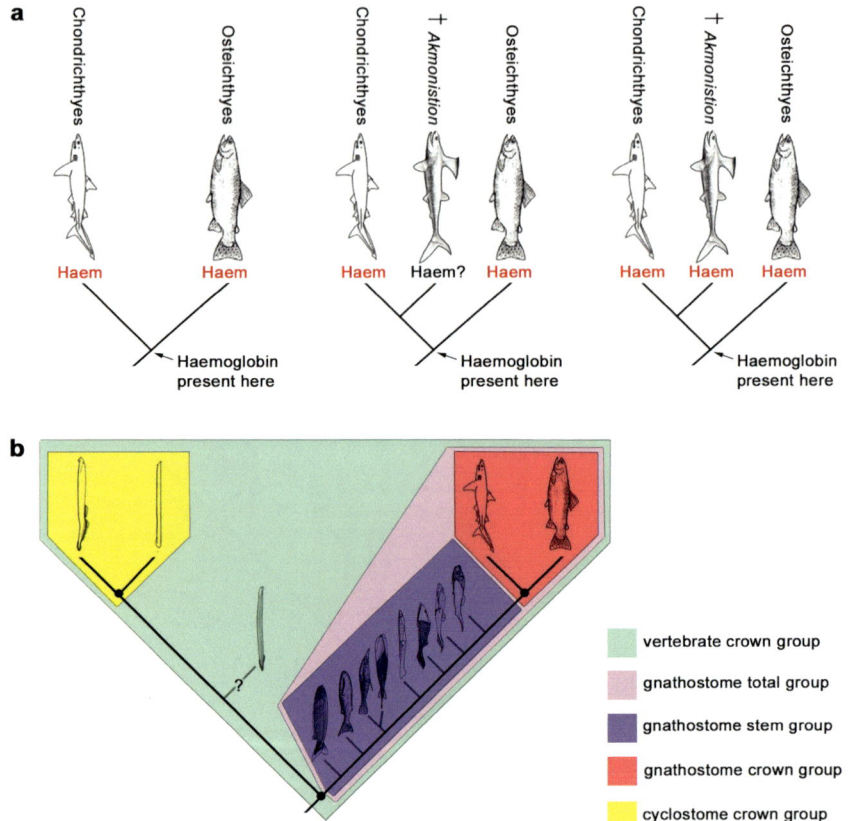

FIGURE 8.3   **a**, diagram illustrating the principle of the extant phylogenetic bracket. **b**, vertebrate phylogeny from Figure 8.2 showing position and extent of vertebrate crown group, gnathostome total group, gnathostome stem group, gnathostome crown group, and cyclostome crown group. (Thumbnail images of animals from Miyashita (2016), except *Akmonistion* from Coates & Sequeira (2001).)

stem group extends from the last common ancestor with cyclostomes up to the last common ancestor of chondrichthyans and osteichthyans. Note that these groups are arranged hierarchically, such that a higher-order crown group will contain lower-order crown groups, total groups and stem groups (Figure 8.3b). The crown group Vertebrata contains the total-groups Cyclostomata and Gnathostomata, which contain further crown and stem groups. A single taxon can thus have multiple memberships of crown and stem groups. For example, a teleost such as the living zebrafish *Danio rerio* is a crown actinopterygian, crown osteichthyan, crown gnathostome, and crown vertebrate; the Silurian (423 million year old) fish *Andreolepis hedei* (Chen et al. 2016) is a stem osteichthyan, crown gnathostome, and crown vertebrate (see Figure 1 in Qu et al. 2015 for respective phylogenetic positions of these species). Extant phylogenetic brackets always map to crown groups, as they are defined by the living members of a clade. EPBs will thus nest hierarchically within each other, and a stem group by definition spans the gap between a more inclusive and a less inclusive EPB. We will explore the significance of this, with specific reference to the neural crest, in the remaining sections.

8.4 EXTANT PHYLOGENETIC BRACKETS FOR THE NEURAL CREST

In the case of the neural crest, three main EPBs are relevant: the EPB of all vertebrates (cyclostomes + gnathostomes), the EPB of all cyclostomes (hagfish + lampreys), and the EPB of all gnathostomes (chondrichthyans + osteichthyans) (Figure 8.4). The robustness of the vertebrate and cyclostome EPBs has been greatly strengthened in recent years by the investigation of hagfish development at the Kuratani lab (Ota et al. 2007, 2011; Kuratani & Ota 2008; Oisi et al. 2013a,b; Kuratani et al. 2016), which has filled in crucial missing data in this part of the tree. In essence, the pattern that emerges is one of contrast between conservation of early-stage crest migration and phylogenetic divergence of later-stage crest fates. This is particularly striking in the craniopharyngeal region, which also undergoes some of the most striking evolutionary transformations among vertebrates.

The organization of the migrating cranial neural crest into distinct trigeminal, hyoid, and branchial (circumpharyngeal) streams is conserved between cyclostomes and gnathostomes (Kuratani et al. 2001, 2016; Ota et al. 2007, 2011; Kuratani & Ota 2008; Oisi et al. 2013a; Kuratani & Ahlberg 2018). The same holds true for the spatial relationship between these streams and neighboring structures in the developing head. Notably, the anterior part of the trigeminal stream always subdivides into supraoptic (=preoptic) and infraoptic (=postoptic) parts, which migrate forwards above and below the optic vesicle, while the posterior part migrates into the first (=mandibular) pharyngeal arch, immediately anterior to the first (=spiracular) pharyngeal pouch. The hyoid stream migrates into the hyoid arch between the first and second pharyngeal pouches, the branchial stream into the more posterior pharyngeal arches. Other major aspects of the molecular and cell-population architecture of the

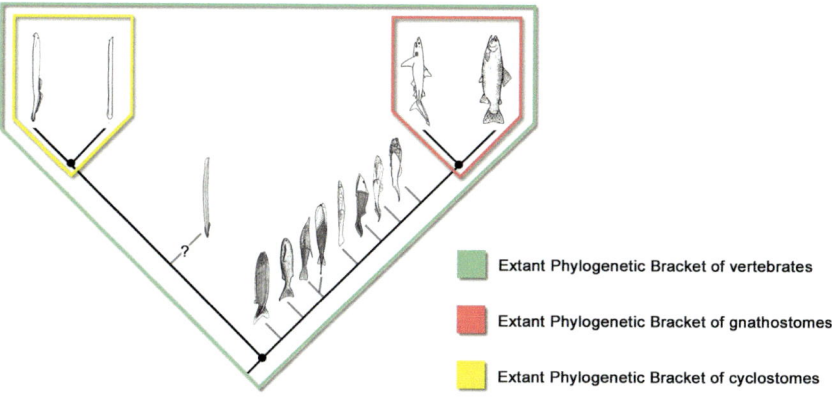

FIGURE 8.4 Diagram showing the extant phylogenetic brackets of vertebrates, gnathostomes, and cyclostomes.

head, such as the gene expression domains in the developing brain (Murakami et al. 2002) and the placodal origin of the nasal sacs and adenohypophysis (Kuratani et al. 2001; Oisi et al. 2013a), are also conserved across all living vertebrates.

The importance of this EPB for the study of fossil vertebrates cannot be overstated. Of all the putative vertebrates known from the fossil record, only the soft-bodied Cambrian genera *Myllokunmingia*, *Haikouichthys*, and *Metaspriggina* are likely to belong to the vertebrate stem group (Shu et al. 1999, 2003; Zhang & Hou 2004; Conway Morris & Caron 2014). All other fossil vertebrates, from the strange armored fishes of the Devonian to family favorites like *Tyrannosaurus rex*, fall within the vertebrate crown group and are thus encompassed by the EPB. This means that their major anatomical landmarks, such as the positions of the inner ears, nasal sacs, hypophysial fossa, cranial nerve openings, and palatoquadrate, can be used with confidence to infer cell population identities and boundaries in the craniopharyngeal region.

In contrast to this strongly conserved early ontogeny, the later fate of the cranial neural crest differs greatly between extant cyclostomes and gnathostomes, creating strongly contrasting cyclostome and gnathostome EPBs. In terms of histology, the major difference between cyclostomes and gnathostomes is the presence of biomineralized tissues in the latter. Neural crest cells make craniofacial cartilages in all extant vertebrates, but only in gnathostomes do they also make bone, dentine, and calcified cartilage. The morphological differences between cyclostomes and gnathostomes are in some respects more subtle but no less important. They principally concern the fate of the trigeminal and hyoid streams, and their interactions with the nasal and hypophysial placodes (Oisi et al. 2013a).

In cyclostomes, there is a single midline nasohypophysial placode that gives rise to the nasal sacs and adenohypophysis. The anterior infraoptic crest-derived ectomesenchyme grows forward around the sides of this placode, forming two lobes that eventually meet in the anterior midline (Figure 8.5a). In lamprey, this tissue creates the anterior part of the "hood" and upper lip, while in hagfish it creates the horizontal septum between the mouth and nasal opening, as well as the tentacles (Oisi et al. 2013a). The result in both cases is a face pierced by a midline cavity into which open both the nasal sacs and the adenohypophysis. The supraoptic ectomesenchyme creates the posterior margin of this midline cavity. The mandibular arch component of the trigeminal stream develops into the lower lip of the mouth and the rasping tongue with its distinctive piston cartilage. Further back, the hyoid arch crest forms a minor component of the cranial endoskeleton, ventral to the otic capsule and anterior to the branchial basket (Kuratani et al. 2016), and the spiracular pouch never develops into a spiracle.

Gnathostomes, by contrast, have separate left and right nasal placodes and a midline hypophysial placode. The anterior infraoptic ectomesenchyme grows forward around the sides of the hypophysial placode (presumably a conserved pattern shared with cyclostomes), but then converges on the midline and grows forward below and between the nasal placodes, forming the trabecular region of the braincase floor. Real trabeculae are absent in cyclostomes (Kuratani & Ahlberg 2018). The nasal sacs open separately to the outside on the face, while the hypophysis opens (if at all) onto the palate through a buccohypophysial foramen; the geometry of the face is

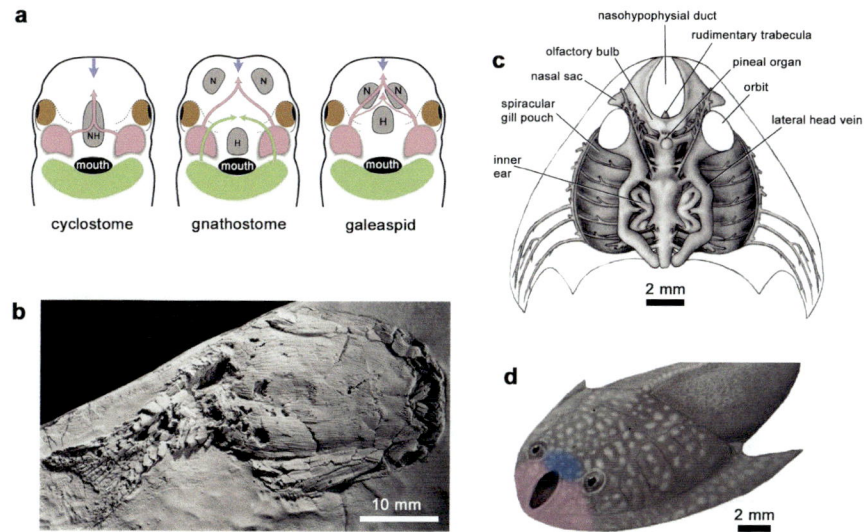

FIGURE 8.5 **a**, migration paths of mandibular (green), infraoptic premandibular (pink), and supraoptic (blue) ectomesenchyme from the trigeminal neural crest stream in cyclostomes (left) and gnathostomes (center). Modified from Kuratani et al. (2001), colors matching those used in Oisi et al. (2013). Nasal (N), hypophysial (H), and nasohypophysial (NH) placodes indicated in gray. On right, inferred placode layout and ectomesenchyme migration paths of the galeaspid *Shuyu*. **b**, specimen of the heterostracan *Athenaegis* in ventral view; the body is completely covered in plates and scales of acellular dermal bone (*aspidin*) ornamented with parallel dentine ridges formed from dermal odontodes. From Soehn & Wilson (1990). **c**, reconstruction of the endocranial spaces (cranial cavity, inner ear cavities, canals for vessels and nerves, roof of pharynx) in the galeaspid *Shuyu*, based on a synchrotron microtomography scan. From Gai et al. (2011). **d**, life reconstruction of *Shuyu* in anterodorsolateral view, from Gai et al. (2011), recolored to show inferred approximate distribution of infraoptic premandibular ectomesenchyme and supraoptic ectomesenchyme. Note that the midline opening in the face is the nasohypophysial duct, not the mouth.

thus fundamentally different from that of cyclostomes. Supraoptic ectomesenchyme creates the nasal capsules. The mandibular arch component of the trigeminal crest forms paired upper and lower jaws, the upper jaw lying alongside the trabecular region and articulating with it. The hyoid arch develops into a large and important structure with dorsal and ventral components; the spiracular pouch often develops into a persistent cavity in the adult, either as an open spiracle (in sharks, rays, and primitive actinopterygians) or as a middle ear cavity (in most tetrapods).

8.5 FOSSIL EVIDENCE FOR THE EVOLUTIONARY HISTORY OF THE NEURAL CREST

The fossils that are most informative about the evolution of neural crest–derived tissues and structures are the members of the gnathostome stem group (Figure 8.2 and 8.3b). Because the gnathostome lineage evolved biomineralization early in its history, it is relatively well represented in the fossil record. By contrast, the cyclostome fossil record is decidedly poor because of the limited preservation potential of soft-bodied animals (Miyashita et al. 2019). The earliest fossil lamprey, *Priscomyzon*, comes from the Devonian of South Africa and is approximately 360 million years old (Gess et al. 2006); the earliest hagfish, *Myxinikela*, dates to the Carboniferous and is thus slightly younger (Miyashita et al. 2019). These fossil forms do not differ radically from modern lampreys and hagfishes and are thus not very informative about the evolution of the cyclostome body plan.

More interesting in this context are the enigmatic fossils known as conodont animals. "Conodont elements" are small (millimeter-scale), tooth-like, biominer-alized structures that are among the most abundant of all fossils during the period 500–200 million years ago, occurring in their billions in marine sediments. Until fairly recently, it was not known what kind of animal they represented, though it had been noted that they were composed of hydroxyapatite, similar to verte-brate biomineralized tissues. It is now clear, following the discovery of complete specimens with soft-tissue preservation (Aldridge et al. 1993), that they are the oropharyngeal "teeth" of elongate soft-bodied animals that appear to be chordates of some kind. The phylogenetic position of this group is obviously of great inter-est in relation to the evolution of vertebrates in general and neural crest–derived tissues in particular, because if they are interpreted under a vertebrate model, the conodont elements lie within the crest-dominated oropharyngeal region. Some recent analyses place conodont animals in the cyclostome stem group (Miyashita et al. 2019), possibly suggesting that the conodont elements are biomineralized precursors of the keratinous "teeth" found on the rasping tongues of hagfish and lampreys. However, this phylogenetic position is not strongly supported as the anatomy is still poorly understood, and the homology between the biomineralized tissues of conodonts and vertebrates is disputed (Murdock et al. 2013; Donoghue & Rücklin 2016).

It is widely agreed that all fossil vertebrates that possess dentine and/or bone belong to the gnathostome total group. This of course implies that the absence of biomineralized tissues in cyclostomes is primary, not a secondary loss. As can be seen from the conodont discussion, there is actually some uncertainty around this, and one recent analysis (Miyashita et al. 2019) places the bony jawless anas-pids in the cyclostome stem group. However, the current consensus view is that unambiguous vertebrate hard tissues such as dentine were never present in the cyclostome lineage. The gnathostome stem group is populated in its lower part by biomineralized jawless vertebrates, collectively known as *ostracoderms*, and in its upper part by primitive jawed vertebrates known as *placoderms*. Both ostraco-derms and placoderms are paraphyletic relative to the gnathostome crown group (Figure 8.2). They have a rich fossil record, spanning from the first unambiguous fragments of vertebrate hard tissues in the Late Cambrian (about 490 million years ago) to the extinction of the last placoderms at the end of the Devonian, 360 million years ago. Together they provide a surprisingly detailed picture of the early evolution of the jawed vertebrate lineage. In so doing, they also cast indirect light on the ancestral vertebrate condition and the changes that have occurred in the cyclostome lineage.

8.6 ANATOMY AND HISTOLOGY AT THE BASE OF THE VERTEBRATES

The EPB of all vertebrates (Figure 8.4) indicates that the last common vertebrate ancestor had separate trigeminal, hyoid, and branchial crest streams, and, as an adult, a cartilaginous pharyngeal arch skeleton derived from these crest streams. There were probably (but see above) no biomineralized structures in the skin or endoskeleton. The fossil record shows unambiguously that biomineralization first evolved in the skin; the earliest hard tissue fragments are dermal (Repetski 1978; Smith et al. 1996; Donoghue et al. 2006), and most of the principal ostracoderm groups—anaspids, arandaspids, heterostracans, and thelodonts—combine dermal biomineralization with an unmineralized endoskeleton. Interestingly, the first unambiguously identifiable vertebrate hard tissue to appear in the fossil record is dentine, which is present in the fragmentary Cambrian-Ordovician vertebrate *Anatolepis* (Repetski 1978; Smith et al. 1996). *Anatolepis* has tubercles containing numerous tubules that extend out from a central pulp cavity, with the mineralized tissue showing centripetal growth towards the pulp; the gaps between the tubercles are filled by a lamellar acellular tissue pierced by occasional vertical canals. The tubercles are evidently dentine, while the lamellar tissue could be a precursor of dermal bone. The *Anatolepis* fragments are clearly dermal rather than oral, but they are otherwise difficult to locate on the body. However, later ostracoderms show that dentinous odontodes cover the entire external surface of the animal, from the snout to the tip of the tail (Figure 8.5b). These are often, but not invariably, attached to dermal bones. If the chondrichthyan data presented by Gillis et al. (2017) have general validity for dermal odontodes, the trunk neural crest of stem-group gnathostomes had skeletogenic capability. Indeed, the dermal domain may have been the original home of the odontode skeleton, as most ostracoderms seem to lack oropharyngeal odontodes; the only definite exceptions are the thelodonts, in which morphologically distinctive odontodes with fused bases have been described from the oral cavity (Smith & Coates 2001; Rücklin et al. 2011; Donoghue & Rücklin 2016). The presence of a more or less complete body covering of odontodes in sharks (as separate placoid scales), in the coelacanth *Latimeria* (as dentine tubercles of the scales and dermal skull bones), and in the basal actinopterygian *Polypterus* (in modified form, as a thin dentine layer underlying the ganoine of the scales) represents a retention of this ancestral condition.

A remarkable window on the capabilities of odontogenetic tissues in jawless stem gnathostomes is provided by an example of wound healing in the Devonian heterostracan *Psammolepis* (Johanson et al. 2013). A deep puncture wound in a dermal plate from the headshield (which consists of well-vascularised but acellular bone, known as *aspidin*, overlain by several generations of dermal odontodes) has been repaired exclusively by the deposition of secondary dentine. The aspidin shows no sign of repair or resorption. Indeed, the reparative dentine encloses not only jagged splinters of aspidin but also grains of quartz sand that presumably entered the wound at the time it was made. It appears that the resorptive and reparative processes of dermal bone had not yet evolved; only the (presumably) crest-derived ectomesenchyme was able to initiate wound healing by differentiating into odontoblasts, which covered up the whole sorry mess with a protective blanket of reparative dentine.

The major evolutionary changes in the odontode skeleton have been the repeated loss of dermal odontodes in different stem and crown gnathostome groups (for example some placoderms, as well as tetrapods and teleosts) and the evolution of organized teeth in the oropharynx. Teeth are first encountered in placoderms, where they are nonshedding (Donoghue & Rücklin 2016). Shedding evolved independently in osteichthyans, where it involves basal resorption of the dentine; and chondrichthyans, where it does not. The earliest evidence for the osteichthyan type of tooth shedding comes from the Late Silurian (approximately 423 million year old) stem osteichthyans *Andreolepis* from Sweden and *Lophosteus* from Estonia (Chen et al. 2016, 2017). Although both genera are fragmentary, the perfect histological preservation allows buried resorption surfaces underlying the teeth to be imaged in 3D using synchrotron microtomography and used to reconstruct the complete ontogenetic history of the dentition.

The entire vertebrate endoskeleton was initially unmineralized. Mineralization first appears in two relatively derived ostracoderm groups, the galeaspids (where it consists of calcified cartilage) and the osteostracans (where the endoskeleton is covered with perichondral bone) (Wang et al. 2005). In both groups, the mineralization is restricted to the so-called head shield, which in fact incorporates both the head *sensu stricto* and the branchial basket, but within this domain it extends across presumable crest–mesoderm boundaries and is thus not specifically informative about the neural crest *per se*. However, by making the endoskeletal anatomy including the cranial cavity visible to us, it gives us a first look at the morphology of the neurocranium in the basal jawless part of the gnathostome lineage. Some less robust but corroborating evidence is provided by heterostracans and thelodonts, in which the pharynx and cranial cavity are partly represented by natural sediment molds (Jarvik 1980; Donoghue & Smith 2001).

The most striking feature about these animals is their possession of a midline nasohypophysial duct. In osteostracans, it is small and located just in front of the eyes, creating an appearance very similar to a lamprey. In galeaspids, the duct is much larger, reaching almost to the upper lip, and varies in shape from a transverse oval to a longitudinal slit (Figure 8.5c,d). Internally, the osteostracan duct is closed ventrally and forms a bag-shaped space at the anterior end of the cranial cavity. This arrangement has traditionally been compared with that of lampreys (Jarvik 1980; Janvier 1996). However, osteostracans differ fundamentally from lampreys, and indeed all other vertebrates, in having an anteriorly displaced pharynx where the spiracular and first two branchial pouches lie anterior to the eyes (Kuratani & Ahlberg 2018). Such a geometry precludes a pharyngeal opening of the nasohypophysial duct, suggesting that the osteostracan and lamprey conditions are convergent. The galeaspid duct as a skeletal structure opens through to the oral cavity, though it is of course possible that it was closed off by a soft-tissue septum. Heterostracans and probably thelodonts appear to have had nasohypophysial ducts that opened anteriorly, somewhat in the manner of hagfish (Jarvik 1980; Donoghue & Smith 2001).

The presence of a nasohypophysial duct in these taxa suggests that the migration pattern of the premandibular infraoptic ectomesenchyme was similar to that in extant cyclostomes (Oisi et al. 2013a). However, there are also important differences. Cyclostomes are characterized by their single nasohypophysial placode, which gives rise to a single midline olfactory organ innervated by separate left and right olfactory nerves (Green et al. 2013) and a closely associated adenohypophysis. The condition in osteostracans is difficult to determine because of their very small and featureless nasohypophysial duct. However, in galeaspids, the recesses for the left and right nasal sacs are widely separated on either side of the nasohypophysial duct, and there is a discrete opening for the adenohypophysis in the posterior wall of the duct (Gai et al. 2011). This layout, which has skeletal tissue of presumably neural crest origin intervening between the hypophysis and nasal sacs, is not readily compatible with a single nasohypophysial placode; rather, it suggests three separate but tightly clustered placodes, which allowed a subset of the premandibular infraoptic ectomesenchyme to migrate into the space between them and form what is effectively the beginning of a trabecular region (Gai et al. 2011) (Figure 8.5a,c–d). The overall pattern of placodes and crest migration would thus combine what we would from a present-day perspective consider as cyclostome and gnathostome features.

Unfortunately, the lack of endoskeletal mineralization in phylogenetically more basal ostracoderms such as heterostracans, arandaspids, and anaspids makes it difficult to determine the condition of the nasohypophysial region at the very base of the gnathostome stem group. The internal surface of the dorsal dermal-bone headshield of heterostracans shows faint impressions of a variety of endocranial structures, and these include, at the anterior end, what appear to be well-separated nasal sacs flanking an anteriorly directed nasohypophysial duct (Jarvik 1980; Janvier 1996). This could represent a condition similar to that in galeaspids. However, the arandaspid *Saccabambaspis*, the earliest vertebrate for which we have an articulated head region, has an extremely short preorbital face—the eyes point forwards like car headlights—and separate left and right nasal openings without a nasohypophysial duct (Miyashita 2016). This morphology suggests paired nasal placodes and very little if any forward expansion of the infraoptic premandibular ectomesenchyme Unfortunately, the hypophysial region is unknown, but the existence of separate nasal placodes implies by geometrical necessity a separate hypophysial placode as well. To summarize, the majority of ostracoderms (jawless stem gnathostomes) appear to have had midline nasohypophysial ducts, suggesting that this aspect of the cyclostome *bauplan* was in fact primitive for vertebrates. However, there is evidence that galeaspids, heterostracans, and arandaspids had three separate placodes instead of a single nasohypophysial placode, and in fact there is no positive evidence for a nasohypophysial placode in any stem gnathostome. It seems that cyclostome-style migration of the infraoptic premandibular ectomesenchyme is not dependent on the presence of a nasohypophysial placode, as has tended to be assumed (Kuratani et al. 2001).

The other respect in which jawless stem gnathostomes appear to differ substantially from cyclostomes is in the branchial and oral apparatus. Osteostracans, galeaspids, and probably heterostracans all have fully developed spiracular gill pouches, showing that the hyoid arch formed part of the branchial skeleton (Figure 8.5c). The spiracular gill pouch lies just behind the eye in galeaspids and heterostracans, as would be expected from the developmental pattern defined by the EPB for all vertebrates, whereas in osteostracans, the pharynx has been displaced anteriorly. There is no evidence that a rasping tongue was present in these groups. Anaspids, by contrast, have a posteriorized branchial region somewhat like that of cyclostomes (Janvier 1996); presence of a rasping tongue is plausible in this group, but direct evidence is lacking.

8.7   THE ORIGIN OF JAWS

The transition from jawless to jawed vertebrates involved a great deal more than just the origin of jaws. In the head region, there was a cascade of apparently correlated changes affecting different parts of the skull and pectoral gridle, including the relationship between the dermoskeleton and endoskeleton; in the postcranium, the changes included the first appearance of the pelvis and pelvic fins. These transformations involved various combinations of neural crest, placode derivatives, unsegmented head mesoderm, somites, and lateral plate mesoderm. Here we will focus on the origin of jaws and the reorganization of the preorbital face, which largely involved derivatives of the trigeminal crest stream, but it is important to understand that these changes did not occur in isolation.

The fossil evidence for the transition is provided by a comparison of jawless and jawed members of the gnathostome stem group. As noted above, the jawed stem gnathostomes are known collectively as *placoderms* (e.g. Janvier 1996). They all possess a suite of derived characters shared with the gnathostome crown group, including jaws, absence of a nasohypophysial duct, a hypophysial fossa that opens onto the palate through a buccohypophysial foramen, nasal sacs that open onto the face through nostrils, a pectoral girdle that carries the pectoral fins and is separated from the skull by an occipital plane (among the jawless ostracoderms, only osteostracans have a pectoral girdle, and this is confluent with the braincase), and an external semicircular canal in the inner ear. In addition to these characteristics, all placoderms have a dermal skeleton on the head and pectoral girdle consisting of large bones with stable identities. The dermal pectoral girdle always forms a complete loop around the body and is often developed into a long trunk armor (Janvier 1996).

Until recently, there was a near-universal consensus that placoderms formed a clade positioned as the sister group to crown gnathostomes (Janvier 1996). The extensive similarities in pattern and structure between the dermal skeletons of placoderms and osteichthyans were dismissed as convergent, and the basal condition for the gnathostome crown group was assumed to be a broadly shark-like vertebrate covered with small dentine scales. During the past decade, a series of high-profile papers have overthrown this phylogenetic hypothesis (Brazeau 2009; Davis et al. 2012; Zhu et al. 2013; Brazeau & Friedman 2014; Dupret et al. 2014; Giles et al. 2015; Zhu et al. 2016; 2018). It is now clear that the dermal skeletons of placoderms and osteichthyans are homologous, and that placoderms are not a clade but a paraphyletic grade of primitive jawed vertebrates. The distinctive characters that constitute the placoderm *gestalt* are primitive for jawed vertebrates as a whole. Modern chondrichthyans are not in any sense living exemplars of primitive gnathostomes; rather, the evolutionary trajectory of gnathostomes can be described as a "placoderm–osteichthyan continuum" of gradual change, with the chondrichthyans forming a very specialized side branch. Derived characters of crown chondrichthyans include more or less complete loss of the ability to form dermal and endoskeletal bone (Ryll et al. 2014;

but see Eames et al. 2007), and acquisition of five or six tandemly duplicated copies of the *Col10a1* gene, which encodes a nonfibrillar network-forming collagen with a skeletogenic role (Debiais-Thibaud et al. 2019). These points need to be stressed, because there is still a widespread belief in the developmental community that chondrichthyans retain substantial parts of the primitive gnathostome character complement and are therefore appropriate model organisms for investigating the origin of gnathostome systems such as the dentition (e.g. Reif 1982; Rasch et al. 2016). Data from chondrichthyans are likely to be misleading if they are interpreted uncritically as representing the primitive gnathostome condition.

Because placoderms constitute a paraphyletic segment of the gnathostome stem, they have potential to illuminate the morphological transition from jawless to jawed vertebrates. Of particular interest in this context is the origin of jaws and the attendant reconfiguration of the preorbital face, two linked transformations involving the derivatives of the trigeminal neural crest stream. Such a fossil-based analysis requires a reasonably well-resolved phylogeny that allows sequences of character change within the placoderm branch segment to be determined. There is still considerable uncertainty about the details of placoderm interrelationships (Brazeau 2009; Davis et al. 2012; Zhu et al. 2013, 2016; Brazeau & Friedman 2014; Dupret et al. 2014; Giles et al. 2015; Qiao et al. 2016; Coates et al. 2018), but almost all recent phylogenetic analyses recover two groups—the antiarchs and acanthothoracids—as basal to all other jawed vertebrates (Figure 8.2). Both groups, like all placoderms, have fully developed jaws with a palatoquadrate that articulates with the braincase, contrasting fundamentally with the jawless osteostracans and galeaspids in which jaws are absent and the (probable) mandibular arch contribution to the neurocranium cannot be clearly demarcated (Figure 8.5c); in other words, the fossil record does not at present cast any direct light on the transformation of the mandibular arch. A useful overview of past and current hypotheses about the nature and causes of this transformation, including a new "mandibular confinement hypothesis", is given by Miyashita (2016). However, with regard to the loss of the nasohypophysial duct and the evolutionary fate of the premandibular infraoptic ectomesenchyme, the acanthothoracids in particular do seem to offer an intriguing glimpse of the transformation process.

Acanthothoracids and antiarchs share a peculiar anatomical feature, which is also present in the stand-alone genus *Brindabellaspis* (sometimes regarded as an acanthothoracid), but is absent from all other fossil and living jawed vertebrates. The nasal capsules, rather than being terminal on the snout, occupy a posterior position immediately in front of the eyes. Anteroventral to the nasal capsule lies a large dermal bone known as the *premedian plate* (Janvier 1996; Dupret et al. 2014, 2017). In antiarchs, the braincase is not ossified, limiting our ability to understand their cranial anatomy, but in acanthothoracids the braincase is often well ossified and highly informative. A recent detailed investigation of the braincase of the acanthothoracid *Romundina*, from the earliest Devonian of Canada (Ørvig 1975), was undertaken with the principal aim of understanding this curious facial anatomy (Figure 8.6) (Dupret et al. 2014, 2017; Kuratani & Ahlberg 2018).

The condition revealed by this investigation is both remarkable and informative. As in many placoderms, the nasal capsule is demarcated from the rest of the braincase by a complete fissure that passes through the optic foramen. Ventral to the nasal

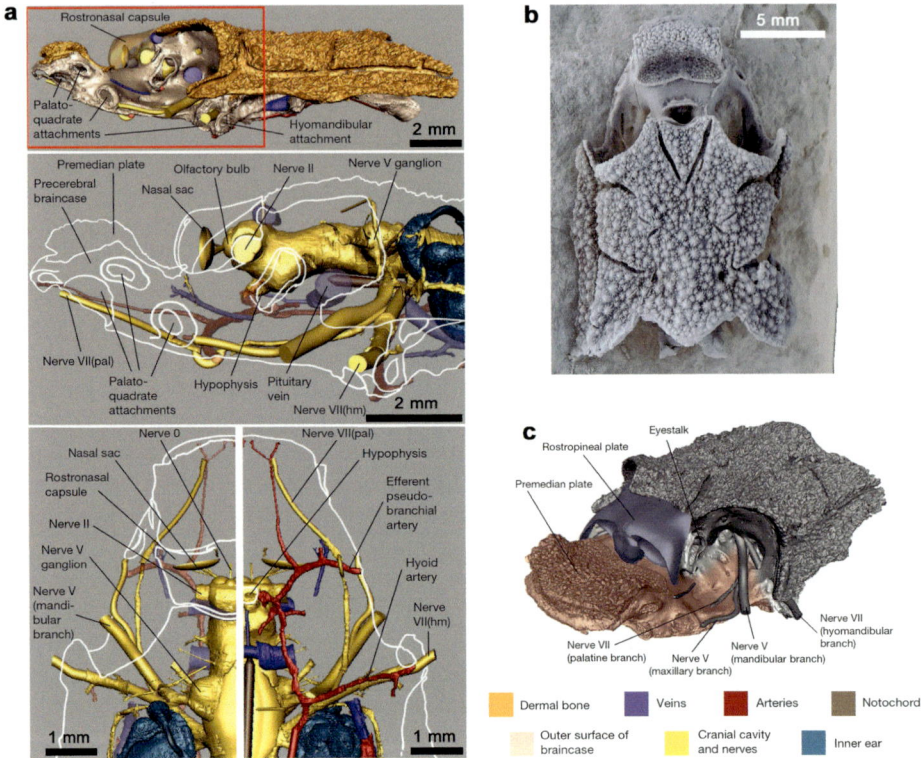

FIGURE 8.6 The neurocranium of the primitive jawed vertebrate *Romundina*. **a**, neurocranium and intracranial spaces modelled from propagation phase contrast synchrotron microtomography scan. Top, neurocranium in left lateral view, anterior to the left; middle, transparent rendering of the anterior part of the neurocranium (indicated with red box in top image) showing cranial cavity and cranial nerves in gold, arteries in red, veins in blue and inner ears in blue-green; bottom, same region in dorsal (left) and ventral (right) views. Color guide below image **c**. From Dupret et al. (2014). **b**, photo of scanned specimen in dorsal view. **c**, inferred distribution of supraorbital (blue) and premandibular infraoptic (pink) ectomesenchyme in the neurocranium of *Romundina*, seen in anterodorsolateral view, from Dupret et al. (2014). Compare with Figure 8.5d.

capsule, the braincase is expanded into a broad plate that extends anteriorly and carries the premedian plate on its dorsal surface (Figure 8.6). Attachment points for the palatoquadrate are found on the lateral margins of this plate, anteriorly beneath the edge of the premedian plate and posteriorly just in front of the exit foramen for the hyomandibular branch of nerve VII. The cranial cavity shows that the telencephalon and olfactory tracts were extremely short; the hypophysial fossa diverges anteroventrally from the cranial cavity, only a short distance behind the level of the optic nerve, with the result that the overall appearance is of the nasal sacs and hypophysis forming a tight triple cluster at the anterior end of the brain. Apart from the hypophysial fossa, no part of the cranial cavity enters the ventral plate of the braincase.

The existence of a number of landmarks, the significance of which is securely anchored within the EPB of all extant vertebrates (Figure 8.3b), makes it possible to reconstruct the cell population map of this anatomy with some confidence. In lateral view, the hyomandibular branch of nerve VII marks the divide between the hyoid and trigeminal crest streams, whereas the optic nerve marks the divide between the supraoptic and infraoptic branches of the trigeminal stream. Ventrally, the hypophysial fossa marks the posterior end of the trabeculae. It is clear from these considerations that the entire face of *Romundina* anterior to the rear wall of the orbit must be composed of trigeminal neural crest. The nasal capsule should be formed from supraoptic neural crest cells, judging from its position above the optic nerve foramen and the fact that this is the composition of the nasal capsule in extant gnathostomes (Oisi et al. 2013; Kuratani & Ahlberg 2018). Remarkably, the position of the ventral plate below the cranial cavity and nasal capsule, its incorporation of the hypophysial fossa, and the palatoquadrate articulations along its sides all indicate that this structure in its entirety is composed of premandibular infraoptic neural crest cells (Figure 8.6a,c). It can be considered a hypertrophied trabecular region, much wider and longer than the corresponding structure in any living gnathostome (Kuratani & Ahlberg 2018). Only the small part that underlies the miniscule forebrain cavity corresponds positionally to the trabeculae of a modern jawed vertebrate. The projecting anterior part of the plate does *not* correspond to a rostrum of the kind seen in sharks and certain other gnathostomes; a rostrum is anterodorsal to the nasal capsules, does not carry attachments for the palatoquadrate, and is presumably always formed from supraoptic neural crest cells.

A comparison with jawless stem gnathostomes offers a possible explanation for the strangely proportioned acanthothoracid face. Galeaspids and osteostracans, the two ostracoderm groups placed closest to jawed vertebrates in all recent analyses, both have neurocrania that extend a considerable distance anterior to the cranial cavity (Figure 8.5b). In both cases, this precerebral region of the neurocranium is pierced by a nasohypophysial duct, and must thus be formed substantially from premandibular infraoptic ectomesenchyme, as discussed here (Figure 8.5a,c–d). (Whether there was also a mandibular arch contribution to the flanks of the precerebral region is unknowable in the absence of landmarks.) The galeaspid cranial cavity (Gai et al. 2011) is remarkably similar to that of *Romundina* (Dupret et al. 2014, 2017), showing the same spatial relationship of the nasal sacs and hypophysis to the short forebrain, and the overall size and shape of the precerebral region is comparable to the ventral plate of the acanthothoracid braincase. This raises the intriguing possibility that the acanthothoracid condition could have arisen directly from a jawless ancestor similar to a galeaspid, by way of two steps: a rerouting of a greater part of the premandibular infraoptic ectomesenchyme into the space between the hypophysial and nasal placodes, obliterating the nasohypophysial duct and turning the precerebral region into a solid plate; and the development of the dorsal half of the mandibular arch into a discrete palatoquadrate lying alongside, and articulating with, this plate. The great length and breadth of the ventral plate compared to the trabecular region of modern gnathostomes suggests, firstly, that the overall degree of proliferation of the premandibular ectomesenchyme remained similar to that in jawless vertebrates, and secondly, that a significant part of this ectomesenchyme still migrated lateral to the nasal placodes.

In more advanced placoderms that are closer to the gnathostome crown group, such as arthrodires and the "maxillate placoderms" *Entelognathus* and *Qilinyu*, the nasal capsules are terminal and there is no projecting anteroventral premedian region (Zhu et al. 2013, 2016; Kuratani & Ahlberg 2018). This transformation has been achieved, not by lengthening the forebrain or olfactory tracts to push the nasal sacs forwards, but by shortening the ventral plate so that it no longer projects beyond the nasal capsules; as a result, these placoderms tend to have very short faces. Lengthening of the forebrain occurs later, possibly independently in different crown gnathostome lineages. It thus appears that the transformation of the fate of the premandibular infraoptic ectomesenchyme from jawless to jawed vertebrates happened in two distinct stages, the first centering on a change in migration path (to obliterate the nasohypophysial duct) and the second on a reduction in proliferation (to shrink the resulting solid structure from a large projecting plate to the rather modest trabeculae of modern gnathostomes).

8.8   CONCLUSION

As we have tried to show with the examples given here, it is possible to investigate the role of neural crest–derived tissues and structures in fossils in a methodologically robust manner by placing the fossil data in appropriate comparative contexts within a phylogenetic framework. Nor is the potential for such analyses limited to the cases we have presented. To take another example, melanocytes (which are always of neural crest origin) are not uncommon in fossils that show soft-tissue preservation, because melanin has high preservation potential (Lindgren et al. 2012, 2018). A more pertinent question is whether the data from fossils add anything unique to our understanding of the evolution of the neural crest. In a sense, investigating neural crest evolution in fossils is easy; for example, all evolutionary changes in the mammalian dentition are by definition "neural crest evolution", because teeth and jaws are neural crest–derived structures. However, these particular changes can be explored quite adequately by applying general concepts from natural selection and functional morphology, and there is nothing about them to suggest that they have been driven specifically by underlying changes in the characteristics or capabilities of the mammalian neural crest.

By contrast, we argue that fossil evidence is directly relevant where it touches on the early evolutionary history of the neural crest—and nowhere more so than in the gnathostome stem group. The behavior and capabilities of the neural crest in early vertebrates can be inferred for two important phylogenetic nodes by means of extant phylogenetic brackets. The vertebrate EPB defines a set of attributes for the neural crest in the last common ancestor of all living vertebrates, while the gnathostome EPB does the same for the last common ancestor of the gnathostome crown group. But apart from the early migratory behavior of the crest, which remains essentially unchanged, these brackets specify very different behaviors and fates. The neural crest cells of crown-group gnathostomes are capable of producing a range of biomineralized tissues, one of which (dentine) appears to be crest-specific; by contrast, the crest cells of the last common vertebrate ancestor may have lacked biomineralizing capability altogether. In the common ancestor of vertebrates, the trigeminal crest stream created a face with a nasohypophysial duct; in gnathostomes, it creates a solid trabecular region and paired jaws.

Fossils cast light on both these transformations in a way that data from living vertebrates cannot do on their own. With regard to the origin of biomineralization, they reveal that dentine was one of the first, possibly *the* first, mineralized tissue to evolve in the gnathostome lineage. This raises the interesting possibility that biomineralization in vertebrates was originally specific to the neural crest before being coopted to the mesoderm. Furthermore, dentine seems to have originated as a dermal tissue that covered the entire body, suggesting that the trunk neural crest is primitively skeletogenic in gnathostomes. These is evidence that wound repair in the dermal skeleton initially relied on production of reparative dentine, before the evolution of resorptive and reparative processes in dermal bone.

With regard to the anatomical transformation of the face, the fossils show that a nasohypophysial duct is not necessarily linked to possession of a single nasohypophysial placode. They also show that the—quite drastic—reconfiguration of the mandibular and premandibular crest populations was not accompanied by any significant change in the disposition of the brain, hypophysis, and nasal sacs, and further suggest that the migration and proliferation patterns of the premandibular crest were not linked but changed independently at different times. All these insights have potential implications for the molecular patterning and regulation of the neural crest, which can be pursued experimentally. The fossil evidence also impacts on the experimental domain by strongly cautioning against using living chondrichthyans as models for primitive gnathostomes. We feel confident in predicting that future fossil discoveries bearing on the origin of jaws and biomineralized tissues, coupled with molecular-developmental discoveries that test the inferences from fossils and add new information to the extant phylogenetic brackets, will continue to expand our understanding of the role of the neural crest in early vertebrate evolution.

REFERENCES

Aldridge RJ, Briggs DEG, Smith MP, Clarkson ENK, Clark NDL. 1993. The anatomy of conodonts. *Philosophical Transactions of the Royal Society of London B* 340(1294):406–420.

Brazeau MD. 2009. The braincase and jaws of a Devonian 'acanthodian' and modern gnathostome origins. *Nature* 457(7227):305–308.

Brazeau MD, Friedman M. 2014. The characters of Paleozoic jawed vertebrates. *Zoological Journal of the Linnean Society* 170(4):779–821.

Chai Y, Jiang X, Ito Y, Bringas P Jr, Han J, Rowitch DH, Soriano P, McMahon AP, Sucov HM. 2000. Fate of the mammalian cranial neural crest during tooth and mandibular morphogenesis. *Development* 127(8):1671–1679.

Chen D, Blom H, Sanchez S, Tafforeau P, Ahlberg PE. 2016. The stem osteichthyan *Andreolepis* and the origin of tooth replacement. *Nature* 539(7628):237–241.

Chen D, Blom H, Sanchez S, Tafforeau P, Märss T, Ahlberg PE. 2017. Development of cyclic shedding teeth from semi-shedding teeth: The inner dental arcade of the stem osteichthyan *Lophosteus*. *Royal Society Open Science* 4(5):161084.

Coates MI, Finarelli JA, Sansom IJ, Andreev PS, Criswell KE, Tietjen K, Rivers ML, La Riviere PJ. 2018. An early chondrichthyan and the evolutionary assembly of a shark body plan. *Proceedings of the Royal Society of London, Series B* 285:20172418.

Coates MI, Sequeira SEK. 2001. A new stethacanthid chondrichthyan from the Lower Carboniferous of Bearsden, Scotland. *Journal of Vertebrate Paleontology* 21(3):438–459.

Conway Morris S, Caron J-B. 2014. A primitive fish from the Cambrian of North America. *Nature* 512(7515):419–422.

Davis SP, Finarelli JA, Coates MI. 2012. *Acanthodes* and shark-like conditions in the last common ancestor of modern gnathostomes. *Nature* 486(7402):247–250.

Debbache J, Parfejevs V, Sommer L. 2018. Cre-driver lines used for genetic fate mapping of neural crest cells in the mouse: An overview. *Genesis* 56(6–7):e23105.

Debiais-Thibaud M, Simion P, Ventéo S, Muñoz D, Marcellini S, Mazan S, Haitina T. 2019. Skeletal mineralization in association with type X collagen expression is an ancestral feature for jawed vertebrates. *Molecular Biology Evolution* pii: msz145. doi: 10.1093/molbev/msz145.

Donoghue PCJ, Rücklin M. 2016. The ins and outs of the evolutionary origin of teeth. *Evolution & Development* 18(1):19–30.

Donoghue PCJ, Smith MP. 2001. The anatomy of *Turinia pagei* (Powrie), and the phylogenetic status of the Thelodonti. *Earth & Environmental Science Transactions of the Royal Society of Edinburgh* 92(1):15–37.

Donoghue PCJ, Sansom IJ, Downs JP. 2006. Early evolution of vertebrate skeletal tissues and cellular interactions, and the canalization of skeletal development. *Journal of Experimental Zoology (Mol. Dev. Evol.)* 306B:278–294.

Dupret V, Sanchez S, Goujet D, Tafforeau P, Ahlberg PE. 2014. A primitive placoderm sheds light on the origin of the jawed vertebrate face. *Nature* 507(7493):500–503.

Dupret V, Sanchez S, Goujet D, Ahlberg PE. 2017. The internal cranial anatomy of *Romundina stellina* Ørvig, 1975 (Vertebrata, Placodermi, Acanthothoraci) and the origin of jawed vertebrates—Anatomical atlas of a primitive gnathostome. *PLOS ONE* 12(2):e0171241.

Eames BF, Allen N, Young J, Kaplan A, Helms JA, Schneider RA. 2007. Skeletogenesis in the swell shark *Cephaloscyllium ventriosum*. *Journal of Anatomy* 210(5):542–554.

Erickson CA, Weston JA. 1983. An SEM analysis of neural crest migration in the mouse. *Development* 74:97–118.

Gai Z, Donoghue PCJ, Zhu M, Janvier P, Stampanoni M. 2011. Fossil jawless fish from China foreshadows early jawed vertebrate anatomy. *Nature* 476(7360):324–327.

Gess RW, Coates MI, Rubidge BS. 2006. A lamprey from the Devonian period of South Africa. *Nature* 443(7114):981–984.

Giles S, Friedman M, Brazeau MD. 2015. Osteichthyan-like cranial conditions in an early Devonian stem gnathostome. *Nature* 520(7545):82–85.

Gillis JA, Alsema EC, Criswell KE. 2017. Trunk neural crest origin of dermal denticles in a cartilaginous fish. *PNAS* 114(50):13200–13205.

Green WW, Basilious A, Dubuc R, Zielinski BS. 2013. The neuroanatomical organization of projection neurons associated with different olfactory bulb pathways in the sea lamprey, *Petromyzon marinus*. *PLOS ONE* 8(7):e69525.

Hawthorn JR, Wilson MVH, Falkenberg AB. 2008. Development of the dermoskeleton in *Superciliaspis gabrielsei* (Agnatha: Osteostraci). *Journal of Vertebrate Paleontology* 28(4):951–960.

Janvier P. 1996. *Early Vertebrates*. Clarendon Press: Oxford.

Jarvik E. 1980. *Basic Structure and Evolution of Vertebrates, Vol. 1*. Academic Press: London.

Kuratani S, Ahlberg PE. 2018. Evolution of the vertebrate neurocranium: Problems of the premandibular domain and the origin of the trabecula. *Zoological Letters* 4(1). doi: 10.1186/s40851-017-0083-6.

Kuratani S, Nobusada Y, Horigome N, Shigetani Y. 2001. Embryology of the lamprey and evolution of the vertebrate jaw: Insights from molecular and developmental perspectives. *Philosophical Transactions of the Royal Society of London B* 356(1414):1615–1632.

Kuratani S, Oisi Y, Ota KG. 2016. Evolution of the vertebrate cranium: Viewed from hagfish developmental studies. *Zoological Science* 33(3):229–238.

Kuratani S, Ota KG. 2008. Hagfish (Cyclostomata, Vertebrata): Searching for the ancestral developmental plan of vertebrates. *BioEssays* 30(2):167–172.

Lindgren J, Sjövall P, Thiel V, Zheng W, Ito S, Wakamatsu K, Hauff R, Kear BP, Engdahl A, Alwmark C, Eriksson ME, Jarenmark M, Sachs S, Ahlberg PE, Marone F, Kuriyama T, Gustafsson O, Malmberg P, Thomen A, Rodriguez-Meizoso I, Uvdal P, Ojika M, Schweitzer MH. 2018. Soft-tissue evidence for homeothermy and Crypsis in a Jurassic ichthyosaur. *Nature* 564(7736):359–365.

Lindgren J, Uvdal P, Sjövall P, Nilsson DE, Engdahl A, Pagh Schultz B, Thiel V. 2012. Molecular preservation of the pigment melanin in fossil melanosomes. *Nature Communications* 3:824.

Matsuoka T, Ahlberg PE, Kessaris N, Iannarelli P, Dennehy U, Richardson WD, McMahon AP, Koentges G. 2005. Neural crest origins of the neck and shoulder. *Nature* 436(7049):347–355.

Miyashita T. 2016. Fishing for jaws in early vertebrate evolution: A new hypothesis of mandibular confinement. *Biological Reviews* 91(3):611–657.

Miyashita T, Coates MI, Farrar R, Larson P, Manning PL, Wogelius RA, Edwards NP, Anné J, Bergmann U, Palmer RA, Currie PJ. 2019. Hagfish from the Cretaceous Tethys Sea and a reconciliation of the morphological–molecular conflict in early vertebrate phylogeny. *PNAS* 116(6):2146–2151.

Murakami Y, Ogasawara M, Satoh N, Sugahara F, Myojin M, Hirano S, Kuratani S. 2002. Compartments in the lamprey embryonic brain as revealed by regulatory gene expression and the distribution of reticulospinal neurons. *Brain Research Bulletin* 57(3–4):271–275.

Murdock DJE, Dong X-P, Repetski JE, Marone F, Stampanoni M, Donoghue PCJ. 2013. The origin of conodonts and of vertebrate mineralized skeletons. *Nature* 502(7472):546–549.

Oisi Y, Ota KG, Kuraku S, Fujimoto S, Kuratani S. 2013a. Craniofacial development of hagfishes and the evolution of vertebrates. *Nature* 493(7431):175–180.

Oisi Y, Ota KG, Fujimoto S, Kuratani S. 2013b. Development of the chondrocranium in hagfishes, with special reference to the early evolution of vertebrates. *Zoological Science* 30(11):944–961.

Ota KG, Kuraku S, Kuratani S. 2007. Hagfish embryology with reference to the evolution of the neural crest. *Nature* 446(7136):672–675.

Ota KG, Kuraku S, Kuratani S. 2011. Identification of vertebra-like elements and their possible differentiation from sclerotomes in the hagfish. *Nature Communications* 2:373.

Ørvig T. 1975. Description, with special reference to the dermal skeleton, of a new Radotinid arthrodire from the Gedinnian of Arctic Canada. Extrait des Colloques internationaux du Centre National de la Recherche Scientifique–Problèmes actuels de Paléontologie–Evolution des Vertébrés. 1975 218:41–71.

Qiao T, King B, Long JA, Ahlberg PE, Zhu M. 2016. Early gnathostome phylogeny revisited: Multiple method consensus. *PLOS ONE* 11(9):e0163157.

Qu Q, Haitina T, Zhu M, Ahlberg PE. 2015. New genomic and fossil data illuminate the origin of enamel. *Nature* 526(7571):108–111.

Rasch LJ, Martin KJ, Cooper RL, Metscher BD, Underwood CJ, Fraser GJ. 2016. An ancient dental gene set governs development and continuous regeneration of teeth in sharks. *Developmental Biology* 415(2):347–370.

Reif W-E. 1982. Evolution of dermal skeleton and dentition in vertebrates: The odontoderegulation theory. *Evolutionary Biology* 15:287–368.

Repetski JE. 1978. A fish from the Upper Cambrian of North America. *Science* 200(4341):529–531.

Rinon A, Lazar S, Marshall H, Büchmann-Møller S, Neufeld A, Elhanany-Tamir H, Taketo MM, Sommer L, Krumlauf R, Tzahor E. 2007. Cranial neural crest cells regulate head muscle patterning and differentiation during vertebrate embryogenesis. *Development* 134(17):3065–3075.

Rücklin M, Giles S, Janvier P, Donoghue PCJ. 2011. Teeth before jaws? Comparative analysis of the structure and development of the external and internal scales in the extinct jawless vertebrate *Loganellia scotica*. *Evolution & Development* 13(6):523–532.

Ryll B, Sanchez S, Haitina T, Tafforeau P, Ahlberg PE. 2014. The genome of *Callorhinchus* and the fossil record: A new perspective on SCPP gene evolution in gnathostomes. *Evolution & Development* 16(3):123–124.

Saldivar JR, Sechrist JW, Krull CE, Ruffins S, Bronner-Fraser M. 1997. Dorsal hindbrain ablation results in rerouting of neural crest migration and changes in gene expression, but normal hyoid development. *Development* 124(14):2729–2739.

Sanchez S, Ahlberg PE, Trinajstic KM, Mirone A, Tafforeau P. 2012. Three-dimensional synchrotron virtual paleohistology: A new insight into the world of fossil bone microstructures. *Microscopy & Microanalysis* 18(5):1095–1105.

Schoch RR. 2004. Skeleton formation in the Branchiosauridae: A case study in comparing ontogenetic trajectories. *Journal of Vertebrate Paleontology* 24(2):309–319.

Serbedzija GN, Bronner-Fraser M, Fraser SE. 1992. Vital dye analysis of cranial neural crest cell migration in the mouse embryo. *Development* 116(2):297–307.

Shimeld SM, Donoghue PCJ. 2012. Evolutionary crossroads in developmental biology: Cyclostomes (lamprey and hagfish). *Development* 139(12):2091–2099.

Shu D-G, Conway Morris S, Han J, Zhang Z-F, Yasui K, Janvier P, Chen L, Zhang X-L, Liu J-N, Li Y, Liu H-Q. 2003. Head and backbone of the early Cambrian vertebrate *Haikouichthys*. *Nature* 421(6922):526–529.

Shu D-G, Luo H-L, Conway Morris S, Zhang X-L, Hu S-X, Chen L, Han J, Zhu M, Li Y, Chen L-Z. 1999. Lower Cambrian vertebrates from South China. *Nature* 402(6757):42–46.

Smith MP, Sansom IJ, Repetski JE. 1996. Histology of the first fish. *Nature* 380(6576):702–704.

Soehn KL, Wilson MVH. 1990. A complete, articulated heterostracan from Wenlockian (Silurian) beds of the Delorme group, Mackenzie Mountains, Northwest Territories, Canada. *Journal of Vertebrate Paleontology* 10(4):405–419.

Steffek AJ, Mujwid DK, Johnston MC. 1979. Scanning electron microscopy (SEM) of cranial neural crest migration in chick embryos. *Birth Defects Original Article Series* 15(8):11–21.

Wang NZ, Donoghue PCJ, Smith MM, Sansom IJ. 2005. Histology of the galeaspid dermoskeleton and endoskeleton, and the origin and early evolution of the vertebrate cranial endoskeleton. *Journal of Vertebrate Paleontology* 25(4):745–756.

Witmer LM. 1995. The extant phylogenetic bracket and the importance of reconstructing soft tissues in fossils. In: Thomason, JJ. (ed). *Functional Morphology in Vertebrate Paleontology*. Cambridge University Press: New York. pp:19–33.

Yoshida T, Vivatbutsiri P, Morriss-Kay G, Saga Y, Iseki S. 2008. Cell lineage in mammalian craniofacial mesenchyme. *Mechanisms of Development* 125(9–10):797–808.

Zhang X-G, Hou X-G. 2004. Evidence for a single median fin-fold and tail in the Lower Cambrian vertebrate, *Haikouichthys ercaicunensis*. *Journal of Evolutionary Biology* 17(5):1162–1166.

Zhu M, Wu X, Ahlberg PE, Choo B, Lu J, Qiao T, Zhao W, Jia L, Blom H, Zhu Y. 2013. A Silurian placoderm with osteichthyan-like marginal jaw bones. *Nature* 302:188–193.

Zhu M, Ahlberg PE, Pan Z, Zhu Y, Qiao T, Zhao W, Jia L, Lu J. 2016. A new Silurian maxillate placoderm illuminates jaw evolution. *Science* 354(6310):334–336.

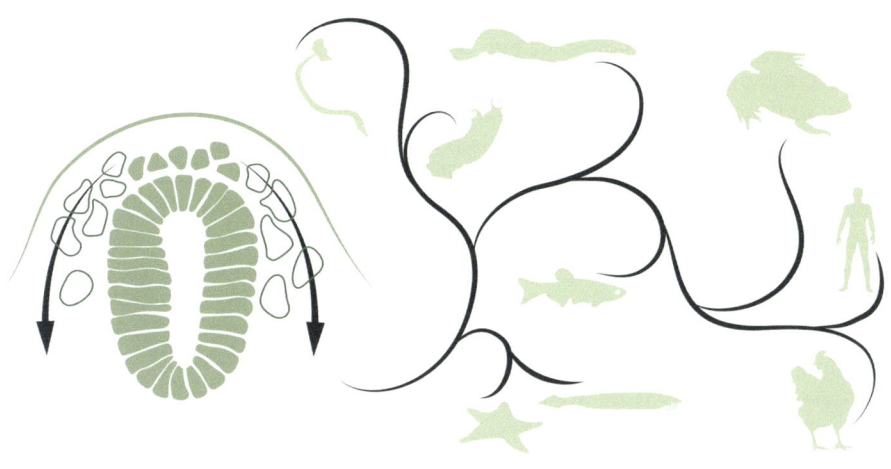

9 Evolving Neural Crest Cells

Hopes for Present and Future Understanding

Igor Adameyko and Brian F. Eames

CONTENTS

9.1   INTRODUCTION

Explaining the origins and subsequent evolution of neural crest cells relies upon what might appear to be quite fragmented cellular, molecular, and paleontological evidence, but the modern synthesis of ideas that emerges is as complex and rich as

265

these fascinating cells themselves. We hope that each chapter of this book inspires the reader's power of knowledge and imagination to connect more dots, building a holistic and up-to-date view on neural crest cell development and evolution.

Indeed, neural crest cells offer a tantalizing puzzle in evolutionary cell biology. When we started to compile this book, 150 years had passed since the initial discovery of neural crest by Wilhelm His (His, 1868), a Swiss-born anatomist and embryologist who also is famous for inventing the microtome. Since His's discovery, researchers have gradually uncovered a wealth of knowledge about many cell biological features of neural crest, including mechanisms of induction, delamination, migration, and differentiation. However, exactly how these fundamental neural crest features appeared and became modified during evolution did not receive as much attention and, thus, remained rather elusive.

Often referred to as the fourth germ layer (in addition to endoderm, mesoderm, and ectoderm), the transient embryonic population of neural crest cells are multipotent, migratory, and plastic—quite the bag of tricks upon which to draw when selective pressures might favor evolutionary innovations. Neural crest cells (or at least neural crest–like cells) arise at the dawn of chordates and ultimately provide ancestral vertebrates with key evolutionary advantages, such as a jawed head, improved hormonal control, pigmentation for display and camouflage, enriched sensory capacities, and a protective bony shield. The life of a neural crest cell involves specification and commitment at the embryonic neural plate, delamination from an epithelium, complex migration patterns throughout the body axis, and differentiation into one of many potential cell fates. Granted, these varied and numerous traits were unlikely to evolve all at once, but perhaps the fact that neural crest cells are a relatively recent innovation gives us a chance to solve the mystery of such a grand evolutionary transition. By contrast, the emergence of mesoderm, which is a much older cell lineage, would be harder to trace, especially since animals without mesoderm are rare, restricting an ability to use outgroups to test hypotheses about the origins of mesoderm.

The question of neural crest evolution demands truly interdisciplinary approaches, integrating such disparate fields as comparative morphology and anatomy, developmental biology, paleontology, and evolutionary genomics. Each chapter in this book combines evidence from at least two of these disciplines, especially taking advantage of the classical triad of comparative embryology, molecular phylogenetics, and paleontology. In total, the authors dissect the evolving neural crest at multiple levels, from gross morphologies, tissues, and cell types down to the molecular control of gene regulatory networks. Furthermore, many chapters show how specific experimental approaches, such as cell transplantation and analyses of gene expression and function, have contributed to our current understanding. For example, the chapter by Alkobtawi and Monsoro-Burq provides a nice outline of the most essential properties of neural crest and touches upon the great leaps in neural crest research that resulted from blending classical and newly emerging technical approaches. We expect that future techniques will expand even further our knowledge of neural crest cell origins and subsequent evolution.

Current understanding of the evolving neural crest must incorporate not only the acquisition of specific traits, but also the molecular mechanisms underlying these evolutionary transitions. Highlighting specific chapters in this book, we follow the developmental life of a neural crest cell, explaining insights into each step during evolution. We finish by discussing some important points to be addressed in future research.

9.2 PRESENT HOPES FOR UNDERSTANDING

9.2.1 SPECIFICATION

The specification of neural crest cells occurs through a series of molecular interactions involving signaling pathways and specifier transcription factors at the neural plate border. This process appears to be relatively conserved, which makes it very difficult to understand how the neural plate border evolved. The evolutionary and mechanistic aspects of neural crest induction is well-summarized in chapters by Alkobtawi and Monsoro-Burq and Yu and Su. These authors revealed how cascades of transcriptional regulators provide for neural crest induction and downstream events, not only in classical animal models but also across the phylogenetic tree of vertebrates. However, the basic evolution of the molecular controls regulating the emergence and specification of the neuroectoderm and neural plate border (corresponding to the future neural crest) is far from being understood, as discussed in the chapter by Yu and Su. These authors argue that neural plate border formation appears to have significant yet moderate conservation of molecular traits across deuterostome clades.

One line of research explores the intricate evolutionary relationship between specification of both neural crest and neurogenic placodes at the neural plate border. As discussed in a recent review by Gerhard Schlosser, two major concepts attempt to explain the emergence of neural crest and placodes (Schlosser et al., 2014). According to the "neural plate border state model", cells that belong to a common neural plate border split into the neural crest and placodal regions. However, according to the "binary competence model", panplacodal primordium develops exclusively from nonneural territory, whereas the neural crest originates from a neural competence region (Patthey et al., 2014; Schlosser et al., 2014). Furthermore, many cranial ganglia, such as the trigeminal, originate from both neural crest–derived and placodal neurons (Steventon et al., 2014). Horie et al. took advantage of single-cell transcriptomics in a powerful combination with lineage tracing and functional studies to challenge the cell fates of anterior lateral plate ectoderm in *Ciona instestinalis* larvae (Horie et al., 2018). These authors transformed a putative protoneural crest cell type, a bipolar tail neuron, into a protoplacodal sensory cell type. These data suggested that the lateral plate ectoderm compartmentalized before the advent of vertebrates and was a common origin for cranial placodes and neural crest (Horie et al., 2018). Further research should clarify the developmental and evolutionary relationships between placodes, neural crest emerging from the neural plate border, and cells of the neural tube. Even the most general concepts proposed to explain the evolution of the central nervous system, which develops as a folding epithelial plate that transforms into a tube, appear grossly insufficient. Therefore, current understanding of the origin of the chordate CNS is extraordinarily enigmatic.

9.2.2 DELAMINATION/MIGRATION

Following specification, two particularly fascinating traits of neural crest cells are delamination from an epithelium and various subsequent migration patterns, permitting them to influence cell types and morphologies throughout the vertebrate body plan. Indeed, dramatic changes in lifestyle, body size, and the increasing complexity of inner organs might have been the driving force and selective pressure that would favor emigration of CNS progenitors outside of the CNS to build adapted sensory inputs, pigmentation, and local organ control. Interspecies transplantation studies (pioneered by Nicole LeDouarin in the 1970s) and other modern cell lineage–tracing strategies elucidated the spectrum of migration patterns and cell types derived from neural crest across vertebrate clades—from lamprey to mouse. The chapter by York, Zehnder, and McCauley tackled molecular aspects of neural crest epithelial-to-mesenchymal transformation (EMT) and migration. These authors explain the evolutionary landscape for employing the preexisting EMT and cell migration modules for efficient delamination of the neural crest towards the periphery and subsequent routing toward specific target destinations within the developing body.

Interestingly, the modes of such migration might be drifting between groups of animals, ranging from solitary migration to navigation in chains to variations of a crowd movement. In the chapter by Adameyko, a special focus is dedicated to peripheral nerve-assisted navigation. Two specific discoveries resulted in a number of peculiar ideas. First, a close lineage relationship was revealed between pigment cells within the CNS and progenitors migrating outside the CNS. Second, nerve-associated cells were observed to give rise to pigmentation and other cell types of the "neural crest spectrum". As a consequence, some studies suggested that neural crest could arise from progenitors of a pigment-containing intra-CNS photosensory system, and that the migration of a protoneural crest population could be facilitated by sensory and motor nerves traversing the bodies of early vertebrates (Ivashkin and Adameyko, 2013), as discussed in a chapter by Adameyko.

Variations on neural crest migration patterns might confound an ability to highlight clearly an ancestral condition. Of course, neural crest can follow different routes in the trunk and head regions, perhaps given different cues from the local adjacent embryonic populations in these different regions that might influence migration patterns. However, these respective patterns (i.e., head vs. trunk) are very similar across vertebrates, as are the numerous premandibular migration patterns of the head neural crest (e.g., supraorbital, infraorbital), highlighted in the chapters of Ahlberg and Haitina, as well as that of Kuratani. Of course, extinct agnathan clades might have evolved unique migration patterns. However, since the extant agnathans lamprey and hagfish appear to have the same relative migration patterns as gnathostomes, inferences can be made about crest migration underlying the evolving head morphologies in critical fossil agnathan clades. Ahlberg, Haitina, and Kuratani extensively discuss comparative aspects and diversity of extinct and extant phenotypes in the lineage of vertebrates. The necessity and the timeliness of such a compendium of ideas and discussions is dictated by the recent surge of discoveries in the field of paleontology, particularly the use of synchrotron-mediated microCT to reveal fossil morphologies noninvasively.

9.2.3 DIFFERENTIATION

One of the trajectories of neural crest cells with perhaps the most promise for experimental resolution is the evolution of cell fates. This knowledge is outlined in chapters by Alkobtawi and Monsoro-Burq; Eames, Gómez-Picos, and Jandzik; and Adameyko. The overall conclusion of these authors from analyses of neural crest derivatives in different animals is that cell fates might have evolved incrementally, especially when it comes to skeletogenic and odontogenic cell types. Importantly, the absence of some key components of the regulatory circuitry utilized by neural crest cells during differentiation does not necessarily mean that the corresponding cell type is not present. Thus, the "new head hypothesis" vividly outlined by Gans and Northcutt in 1983 (Gans and Northcutt, 1983) still stays enigmatic, particularly when attempts are made to reveal the cellular and molecular mechanisms involved in evolution of the cranial skeletogenic population. Chapters written by Abitua, Adameyko, and Eames with their coauthors focus on cellular aspects of neural crest evolution, specifically the possible mechanistic scenarios causing cell types to transform. For example, introduction of the transcription factor Twist into the overall picture, given its capacity to "mesenchymize" cell populations in *Ciona*, led to an interesting conclusion regarding the molecular mechanisms that might have introduced skeletogenic potential into neural crest. In tunicates, observations of cell types demonstrating neural crest–like behavior and multiple fate potentials led to a new hypothesis about the role of specific CNS progenitors in neural crest evolution (as discussed in chapters by Abitua and Yu with coauthors).

Although some dots seemingly connect quite well, lots of important information is still missing, especially when it comes to genomic and epigenetic mechanisms of cell lineage transformations. For instance, evolution of cell differentiation and cell fate choices leaves almost no traces, especially if extant forms representing ancestral conditions are scarce. Nevertheless, direct experimentation in nonvertebrate chordates (discussed in chapters by Abitua and Adameyko) might pave a way towards better understanding of molecular aspects of cell type evolution, cooptions of fates, and molecular mechanism enabling blending the properties of several cell lineages. It is plausible to suggest that neural crest took advantage of these evolutionary processes ever since their first appearance.

9.3   FUTURE HOPES FOR UNDERSTANDING

Having summarized some of the main points addressed by the authors of this book, we now turn to a prospective on tantalizing future areas of neural crest research. Although many comparative evolutionary studies already have advanced our understanding of neural crest evolution, both at the level of genomes and morphologies, there are still many questions left unanswered. For instance, mechanisms of evolutionary segregation of cell fate potential in cranial, vagal, cardiac, trunk, and sacral neural crest cells remain largely unknown. With new methods addressing evolutionary changes in enhancers and other regulatory elements, and especially by taking advantage of the dawn of single-cell technologies, we might have a fair chance to address the emergence of distinct control mechanisms biasing neural crest cells towards specific ranges of fates at different axial levels. This, in turn, will help us to understand the mechanisms behind the diversity of neural crest–specific adaptation, ranging from a putative contribution to the turtle carapace (Gilbert et al., 2007) to participation in ruminant headgear, such as antlers and horns (Wang et al., 2019).

Molecular genomic approaches have shined a light on evolutionary changes in the regulation of neural crest multipotency. The conservation of major gene regulatory networks suggests that core components and regulatory loops are evolutionarily stable. On the other hand, later events during the life of a neural crest cell might be a convenient evolutionary substrate to introduce adaptions to a particular lifestyle and to drive diversification of species. A particular point of emphasis for future research on the origins of neural crest and the ancestral state of neural crest is the importance of outgroup comparisons. We need more and more experiments with appropriate vertebrate outgroup controls, including species in various urochordate, hemichordate, cephalochordate, and echinoderm clades, as highlighted in chapters by Yu, Abitua, and York with coauthors.

As mentioned earlier, a major conundrum perplexing generations of neural crest researchers is the apparently divergent repertoire of cell fates attained by cranial versus trunk neural crest cells. Specifically, cranial neural crest cells can form skeletal tissues, such as dentine, bone, and cartilage, while trunk neural crest cells were argued to lack this ability. However, this story might be another example of biased results stemming from a limited phylogenetic sampling. While many experiments have demonstrated that trunk neural crest cells in amniotes, such as mouse or chick, do not form skeletal tissues, cell labelling data from bony and cartilaginous fishes indeed suggest that trunk neural crest cells primitively formed skeletal tissues (Gillis et al., 2017; Smith et al., 1994). Of course, the dermal armor of agnathan fossil species always was assumed to have had trunk neural crest origins, again arguing for a primitive skeletogenic role for all neural crest at the base of the vertebrate clade. However, experiments to verify this conjecture currently are impossible to conduct on extinct animals.

Modern molecular techniques might finally shed light on classical questions regarding ancestral features of vertebrate neural crest cells. Generally, the new technique of single-cell transcriptomics allows a bioinformatic dissection of various stages of neural crest cell specification and differentiation. Comparative approaches using this technique can reveal how developmental processes of neural crest cells changed during evolution. Specifically, comparative analyses of neural crest gene regulatory networks in extant vertebrates and closely related deuterostomes can provide insights into the primitive state of neural crest–like progenitors. Along these lines, Marianne Bronner and coauthors argue that molecular circuits of lamprey cranial neural crest are more similar to trunk neural crest of jawed vertebrates than the cranial neural crest of those same jawed animals (Martik et al., 2019). Thus, trunk neural crest of jawed vertebrates might represent better primitive features of neural crest that were present in the common ancestor of cyclostomes and gnathostomes. If this conclusion were true, then cranial neural crest cells might have evolved additional gene-regulatory modules and components on top of the primitive neural crest condition. Of course, the fact that lamprey do not form bone or dentine (and never did) might bias these results, since they likely do not represent well neural crest from most agnathan clades. Also, gene regulatory networks underlying the development of cranial neural crest–specific features in lamprey might have evolved independently from other vertebrates, but similar comparative approaches can address this directly in further research.

How do variations in neural crest drive morphological diversity? This question has received an almost disproportionate amount of experimental research, traditionally using model organisms, but also more recently using nonmodel organisms. While this book did not focus on such experiments, they reveal that neural crest are major drivers of morphological diversity, especially in the craniofacial complex of vertebrates (Sanchez-Villagra et al., 2016; Usui and Tokita, 2018). Despite the data showing that neural crest might be the primary determinants of facial morphology, these specific traits and many others are dependent upon a series of tissue interactions between neural crest–derived mesenchyme and associated epithelia (Eames and Schneider, 2005, 2008; Hu et al., 2003; Thesleff, 2003). Accordingly, when seeking to understand complex developmental traits that result from tissue interactions, a focus on neural crest alone is short-sighted.

The evolution of neural crest and neural crest–derived structures is not finished and continues today. In humans, for example, the set of specific facial features seen in various populations around the world results from currently unknown evolutionary tinkering, likely involving cranial neural crest. Recently, DNA sequencing data has even been used to predict facial morphology (Lippert et al., 2017; Sero et al., 2019), linking genomic techniques to comparative anatomy, propelling designer baby faces from perhaps Isaac Asimov's creative mind to potential real-life scenarios. From a clinical perspective, the range of healthy variations in human facial morphologies is likely connected with the incidence rates of discrete and specific craniofacial abnormalities. Accordingly, the distribution of specific alleles in a human population balances the diversity of "normal" morphologies with a plethora of pathological conditions. Thus, further evolution of the cranial neural crest lineage will be driven by complex dynamics of a fast-growing human population. In particular, modern migration patterns involve people moving literally around the world, due to war, famine, or even relatively simple job opportunities, increasing breeding between genetically diverse populations and putting together unique combinations of gene allele frequencies that might impact neural crest cell development.

With respect to nonhuman vertebrate species, they also keep diversifying to occupy new ecological niches, especially in light of expanding anthropogenic habitats and climate change. Morphology of mouthparts and the position and development of sensory organs are key features that enable animals to adapt to such ecological changes. Since these traits are likely to be controlled by cranial neural crest, the corresponding underlying gene regulatory networks likely serve as an evolutionary substrate at the molecular level. It would be good to know how modern species continue to transform and adapt different phases of neural crest development, starting from induction and migration to elaboration of neural crest–derived cell types. Without a doubt, evolutionary modulations of the cranial neural crest lineage stand behind morphological diversity of craniofacial arrangements in nature—including a variation of beaks in famous Darwinian finches (Lamichhaney et al., 2015) or a stunning repertoire of extinct cranial designs discussed and interpreted in chapters by Ahlberg, Haitina, and Kuratani.

With the expansion of systems-level analyses in the field of cognition, an intriguing future area of research is the neural crest basis of behavior. For example, neural crest cells contribute to the adrenal medulla and, as a result, regulate hormonal control. By extension, neural crest might be connected to the evolution of behavioral traits, such as aggression. Indeed, changes in hormonal control were attributed to the so-called "domestication syndrome", a hypothesis stating that selection of alleles affecting neural crest cells causes domesticated behavioral phenotypes that are less aggressive and more social (Wilkins et al., 2014). Interestingly, such traits coincided with a specific set of craniofacial morphologies. When it comes to human evolution, a "self-domestication" hypothesis (Calvey, 2019; Niego and Benitez-Burraco, 2019; Thomas and Kirby, 2018) attempts to explain modern human transitions towards increased sociality and development of fine, baby-like facial features.

As today's nascent era of synthetic biology takes off, we now possess the ability to engineer new molecular circuits, expanding our capacity to investigate how various traits can be modified in neural crest. For example, by making targeted, functional perturbations, we can address experimentally how new cell types might have been added to the ancestral spectrum of neural crest–derived cell types. Attempts to engineer new neural crest properties, such as migratory routes and morphologies, should result in deep insights into the organization and stability of gene regulatory networks. A major evolutionary mechanism highlighted in chapters by Abitua, Adameyko, and Eames and coauthors is the cooption of gene regulatory networks by neural crest cells. New molecular manipulation techniques can challenge how neural crest cooption of gene regulatory networks from other cell lineages can proceed without unresolvable conflicts with the previously established regulatory system. In parallel, experiments with natural and artificial selection might result in better understanding of the plasticity and evolution of multigenic neural crest–associated traits. For example, the natural variations of craniofacial morphologies in wild cichlid fishes or the diversity of skull designs in domesticated dogs and cats can become excellent model systems to study the molecular basis of many parameters of an evolutionary process, including selection, mutation landscapes, population dynamics, and speciation, all with neural crest likely playing a central role.

If the reader is pessimistic, then perhaps understanding the evolution of neural crest in the distant past is an unresolvable challenge. However, the set of principles uncovered from studies of neural crest evolution might still operate at present, providing a feasible aim with an incredible practical importance. For instance, neural crest–derived cancers, such as neuroblastoma, pheochromocytoma, paraganglioma, neurofibromatosis, melanoma, and others, remain major threats to human health. The molecular mechanisms and developmental strategies that were employed during neural crest evolution might be applied to prevent high incidence rates of these cancers, as well as the other neurocristopathies, representing a major biomedicine-oriented interest.

In closing, we hope that the reader, by traversing the peaks and valleys of this book, will not only learn from the experts of the field, but will be entertained by a set of bizarre and exquisite evolutionary ideas about the emergence of a cell lineage that largely shaped our own existence.

ACKNOWLEDGMENTS

We wish to thank Dan Medeiros for roping us into editing this book after he realized that his constantly updated familial obligations precluded his active participation in this project after he had recruited and confirmed participation of many of the authors in the book. We also thank Dr. Olga Kharchenko for providing the artistic illustrations throughout this book. Finally, we wanted to send a heartfelt thanks for the inspirational and motivating force of Dr. Brian Hall. Not only did Brian champion this book, along with many others in the book series *Evolutionary Cell Biology* that he organizes, he was also the main reason that we decided to put all of our efforts into this book. We hope that it can serve to motivate the next generation of neural crest researchers, similar to how Brian indeed motivated our work on these fascinating cells.

REFERENCES

Calvey, T., 2019. Human self-domestication and the extended evolutionary synthesis of addiction: How humans evolved a unique vulnerability. *Neuroscience* 419, 100–107.

Eames, B.F., Schneider, R.A., 2005. Quail-duck chimeras reveal spatiotemporal plasticity in molecular and histogenic programs of cranial feather development. *Development* 132(7), 1499–1509.

Eames, B.F., Schneider, R.A., 2008. The genesis of cartilage size and shape during development and evolution. *Development* 135(23), 3947–3958.

Gans, C., Northcutt, R.G., 1983. Neural crest and the origin of vertebrates: A new head. *Science* 220(4594), 268–273.

Gilbert, S.F., Bender, G., Betters, E., Yin, M., Cebra-Thomas, J.A., 2007. The contribution of neural crest cells to the nuchal bone and plastron of the turtle shell. *Integr Comp Biol* 47(3), 401–408.

Gillis, J.A., Alsema, E.C., Criswell, K.E., 2017. Trunk neural crest origin of dermal denticles in a cartilaginous fish. *Proc Natl Acad Sci U S A* 114(50), 13200–13205.

Horie, R., Hazbun, A., Chen, K., Cao, C., Levine, M., Horie, T., 2018. Shared evolutionary origin of vertebrate neural crest and cranial placodes. *Nature* 560(7717), 228–232.

Hu, D., Marcucio, R.S., Helms, J.A., 2003. A zone of frontonasal ectoderm regulates patterning and growth in the face. *Development* 130(9), 1749–1758.

Ivashkin, E., Adameyko, I., 2013. Progenitors of the protochordate ocellus as an evolutionary origin of the neural crest. *EvoDevo* 4(1), 12.

Lamichhaney, S., Berglund, J., Almen, M.S., Maqbool, K., Grabherr, M., Martinez-Barrio, A., Promerova, M., Rubin, C.J., Wang, C., Zamani, N., Grant, B.R., Grant, P.R., Webster, M.T., Andersson, L., 2015. Evolution of Darwin's finches and their beaks revealed by genome sequencing. *Nature* 518(7539), 371–375.

Lippert, C., Sabatini, R., Maher, M.C., Kang, E.Y., Lee, S., Arikan, O., Harley, A., Bernal, A., Garst, P., Lavrenko, V., Yocum, K., Wong, T., Zhu, M., Yang, W.Y., Chang, C., Lu, T., Lee, C.W.H., Hicks, B., Ramakrishnan, S., Tang, H., Xie, C., Piper, J., Brewerton, S., Turpaz, Y., Telenti, A., Roby, R.K., Och, F.J., Venter, J.C., 2017. Identification of individuals by trait prediction using whole-genome sequencing data. *Proc Natl Acad Sci U S A* 114(38), 10166–10171.

Martik, M.L., Gandhi, S., Uy, B.R., Gillis, J.A., Green, S.A., Simoes-Costa, M., Bronner, M.E., 2019. Evolution of the new head by gradual acquisition of neural crest regulatory circuits. *Nature* 574(7780), 675–678.

Niego, A., Benitez-Burraco, A., 2019. Williams syndrome, human self-domestication, and language evolution. *Front Psychol* 10, 521.

Patthey, C., Schlosser, G., Shimeld, S.M., 2014. The evolutionary history of vertebrate cranial placodes--I: Cell type evolution. *Dev Biol* 389(1), 82–97.

Sanchez-Villagra, M.R., Geiger, M., Schneider, R.A., 2016. The taming of the neural crest: A developmental perspective on the origins of morphological covariation in domesticated mammals. *R Soc Open Sci* 3(6), 160107.

Schlosser, G., Patthey, C., Shimeld, S.M., 2014. The evolutionary history of vertebrate cranial placodes II. Evolution of ectodermal patterning. *Dev Biol* 389(1), 98–119.

Sero, D., Zaidi, A., Li, J., White, J.D., Zarzar, T.B.G., Marazita, M.L., Weinberg, S.M., Suetens, P., Vandermeulen, D., Wagner, J.K., Shriver, M.D., Claes, P., 2019. Facial recognition from DNA using face-to-DNA classifiers. *Nat Commun* 10(1), 2557.

Smith, M.M., Hickman, A., Amanze, D., Lumsden, A., Thorogood, P., 1994. Trunk neural crest origin of caudal fin mesenchyme in the zebrafish Brachydanio rerio. *Proc R Soc Lond B* 256(1346), 137–145.

Steventon, B., Mayor, R., Streit, A., 2014. Neural crest and placode interaction during the development of the cranial sensory system. *Dev Biol* 389(1), 28–38.

Thesleff, I., 2003. Epithelial-mesenchymal signalling regulating tooth morphogenesis. *J Cell Sci* 116(9), 1647–1648.

Thomas, J., Kirby, S., 2018. Self domestication and the evolution of language. *Biol Philos* 33(1), 9.

Usui, K., Tokita, M., 2018. Creating diversity in mammalian facial morphology: A review of potential developmental mechanisms. *EvoDevo* 9, 15.

Wang, Y., Zhang, C., Wang, N., Li, Z., Heller, R., Liu, R., Zhao, Y., Han, J., Pan, X., Zheng, Z., Dai, X., Chen, C., Dou, M., Peng, S., Chen, X., Liu, J., Li, M., Wang, K., Liu, C., Lin, Z., Chen, L., Hao, F., Zhu, W., Song, C., Zhao, C., Zheng, C., Wang, J., Hu, S., Li, C., Yang, H., Jiang, L., Li, G., Liu, M., Sonstegard, T.S., Zhang, G., Jiang, Y., Wang, W., Qiu, Q., 2019. Genetic basis of ruminant headgear and rapid antler regeneration. *Science* 364(6446).

Wilkins, A.S., Wrangham, R.W., Fitch, W.T., 2014. The "domestication syndrome" in mammals: A unified explanation based on neural crest cell behavior and genetics. *Genetics* 197(3), 795–808.

Index